FOURTH EDITION

Statistics in Kinesiology

William J. Vincent, EdD
Brigham Young University

Joseph P. Weir, PhD
Des Moines University

Human Kinetics

Library of Congress Cataloging-in-Publication Data

Vincent, William J.
 Statistics in kinesiology / William J. Vincent, Joseph P. Weir. -- 4th ed.
 p. ; cm.
 Includes bibliographical references and index.
 ISBN-13: 978-1-4504-0254-5 (hardcover)
 ISBN-10: 1-4504-0254-2 (hardcover)
 I. Weir, Joseph P., 1965- II. Title.
 [DNLM: 1. Kinesiology, Applied--Statistics. WE 103]

 612.7'6--dc23

2011044106

ISBN-10: 1-4504-0254-2
ISBN-13: 978-1-4504-0254-5

The web addresses cited in this text were current as of October, 2011, unless otherwise noted.

Acquisitions Editor: Myles Schrag; **Developmental Editor:** Kevin Matz; **Assistant Editor:** Steven Calderwood; **Copyeditor:** Amanda M. Eastin-Allen; **Indexer:** Betty Frizzéll; **Permissions Manager:** Dalene Reeder; **Graphic Designer:** Robert Reuther; **Graphic Artist:** Kathleen Boudreau-Fuoss; **Cover Designer:** Keith Blomberg; **Photographs (interior):** © Human Kinetics; **Photo Asset Manager:** Laura Fitch; **Visual Production Assistant:** Joyce Brumfield; **Photo Production Manager:** Jason Allen; **Art Manager:** Kelly Hendren; **Associate Art Manager:** Alan L. Wilborn; **Illustrations:** © Human Kinetics; **Printer:** Sheridan Books

Printed in the United States of America   10 9 8 7 6 5 4 3 2

The paper in this book is certified under a sustainable forestry program.

Human Kinetics
Website: www.HumanKinetics.com

United States: Human Kinetics
P.O. Box 5076
Champaign, IL 61825-5076
800-747-4457
e-mail: humank@hkusa.com

Canada: Human Kinetics
475 Devonshire Road Unit 100
Windsor, ON N8Y 2L5
800-465-7301 (in Canada only)
e-mail: info@hkcanada.com

Europe: Human Kinetics
107 Bradford Road
Stanningley
Leeds LS28 6AT, United Kingdom
+44 (0) 113 255 5665
e-mail: hk@hkeurope.com

Australia: Human Kinetics
57A Price Avenue
Lower Mitcham, South Australia 5062
08 8372 0999
e-mail: info@hkaustralia.com

New Zealand: Human Kinetics
P.O. Box 80
Torrens Park, South Australia 5062
0800 222 062
e-mail: info@hknewzealand.com

E5322

This book is dedicated to my eternal family:
Clarence, Orpha, Jerry, Diana, Steven,
Daniel, Susan, Gail, David, Nancy,
and all who have gone before or will come after.
—*William J. Vincent, EdD*

To Sumiko Inoue Weir (1933-1982)
and Julian Paul Weir (1923-2009).
—*Joseph P. Weir, PhD*

Contents

Preface

This edition of *Statistics in Kinesiology* represents a second phase in the life of this unique book. The most obvious change is the addition of Joe Weir of Des Moines University Osteopathic Medical Center as a coauthor. Bill Vincent brought to the first three editions of this book more than 35 years of experience in teaching statistics. Joe Weir has more than 15 years of experience in teaching research and statistics, primarily to physical therapy students. We both emphasize the practical use of statistics as a tool to help those in the movement sciences (e.g., physical educators, coaches, biomechanists, sport psychologists, exercise physiologists, athletic trainers, and physical therapists) analyze quantitative data. The goal is always to educate students in the proper use of statistical tools that can help them answer questions in their specific disciplines.

In updating this book, we retained all the qualities that made previous editions such a success. Specifically, the examples of statistical procedures still focus on variables in kinesiology so that students can better relate to how the procedures are used and how the procedures can help them answer questions. We retained the use of hand calculations because we think that doing some of the calculations manually on simple data is an important learning tool. Nonetheless, the mathematics shown in the examples involve only basic algebra skills. As with the previous editions, we emphasize topics that are commonly seen in our disciplines, such as repeated measures analysis of variance and the interpretation of interactions in factorial analyses of variance.

We have also made some substantial changes to the content of the book. Some topics have been expanded. For example, we increased coverage of effect sizes and the use of confidence intervals. We now have a separate chapter on analysis of covariance because it is becoming the technique of choice for analyzing the pretest–posttest control group design. Similarly, we expanded the coverage of the quantification of reliability. We also added new content on clinical measures of association, such as relative risk and odds ratios, that are relevant to clinical disciplines in kinesiology.

We hope this fourth edition will be a valuable tool in helping both students and researchers analyze quantitative data in the kinesiological disciplines.

—William J. Vincent , EdD
—Joseph P. Weir, PhD

Acknowledgments

My deepest acknowledgments are to my students who over the years have inspired me and encouraged me to write this book and its subsequent editions. They are the ones who have challenged me and provided the motivation to teach and to write. While administrative responsibilities are rewarding, I must honestly say that the classroom is where I find the greatest joy. It is there that the interaction takes place which inspires teachers and empowers learners.

I would like to acknowledge four people who have had a major impact on my professional life: Dr. Glen Arnett (deceased), who hired me in my first job at California State University, Northridge, Dr. Ruel Barker, Chair, and Dr. Robert Conlee, Dean at Brigham Young University, who hired me as an adjunct professor after retirement from CSUN, and Dr. Larry Hall who continued to support me for many years at BYU.

Special thanks to Kevin Matz and his staff at Human Kinetics for guiding us through this fourth edition with expertise and skill. And to my friend and coauthor, Dr. Joe Weir: thanks Joe for the great contributions you have made to this edition.

Finally I acknowledge my wife Diana, who for 53 years has been my eternal sweetheart and the love of my life. She has supported me all the way.

—*William J. Vincent EdD*

I am indebted to many people who have impacted me both professionally and personally over the years. In particular, I would like to thank Dr. Terry Housh at the University of Nebraska-Lincoln, who was my PhD advisor and remains my friend and mentor, and Dr. Ronald DeMeersman, who shepherded me into the professoriate at Teachers College, Columbia University when I was fresh out of graduate school. In addition, it has been a great pleasure to collaborate with Dr. Bill Vincent on the fourth edition of this book. He is a gentleman of the highest order. Finally, I wish to thank my wife, Dr. Loree Weir, who has been extremely supportive of all my professional pursuits.

—*Joseph P. Weir, PhD*

List of Key Symbols

α	1. Greek letter, lowercase *alpha*
	2. Area for rejection of H_0 on a normal curve
ANCOVA	Analysis of covariance
ANOVA	Analysis of variance
β	Slope of a line; Greek letter, lowercase *beta*
B	Greek letter, uppercase *beta*
χ	Greek letter, lowercase *chi*
χ^2	*Chi*-square
C	1. Column
	2. Number of comparisons
	3. Constant
	4. Cumulative frequency
	5. *Y*-intercept of a line
d	Deviation (the difference between a raw score and a mean)
df	Degrees of freedom
D_1	The 10th percentile
E	Expected frequency
ES	Effect size
f	Frequency
F	Symbol for ANOVA
FW_α	Familywise *alpha*
H	The highest score in a data sheet. Also, the value of Kruskal-Wallis ANOVA for ranked data.
H_0	The null hypothesis
H_1	The research hypothesis
HSD	Tukey's honestly significant difference
i	Interval size in a grouped frequency distribution
I	Scheffé's confidence interval
IQR	Interquartile range
k	Number of groups in a data set
L	The lowest score in a data sheet
MANOVA	Multiple analysis of variance
M_G	The grand mean in ANOVA
MS	Mean square
MS_E	Mean square error
μ	1. Greek letter, lowercase *mu*
	2. Mean of a population

n	Number of scores in a subgroup of the data set
N	Total number of scores in a data set
O	Observed frequency
ω	Greek letter, lowercase *omega*
ω^2	*Omega* squared
p	1. Probability of error
	2. Proportion
P	Percentile
Q_1	The 25th percentile
r	Pearson's correlation coefficient
R	1. Range
	2. Rows
	3. Multiple correlation coefficient
R_1	Intraclass correlation
R_2	Coefficient alpha in intraclass correlation
ρ	1. Greek letter, lowercase *rho*
	2. Spearman's rank order correlation coefficient
SD	Standard deviation (based on a sample)
SE_D	Standard error of the difference
SE_E	Standard error of the estimate
SE_M	Standard error of the mean
σ	1. Greek letter, lowercase *sigma*
	2. Standard deviation (based on a population)
σ_p	Standard error of a proportion
SS	Sum of squares
Stanine	Standard score with middle score = 5 and R = 1 to 9
Σ	1. Greek letter, upper case *sigma*
	2. The sum of a set of data
t	Student's t
T	T score (standard score with $\bar{X}$ = 50 and σ = 10.0)
U	Mann-Whitney U test
V	Variance
X	A single raw score
$\bar{X}$	The mean
X_{mid}	The middle score in an interval of scores
Z	Z score (standard score with $\bar{X}$ = 0 and σ = 1.0)
Z_α	Alpha point

$$H = \left[\frac{12}{18(18+1)}\right]\left[\frac{82.5^2}{6} + \frac{54.5^2}{6} + \frac{34.0^2}{6}\right] - 3(18+1)$$

$$H = [.035][1822.09] - 57$$

$$H = 6.77$$

Measurement, Statistics, and Research

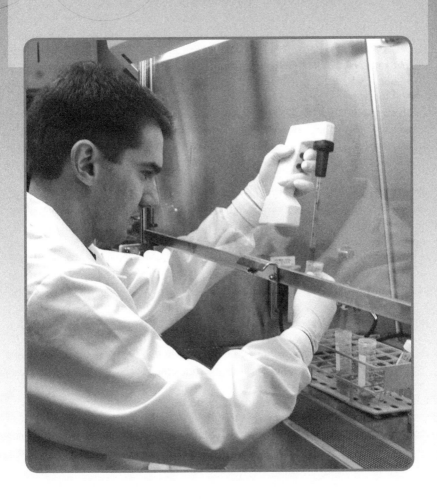

T he most important step in the process of every science is the measurement of quantities . . . . The whole system of civilized life may be fitly symbolized by a foot rule, a set of weights, and a clock.

James Maxwell

Kinesiology is the study of the art and science of human movement. Artistic researchers may study and evaluate the qualitative beauty and form of human movement in dancers, gymnasts, or other artistic movement forms. Scientific researchers are interested in the quantitative characteristics of the human body as it moves. This book addresses the science of human movement and how it is measured, statistically analyzed, and evaluated.

Statistical procedures are the same regardless of the discipline from which the data was collected. The researcher chooses the proper procedure with which to analyze the data in a particular field. In this fourth edition of *Statistics in Kinesiology*, the authors include examples of statistical procedures from a wide ranging list of disciplines in kinesiology including physical education and sports; motor learning; biomechanics; exercise physiology; sport psychology; health, leisure studies, and recreation; athletic training; and physical therapy. From these examples, the reader can choose the appropriate procedures and apply them to data collected in any related field.

All science is based on precise and consistent measurement. Whether or not you consider yourself to be an exercise scientist or a physical education teacher or a physical therapist, you can gain much by learning proper measurement techniques. Exercise scientists measure various attributes of human performance in laboratories, teachers measure students' performances in the field, and physical therapists measure patients' performance in the clinic.

Maxwell's observation indicates that most measurements are of quantitative values—distance, force, and time. However, counting the frequency of occurrence of events is also measurement. These same measurements apply to kinesiology, where we commonly measure *distance* (how tall people are, or how far they can jump), *force* (how much they weigh, or how much they can lift), *time* (how fast they can run, or how long they can run at a given pace on a treadmill), and *frequency* (how many strides it takes to run 100 meters, or how many times the heart beats in a minute). These measurements are sometimes referred to as *objective* because they are made with mechanical instruments, which require minimal judgment on the part of the investigator and reduce investigator bias to a minimum.

Other measurements are classified as qualitative (or subjective), because they require human judgment, and are used to determine the quality of a performance, such as a gymnastics routine or a golf swing, or other factors such as a patient's level of pain.

What Is Measurement?

Put simply, **measurement** is the process of comparing a value to a standard. For example, we compare our own weight (the force of gravity on our body) with the standard of a pound or a kilogram every time we step on a scale. When a physical education teacher tests students in the long jump, the process of measurement is being applied. The standard with which the jumps are compared is distance (meters). When the teacher uses an instrument (in this case, a tape measure) to determine that a student has jumped 5.2 meters, this bit of information is called data.

Data are the result of measurement. When individual bits of data are collected they are usually disorganized. After all of the desired data points are known, they can be organized by a process called statistics. **Statistics** is a mathematical technique by which data are organized, treated, and presented for interpretation and evaluation. **Evaluation** is the philosophical process of determining the worth of the data.

To be useful, the data from measurement must be reproducible—that is, a second measurement under the same conditions should produce the same result as the first measurement. Reproducibility is typically discussed under the heading of reliability. **Reliability** (sometimes referred to as the consistency of the data) is usually determined by the test–retest method, where the first measure is compared with a second or third measure on the same subjects under the same conditions. The quantification of reliability is addressed in chapter 13.

To be acceptable, data must also be valid. **Validity** refers to the soundness or the appropriateness of the test in measuring what it is designed to measure. Validity may be determined by a logical analysis of the measurement procedures or by comparison to another test known to be valid. In kinesiology, we often quantify validity by determining the extent to which a test is correlated to another index. The techniques of correlation and regression analysis are often used for these purposes. For example, the validity of skinfold measures for body fat assessment is quantified by how well the skinfold measures correlate with the results from underwater weighing. Chapters 8 and 9 address correlation and regression procedures that can be used to quantify validity.

Objectivity means that the data are collected without bias by the investigator. Bias can be detected by comparing an investigator's scores against those of an expert or panel of experts. Objectivity is sometimes referred to as inter-rater reliability (Morrow et al., 2000, p. 78).

Process of Measurement

Measurement involves four steps:

1. The object to be measured is identified and defined.
2. The standard with which the measured object will be compared is identified and defined.

3. The object is compared with the standard.

4. A quantitative statement is made of the relationship of the object to the standard.

For example, if we measured the height of a person who is 2 meters tall, we would conclude that the person's height (the object measured) is two times greater (the relationship) than 1 meter (the standard).

The standard used for comparison is critical to the measurement process. If the standard is not consistent, then the data will change each time an object is compared with that standard. In the English system of measurement, the original standard was not consistent. About 600 years ago in England the standard for measuring distance was the length of the king's foot. When the king died and another with a smaller or larger foot took his place, the standard changed and the length of all objects in the kingdom had to be redetermined.

The English system of measuring distance was originally based on anatomical standards such as foot, cubit (distance from the elbow to the finger tips—typically about 1.5 feet), yard (a typical stride for a person), and hand (the spread of the hand from end of the little finger to end of the thumb). Force was based on pounds (7,000 grains of wheat equaled 1 pound), and each pound was further divided into 16 ounces. In years when the rain was adequate the grains of wheat were large, but in drought years the grains were smaller. So the standard changed from year to year.

Eventually these measures were standardized, but the English system is difficult to use because it has no common numerical denominator for all units of measurement. Sixteen ounces make a pound, but 2,000 pounds make a ton; 2 cups make a pint and 2 pints make a quart, but it takes 4 quarts to make a gallon. Twelve inches make a foot, 3 feet make a yard, and 5,280 feet constitute a mile. It is no wonder that children have difficulty learning this system.

The metric system, which is more consistent and easier to understand, was first introduced in France in 1799. It is now used everywhere except in the United States and the countries in the British Empire. In this system, the units of measurement for distance, force, and volume are based on multiples of 10. The metric system uses the following terms:

Prefix	Value
milli	1/1,000
centi	1/100
zero	0
deca	10
hecto	100
kilo	1,000
mega	1,000,000
giga	1,000,000,000

These terms make it easy to convert units of measure. For example, a kilometer is 1,000 meters and a centimeter is 1/100 of a meter. The same terminology is used for force, volume, and distance. The metric system is based on geophysical values that are constant. A meter was formerly defined to be 1/10,000,000 of the distance from the equator to a pole on the earth but is now defined as the length that light travels in a vacuum over 1/299,793,458 of a second. A gram is the force of gravity on 1 cubic centimeter of water at 4 °C (its maximum density).

Fortunately, both the English and metric systems use the same units of measurement for time: seconds, minutes, hours, and so on. Although measurements are not based on multiples of 10 (there are 60 seconds to 1 minute), the relationships are the same in both systems and are common throughout the world.

The metric system is clearly superior to the English system. If it were used everywhere in the world, students would have less confusion and frustration when learning measurement concepts.

Variables and Constants

When we measure human performance, we measure variables. A **variable** is a characteristic of a person, place, or object that can assume more than one value. For example, one characteristic of a person that varies is the time he or she takes to run a mile.

Other examples of variables in kinesiology are weight, height, the number of basketball free throws made out of 10 tries, oxygen consumption and heart rate during a treadmill test, angles at the knee joint during various phases of a gait cycle, scores on manual muscle tests, and placement in a ladder tournament.

Note that different people score differently on the same variable, and a single person may perform differently when measured on the same variable more than once. The data vary between people as well as within people. Because variables change, we must monitor, or measure, them frequently if we need to know the current status of the variable.

A characteristic that can assume only one value is called a **constant**. A constant never changes. Therefore, once we measure it with accuracy, we do not have to measure it again. The number of players on an official baseball team is a constant. It must be nine, no more or no less. The distances in track events are constants. In a 100-meter dash, the distance is always 100 meters. The distances from base to base on a baseball field and from the pitcher's mound to home plate are constants.

Many anatomical characteristics are constants. Generally a person has only one head, one heart, two lungs, and two kidneys. Of course, illness or accidents may disfigure the body and thus change some of these characteristics, but typically the anatomy of the human body is constant.

Probably the most well-known constant in mathematics is pi. Pi is the number of times the diameter of a circle can be laid around its circumference, and it is always 3.14159. . . .

Types of Variables

Variables may be classified as continuous or discrete. A **continuous variable** is one that theoretically can assume any value. Distance, force, and time are continuous variables. Depending on the accuracy of the measuring instrument, distance can be measured to as short as a millionth of a centimeter or as long as a light-year. Time can be measured in milliseconds, days, or centuries.

The value of a **discrete variable** is limited to certain numbers, usually whole numbers or integers. For example, sex is classified as male or female. When counting people, we always use whole numbers because it is impossible to have a fraction of a person. The number of heartbeats in a minute and the number of baskets scored in a basketball game are also examples of discrete variables. One cannot count half of a heartbeat or score part of a basket. Many clinical variables are discrete variables. When making a diagnosis, a person is classified as either having the condition (e.g., torn anterior cruciate ligament) or not.

Classification of Data

Data collected on variables may be grouped into four categories, or scales: nominal, ordinal, interval, and ratio. In **nominal scales,** subjects are grouped into mutually exclusive categories without qualitative differentiation between the categories. Subjects are simply classified into one of the categories and then counted. Data grouped this way are sometimes called **frequency data** because the scale indicates the frequency, or the number of times an event happens, for each category. For example, a teacher classified students as male or female and then counted each category. The results were 17 males and 19 females. The values 17 and 19 represent the frequencies of the two categories, male and female.

Some nominal scales have only two categories, such as male or female, or yes and no. Others, such as an ethnicity scale, have more than two divisions. Nominal scales do not place qualitative value differences on the categories of the variable. However, numbers are assigned to the categories. The size of the number does not indicate an amount, but rather is used to indicate category assignment in a data file. For example, we might assign the number 0 to males and 1 to females. Here, the choice of number is completely arbitrary.

An **ordinal scale,** sometimes called a rank order scale, gives quantitative order to the variables but does not indicate how much better one score is than another. In a physical education class, placement on a ladder tournament is an example of an ordinal scale. The person on top of the ladder tournament has performed better than the person ranked second, but no indication is given of how much better. The top two persons may be very close in skill, and both may be considerably more skilled than the person in third place, but the ordinal scale does not provide that information. It renders only the order of the players, not their absolute abilities. Differences between the positions on an ordinal scale may be unequal. If 10 people are placed in order from short to tall, then numbered 1 (shortest) to 10 (tallest), the values of 1 to 10 would represent ordinal data. Ten is taller than 9, and 9 is taller than 8, but the data do not reveal how much taller. Clinicians often use 0-to-10 ordinal scales to quantify pain. Similarly, in exercise physiology the 6-to-20 Borg scale is an ordinal scale that is used to quantify the rating of perceived exertion during exercise.

An **interval scale** has equal units, or intervals, of measurement—that is, the same distance exists between each division of the scale—but has no absolute zero point. Because zero does not represent the absence of value, it is not appropriate to say that one point on the scale is twice, three times, or half as large as another point.

The Fahrenheit scale for measuring temperature is an example of an interval scale: 60° is 10° hotter than 50° and 10° cooler than 70° (the intervals between the data points are equal), but 100° is not twice as hot as 50°. This is because 0° does not indicate the complete absence of heat. In athletics, interval scores are used to judge performances in sports such as ice skating, gymnastics, diving, and synchronized swimming. A 9.0 in gymnastics is halfway between 10.0 and 8.0, but it is not necessarily twice as good as a 4.5. A score of 0 does not mean the absence of skill; it means that the performance did not contain sufficient skill to be awarded any points.

The most complete scale of measurement is the **ratio scale.** This scale is based on order, has equal distance between scale points, and uses zero to represent the absence of value. All units are equidistant from each other, and proportional, or ratio, comparisons are appropriate. All measurements of distance, force, or time are based on ratio scales. Twenty kilograms is twice the mass of 10 kilograms, and 100 meters is twice as far as 50 meters. A negative score is not possible on a ratio scale. A person cannot run a race in negative seconds, weigh less than 0 kilograms, or score negative points in basketball. In kinesiology, data of the ratio type are often used.

Nominal and ordinal scales are called **nonparametric** because they do not meet the assumption of normality. Interval and ratio scales are classified as **parametric.** These concepts are discussed further in chapter 16.

Research Design and Statistical Analysis

Research design and statistical analysis are both based on measurement principles. They are so intertwined that it is difficult to discuss one without referring to the other. This book presents statistical techniques that are designed to assist in evaluating data. How the data are collected, which instruments are used, and how the variables are controlled are all part of the research design.

Research is a special technique for solving problems. Identifying the problem is a critical part of research. A problem usually begins with such questions as "I wonder if . . . ," "I wonder why . . . ," or "I wonder what . . . ." If we do not find the answer to the question, then we have a problem to be researched. First, we might ask an expert in the field about the question. If the expert knows the answer, our problem is solved (assuming the expert is not wrong!).

If the expert does not know, we might visit the library. If we find the answer in the library, our problem is solved. But when we cannot find the answer by looking in the library or talking to other people, then the only way to solve the problem is to conduct research.

Many people think of an experiment when they hear the word research, but research is not limited to exploring scientific, experimental design problems. The problem may be solved by historical research, descriptive research, or experimental research. **Historical research** is a search through records of the past to determine what happened and why. It is an attempt to solve present problems by learning from the past. This book does not address historical research; however, many available texts on research design do discuss historical research.

Observational research (or **descriptive research**) involves describing events or conditions, which the researcher does not actively manipulate. In this type of study, researchers often examine correlations in the data. A common tool of observational research is the survey. The researcher identifies the events or conditions to be described and seeks information from people or other sources by asking questions, often by using a questionnaire. Statistical techniques are used to organize, treat, and present the data from observational research for evaluation and interpretation.

Experimental research is the research process that involves manipulating and controlling events or variables to solve a problem. Experimental research puts researchers in the strongest position to make cause–effect inferences from the data. The remainder of this chapter discusses research of the experimental design type. **Experimental design** is a process that involves manipulating and controlling certain events or conditions to solve a problem

To begin your research, you need a plan. The research design is the plan that sets out the manner in which the data will be collected and analyzed. Often the plan calls for subproblems to be solved first or for pilot work (preliminary data collection) to be conducted to determine the feasibility of the research design. A common error of beginning researchers is to jump into data collection before doing adequate planning and pilot work.

If the problem has not been solved after the problem has been identified and the library search has been completed, we are ready to make a hypothesis. A **hypothesis** is an educated guess or a logical assumption that is based on prior research or known facts and that can be tested by the experimental design. The hypothesis must be stated in such a way that statistical analysis can be performed on the data to determine the probability (or odds) of obtaining the given results if the hypothesis is true.

Hypothesis Testing

The hypothesis that prompts the research is called the **research hypothesis,** or the **alternate hypothesis,** and is symbolized by H_1. It usually predicts relationships or differences between or among groups of subjects. After all, if we didn't believe that relationships or differences existed, then we would probably not perform the experiment in the first place. However, H_1 is not the hypothesis that is normally tested by statistical analysis.

The hypothesis usually tested statistically is called the **null hypothesis,** or H_0. The null hypothesis predicts no relationship or no difference between the groups. If H_0 is true, then H_1 is false, and vice versa. They are mutually exclusive propositions. Essentially, the null hypothesis states that any relationship or difference that may be observed by measurement is the result of random occurrences.

The statistical analysis reports the probability that the results would occur as we observe them if the null hypothesis is true. For example, if the null hypothesis is true the statistics may indicate the probability that the obtained results could occur by random chance is less than one in 100 ($p = .01$). If the probability is small, we reject the null hypothesis and accept the alternate, or research hypothesis (H_1). If the probability is large, we accept the null hypothesis.

The null hypothesis is normally used to evaluate results because researchers are usually not sure what the results will be when they conduct an experiment. If we knew the results before we started, we would not need to do the experiment. But researchers do begin an experiment with an idea in mind, and this idea is represented by the research hypothesis. Usually the research hypothesis prompts the experiment, but it is the null hypothesis that is statistically tested.

The researcher should test the null hypothesis unless strong evidence—from other research, from related literature, from the opinion of experts, or from other reliable sources—suggests that a significant difference is to be expected. The opinion (perhaps biased) of the investigator is not sufficient reason alone to test the research hypothesis.

We reject H_0 and accept H_1 when differences or relationships among variables are established beyond a reasonable doubt, or at an acceptable level of confidence. Before we collect data, we must establish a level of confidence. For example, we may decide to reject the null hypothesis if the odds against it are better than 95 to 5. Stated another way, p (the probability that we could have gotten the data that we did if the null is true) is less than or equal to .05 ($p \leq .05$). This means that the odds that the null hypothesis is true are less than or equal to 5 in 100. Chapter 7 addresses these processes in more detail.

Independent and Dependent Variables

In an experiment, the **independent variable** refers to the variable that is manipulated or controlled by the researcher. This is sometimes called the treatment. A sports nutrition researcher might randomize subjects to receive either a creatine supplement or placebo. Here we might call this independent variable "supplement condition" and say that it has two levels (creatine vs. placebo). Similarly, a researcher might be interested in examining the effect of stretching duration on range of motion of the lower back and hamstrings. Subjects could be randomized to one of three groups that perform a hamstring stretch for 15, 30, or 60 seconds. The independent variable might be called stretching duration and it has three levels.

A **dependent variable** is dependent on the effects of one or more other variables. In an experimental study, the dependent variable is that which is measured. Under these conditions, the researcher examines the effect of the independent variable on the dependent variable. In the creatine study, the dependent variable might be peak power from a Wingate anaerobic power test. The researcher then might test the hypothesis that creatine supplementation causes an increase in anaerobic power. If the study is well designed and controlled, and if the creatine-supplemented subjects score higher (on average) than the placebo subjects, then the hypothesis is supported. For the stretching example in the previous paragraph, the dependent variable might be lower back and hamstring flexibility as measured by the sit-and-reach test.

The terms independent variable and dependent variable are also used in the context of observational studies. Imagine that a physical education researcher asks athletes how often they practice and then gives them a motor skills test. One might suspect that skill is dependent on practice; therefore, skill would be the dependent variable and practice would be the independent variable. Usually the more one practices, the better one performs. If a graph of practice time (on the X-axis) and skill (on the Y-axis) were constructed, the plot of the relationship would progress from lower left to upper right (see figure 1.1). As practice time increases on the X-axis, skill would also increase on the Y-axis. The independent variable is commonly plotted on the X-axis and the dependent variable is plotted on the Y-axis, and we often refer to the independent variable as the X-variable and the dependent variable as the Y-variable. Note that the researcher did not manipulate practice time. Instead, the amount of practice time was estimated based on self-report by the subjects. Therefore, this study is observational and not experimental.

It is common to refer to the independent variable as the **predictor variable** and to call the dependent variable, which is being studied, the **criterion variable.** If we measured the effect of a diet on weight gain or loss, the diet would be the independent, or predictor, variable, and weight would be the dependent, or criterion, variable. The weight gain or loss is dependent on the effects of the diet. We study the effect of the predictor variable on the criterion variable.

Internal and External Validity

Previously, we noted that validity is the extent to which a test measures what it purports to measure. We also use the term validity in the context of the quality of a research study in that research conducted by experimental design must demonstrate both internal and external validity. **Internal validity** refers to the design of the study itself; it is a measure of the control within the experiment to ascertain that the results are due to the treatment that was applied. Sometimes when people take motor skill tests, they improve their performance simply by taking the test twice. If a pretest–treatment–posttest design is conducted, the subjects may show improvements on the posttest because they learned specific test-taking techniques while taking the pretest. If these improvements are attributed to treatment when in fact they are due to practice on the pretest, an error has been made.

To find out if the changes are due to the treatment or to practice on the test, the researchers could use a control group. The control group would also take both the pre- and posttest but would not receive the treatment; the control subjects may or may not show posttest improvement. Analyzing the posttest differences between the

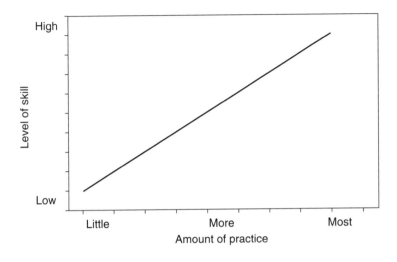

Figure 1.1 Relationship of skill to practice.

experimental group and the control group helps us sort out how much improvement is due to (a) the treatment and (b) the learning effect from the pretest. A design of this type without a control group can be criticized for having weak internal validity.

Other factors, which are not controlled in the experiment, can also reduce the internal validity. These factors are sometimes called **intervening variables,** or extraneous variables; they intervene to affect the dependent variable but are extraneous to the research design. For example, if the posttest is given too soon after treatment, fatigue may affect the results. If fatigue is not one of the variables being studied, and if it is not properly controlled, the results may not reflect the true effect of the treatment. Likewise, a study of the effects of a low-fat diet on body composition that does not control for the variable of exercise may reach erroneous conclusions. To preserve internal validity, all potential intervening variables must be controlled or equated between the groups in the experiment.

Two other factors that may reduce internal validity are instrument error and investigator error. **Instrument error** refers to incorrect data that are due to a faulty instrument. If we use an inaccurate stopwatch to measure time, then the data will be faulty. Complicated instruments, such as oxygen analyzers, respirometers, force plates, and dynamometers, need to be continually calibrated to ensure that they are correctly measuring the variable.

Investigator error occurs when the investigator introduces bias in recording the data. This bias may be intentional or unintentional. Of course, we hope that no one would intentionally falsify data, but it has been known to happen. Usually, however, the bias is unintentional.

For measurements of distance, force, time, or frequency, investigator error is usually minimal. It is fairly easy to read an instrument, especially one with a digital readout. But when judging a skilled performance, the investigator, or judge, can produce biased scores. Other errors occur when the investigator is not skilled in the data collection technique, such as taking skinfold measurements for body composition analysis or measuring joint angle with a hand-held goniometer. The amount of error an investigator introduces can be determined by comparing the investigator's data with that collected on the same subjects by an expert or panel of experts.

External validity refers to the ability to generalize the results of the experiment to the population from which the samples were drawn. If a sample is not random, then it may not represent the population from which it was drawn. It is also possible that other factors (intervening variables) that were controlled in the experiment may not be controlled in the population. The very fact that the experiment was tightly controlled may make it difficult to generalize the results to an actual situation in which these variables are left free to influence performance.

When conducting research, we need to demonstrate that we have designed the experiment properly and have controlled all appropriate variables to ensure internal validity. We also need to show that the results can be applied to the real world. The process of generalizing from a sample to a population is called **statistical inference** and is one of the basic tools of the statistician.

Statistical Inference

Much of the work performed by a statistician involves making predictions about a large group (usually a group of people, but it could be a group of animals, rocks, fish, stars, or almost any object) based on data collected from a small portion of the group. A **population** is any group of persons, places, or objects that have at least one common characteristic. These characteristics are specified by the definition of the population. All of the seventh-grade girls in a given school would be one population. Males between the ages of 31 and 40 who do not participate in regular exercise are another population. In short, a population can be any group, as long as the criteria for inclusion in the group are defined so that it is clear who qualifies as a member.

Usually, the population of interest is quite large, so large that it would be either practically impossible or financially unreasonable to measure all of the members. If it is impossible or impractical to measure all members of a population, then we measure a portion, or fraction, of the population, which is called a **sample.** We assume that the subjects in the sample represent, or have the same characteristics as, the population. Thus, data collected on the sample can be generalized to estimate the characteristics of the population.

The larger the sample, the more accurately it will represent the population. If we want to know the average height of all 15-year-old females in a city, we are more likely to get an accurate picture of the population by measuring 1,000 subjects than by measuring only 10 subjects. The error in the prediction is inversely related to the size of the sample. The larger the sample, the smaller the error, and the smaller the sample, the larger the error.

In a **random sample,** each member of the population has an equal opportunity of being selected into the sample. Any inference made about the population based on a nonrandom sample is of doubtful value. In fact, all of the formulas and calculations that are used to make inferences about populations are based on the assumption of random sampling. Major errors in the estimate of the population can be made if the sample is not random.

For example, suppose we needed to estimate the average percent body fat of all males at a university. A random sample would require that all male students have an equal chance of being selected. If we select only males who enter the gymnasium between 1:00 p.m. and 3:00 p.m., the sample is likely to represent a larger portion of athletes than is in the population because many athletes enter the gymnasium at this time to prepare for practice. Athletes tend to have low body fat, so such a sample would probably underestimate the average percent body fat of the entire population and would be considered biased.

Possibly no one place on campus exists where everyone has an equal chance of being chosen at any given time. If the sample is taken during the day, night students are excluded. If it is taken on Tuesday, then students who are not on campus on Tuesday are excluded. If it is taken near the engineering building, then it is likely that a disproportionate number of engineering majors will be chosen. Probably the only place where a true random sample can be taken at a university is in the files of the administration building. Using a computer to select random identification numbers from a list of all students could ensure that each student has an equal chance of being represented in the sample.

When random selection in a large population is desired and subcategories of the population are of interest, a **stratified sample** may be taken. To do this, we select a certain part of the sample from each subgroup of the population. In the previous example, we may want to ensure that we include male students from each major, some from day classes, some from night classes, some of each academic class (freshmen, sophomores, juniors, seniors, graduate students), and some from each ethnic group. The proportion of students from each subgroup of the sample must be the same as the proportion of that subgroup in the entire population. Statistics may then be compiled on the entire sample or on any subgroup.

A sample cannot accurately represent a population unless it is drawn without bias. **Bias** means that factors operate on the sample to make it unrepresentative of the population. These factors are sometimes very subtle and we may not even be aware of them. But if the sample is totally random and sufficiently large, even unknown factors that may bias the results will be eliminated or distributed within the sample in the same way they are distributed in the population.

Parameters and Statistics

The only way to know the exact characteristics of a population is to measure all of its members. A **parameter** is a characteristic of the entire population. A **statistic** is a characteristic of a sample that is used to estimate the value of the population parameter.

Any estimate of a population parameter based on sample statistics contains some error. The amount of error is never known exactly, but it can be estimated based on the size and variability of the sample. For example, if we measure all males at a university and determine that the average percent body fat in the population is 21%, this value is called a parameter. If we then take a sample and find that the percent fat in the sample is 18%, this value is called a statistic. The difference of 3% between the parameter and the statistic is the result of sampling error. Chapter 7 explains how to estimate the amount of error in a sample.

Probability and Hypothesis Testing

Statistics has been called the science of making educated guesses. Unless the entire population is measured, the only statement that can be made about a population based on a sample is an educated guess accompanied by a probability statement. Statistics allow us to make a statement and then cite the odds that it is correct. For example, if we wanted to know the height of 15-year-old females in a city, it might be stated that based on a random sample of 200 females the average population height is estimated to be 158 centimeters with an error factor of ±5 centimeters and that the odds that this estimate is correct are 95 to 5. In statistical terms, this is called a 95% level of confidence, or a 5% probability of error. This error factor is usually represented as $p < .05$ (p is almost never exact, so we usually say "the probability of error is less than 5%").

Another approach might be to hypothesize (guess or predict) what the average population height is likely to be and then use a sample and a probability statement to test the hypothesis. Suppose we hypothesize that the average height of the population of 15-year-old females in a city is 160 centimeters. Using the data from the sample of 200 described earlier, we infer that a 95% chance exists that the average height of the population is 158 centimeters with an error factor of ±5 centimeters (chapter 7 describes how to predict population parameters using statistical inferences). This means that the odds are 95 to 5 that the true average height of the population of females is between 153 and 163 centimeters. Under these conditions, we accept as true the hypothesis that the height is 160 centimeters because it lies within the limits of the values estimated from the sample. However, a hypothesis that the average or mean height is 165 centimeters would be rejected because that value does not fall within the limits of the estimated population value.

The technique of hypothesizing and then testing the hypothesis by experimentation provides answers to questions that cannot be directly answered in any other way. Herein lies the real value of inferential statistics: the ability to make a statement based on research and estimate the probability that the statement is correct.

Theories and Hypotheses

A **theory** is a belief regarding a concept or a series of related concepts. Theories are not necessarily true or false. They are only productive or nonproductive in producing hypotheses. Fruitful theories produce many hypotheses that can be tested by the scientific process of experimentation. A hypothesis is stated more specifically than a theory and can be tested in an experiment. When the experiment is complete, the odds that the hypothesis is correct or incorrect can be stated. We determine these odds by using statistical procedures.

When a theory produces many hypotheses, most or all of which are deemed to be correct (by original and confirming studies), the theory is accepted as true. Usually this process goes through many stages, and the theory is revised as the process proceeds. The process may take years, decades, or even centuries. Finally the theory is accepted as the most correct interpretation of the data that have been observed. When many or most of the hypotheses produced by a certain theory are rejected or cannot be confirmed by other unbiased scientists, the theory is revised or abandoned.

An example of this process in the field of motor behavior is found in the concept of mental practice, or visualization. For many years, scientists and athletes believed, but could not prove, that the learning of motor skills was best accomplished by a combination of mental and physical processes. Athletes seemed to intuitively know that they could improve performance by visualizing themselves executing the skill. This theory produced many hypotheses, which were tested. One popular hypothesis was that if a person just thought about a physical skill, without ever having performed it physically, his or her performance on the skill would improve.

A few studies in the mid 20th century produced some hints that this may indeed be true. One early researcher in this area (Twinning, 1949) published a study that demonstrated significant differences in physical performance between control subjects and subjects who visualized themselves performing a novel skill. Soon other studies with similar, but modified, hypotheses were added to the knowledge base.

In the 1960s many studies (Egstrom, 1964; Jones, 1965; Oxendine, 1969; Richardson, 1967; Vincent, 1968) based on visualization theory were published, most of which confirmed the conclusion that visualization groups perform significantly better than control groups, but not as well as physical practice groups. The theory, or general belief, produced testable hypotheses that were supported by many experiments. Today, the theory that visualization improves performance of a physical skill is well accepted.

Misuse of Statistics

Advertisements tend to quote statistics freely. We are told that a certain toothpaste is better than another, that "with our magic exerciser, you can lose 10 pounds per week with only 1 minute of exercise per day," or that "8 out of 10 doctors recommend our product." These are examples of statistics that may or may not be true. Such misleading statements usually result from an inadequate definition of terms, the lack of a random sample, too small a sample; or a statement may have no statistical analysis for support or be just an uneducated guess.

One educational institution claimed that 33% of coeds marry their professors. Although this was true at the time for that school (only three women were enrolled in the school and one married a professor), it is an example of an improper generalization from a small sample. It leads the reader to an incorrect conclusion.

Sometimes the word *average* is misleading. For example, it is possible to show that more than half of the population has an annual income lower than the average. If 9 people make $10,000 per year, and one makes $100,000, then 9 out of 10 make less than the average income of $19,000. Perhaps a better description of the income of the population would be the typical income (the 50th percentile). Extreme scores, sometimes called *outliers,* have a disproportionate effect on the statistics. This is somewhat analogous to the joke about the man with his head in the oven and his feet in the freezer who remarked, "On the average, I feel fine."

Statistics is the process of making educated guesses. A statistician attempts to explain or control the random effects that may be operating and to make a probability statement about the remaining real effects. Statistics do not prove a statement. At best, they give us enough information to determine the probability that the statement is true.

We need to be careful when producing or using statistics. We should report only well-documented, valid, reliable, and objective results. And we should be wary of unusual or unique claims, especially if the source of the statistics is likely to benefit financially when we believe the claims.

Summary

This chapter introduced the concepts of measurement, statistics, and research, discussing the essentials of measurement theory and the process of how things are measured. This chapter also discussed the interrelationships among measurement principles, statistics, and research design. Research design and statistical analysis are always interrelated. Research cannot be conducted without statistical analysis, and proper statistical analysis depends on the quality of the research design.

The process of hypothesis testing and the appropriate use of the research and the null hypothesis were introduced. Statistical inference is an essential tool used by the statistician to estimate population parameters from sample statistics.

Problems to Solve

1. Define the following terms and state their relationship to one another: (a) measurement, (b) data, (c) statistics, and (d) evaluation.

2. What is the difference between a variable and a constant? Between independent and dependent variables? Give examples of each.

3. List the four classes of data and provide an example of each one. Explain why your example fits the class in which you placed it.

4. Explain how statistical inference is conducted. Define population, sample, random sample, and stratified random sample.
5. What is the difference between a parameter and a statistic?
6. What is the relationship between a theory and a hypothesis? Give an example of how a hypothesis may be used to solve a problem in kinesiology.
7. Give an example of a misuse of statistics that you have observed.
8. List three journals in the general field of kinesiology in which research and statistical conclusions are presented.

See appendix C for answers to problems.

Key Words

alternate hypothesis
bias
constant
continuous variable
criterion variable
data
dependent variable
descriptive research
discrete variable
evaluation
experimental design
experimental research
external validity
frequency data
historical research
hypothesis
independent variable
instrument error
internal validity
interval scale
intervening variable
investigator error
kinesiology

measurement
nominal scale
nonparametric
null hypothesis
objectivity
ordinal scale
parameter
parametric
population
predictor variable
random sample
ratio scale
reliability
research
research hypothesis
sample
statistic
statistical inference
statistics
stratified sample
theory
validity
variable

$$H = \left[\frac{12}{18(18+1)} \right]\left[\frac{82.5^2}{6} + \frac{54.5^2}{6} + \frac{34.0^2}{6} \right] - 3(18+1)$$

$$H = [.0A5][1822.09] - 57$$

$$H = 6.77$$

Organizing and Displaying Data

Scientists at the 2004 Olympics in Athens, Greece, recorded data on the performances of all athletes in all events. The data were not all of the same type. There were (1) judges' scores, such as those in gymnastics and diving; (2) times in minutes and seconds, such as those in races in track, swimming, and rowing; (3) distances in meters, such as those in throwing and jumping events; (4) forces in kilograms, such as those in weightlifting; and (5) counting, such as points in basketball, runs in softball, or goals in soccer. Many scientists at each of the venues recorded data.

After the Olympics, the scientists placed the results into a single database so that they could compare the data among all the athletes and across all the nations. When the scientists finished entering the data, they found that they had an enormous database that was cumbersome and difficult to read. It had hundreds of lines of data (each line representing an athlete and his or her nation) and multiple scores on each athlete or team. The data were all there, but they were disorganized. How can the data be organized and displayed so that appropriate comparisons can be made?

Organizing Data

In chapter 1, statistics was defined as a mathematical technique by which data are organized, treated, and presented for interpretation and evaluation. Computers are wonderful devices that can organize, analyze, and display data much better than humans. Computers enhance the human brain because they can remember large amounts of data and compute faster. You should use a computer whenever possible. However, computers are only an extension of the human brain. They will not perform well unless someone enters the data correctly and understands the output. Before a computer will be of value to you, you must know what you want it to do and what to expect in the output. This chapter demonstrates techniques for organizing and displaying raw data into a readable and useful format. It explains how to display the data in tables and how to create graphs so that the data can be more easily interpreted. Once you understand the methods involved, you should use a computer to speed up the process.

The data scientists collect contain much information of potential value. But before we can derive information from raw data, we must organize the data into a logical, readable, and usable format called a distribution table. Three types of distribution tables are discussed in this chapter: the rank order distribution, the simple frequency distribution, and the grouped frequency distribution.

Rank Order Distribution

A **rank order distribution** is an ordered listing of the data in a single column. It presents a quick view of the spread or variability of the group. It easily identifies the extreme scores, the highest and the lowest. A rank order distribution is used when the number of data points (N) is relatively small (i.e., ≤ 20). If the ordered listing of the data can fit on one page, then the rank order distribution is appropriate.

The **range** (R) is the distance in numerical value from the highest (H) to the lowest (L) score. It is calculated by subtracting the lowest score from the highest score:

$$R = H - L. \qquad (2.01)$$

The range is not easily determined until the raw data have been ordered. Following is a subset of data taken from a physical education teacher's roll book. The data represent the number of pull-ups performed by 15 eighth-grade boys.

Raw Data: 12, 10, 9, 8, 2, 5, 18, 15, 14, 17, 13, 12, 8, 9, 16.

These data are difficult to interpret because they are not organized. When the data are placed into a rank order distribution, from highest to lowest (see table 2.1), they are easier to interpret. The symbol X represents a variable, in this case pull-ups, and N (the number of scores) = 15.

When the pull-up scores are organized into a rank order distribution, it is easy to see that a large difference exists in ability to perform pull-ups. The spread of the group, or the range, is considerable—from 2 to 18 pull-ups.

TABLE 2.1

Rank Order Distribution of Pull-Up Scores

X
18
17
16
15
14
13
12
12
10
9
9
8
8
5
2

$N = 15$
$H = 18$
$L = 2$
$R = 18 - 2 = 16$

Simple Frequency Distribution

When the number of cases being studied is larger than will fit on one page, it is inconvenient to list them separately because the list would be too long. Larger data sets can be organized into a **simple frequency distribution,** an ordered listing of the variable being studied (X), with a frequency column (f) that indicates the number of cases at each given value of X. (This pattern is referred to as a simple frequency distribution because there is also a more complicated distribution called a *grouped* frequency distribution, which we will consider next.)

Suppose a student majoring in kinesiology wanted to analyze the pull-up scores for all male kinesiology majors at the university ($N = 212$). To avoid a long list, the student could arrange the scores into the simple frequency distribution presented in table 2.2.

TABLE 2.2

Simple Frequency Distribution of Pull-Up Scores

X	f
20	2
19	0
18	3
17	6
16	8
15	10
14	17
13	21
12	25
11	24
10	26
9	19
8	16
7	12
6	10
5	4
4	3
3	2
2	1
1	2
0	1
	212

$N = 212$
$H = 20$
$L = 0$
$R = 20 - 0 = 20$

The left column in table 2.2 labeled *X* represents the variable (pull-up scores), and the right column labeled *f* represents the number of subjects who received a given score. The number of subjects measured (*N*) is represented by the sum of the frequency column (212). Use this type of simple frequency distribution when *N* > 20 and *R* ≤ 20. This will allow you to fit all of the data on one sheet of paper.

Grouped Frequency Distribution

It's fairly easy to organize data into a simple frequency distribution if the range of scores is only 20, but researchers often work with variables that produce a range of more than 20 scores. When physical education teachers looked at their students' aerobic capacity as measured by their scores on the mile run (*N* = 206), they found that the lowest score, or fastest time, was 302 seconds and the highest score, or slowest time, was 595 seconds—a range of 293 seconds (595 − 302 = 293). With such a range, it is impractical to list all of the scores in one line. The teachers needed a method of grouping the data that would reduce the length of the raw data list.

A **grouped frequency distribution** is an ordered listing of, in one column, a variable (*X*) in groups of scores and, in a second column [the frequency column (*f*)], the number of persons who performed in each group of scores. When *N* > 20 and *R* > 20, a grouped frequency distribution should be used.

When setting up a grouped frequency distribution, we must first decide how many groups should be formed. For convenience, we will adopt 15 as the ideal number of groups. Fifteen was chosen because it represents a reasonable number of groups to list on one page of paper, but the decision is an arbitrary one.

When the number of groups becomes as small as 10, the number of scores in each group may become too large. This tends to obscure the data because so many cases are crowded together. Likewise, when there are more than 20 groups, the groups are spread out so far that some groups may lack cases and the list becomes long and cumbersome. For these reasons, it is best to keep the number of groups at about 15, but the number may vary between 10 and 20. In the previous example of the mile-run times, groups of 20 scores each (e.g., 300–319) produce 15 groups (see table 2.3).

The formula for **interval size** (the size of a group of scores on a given variable; *i*) is

$$i = \text{range}/15. \qquad (2.02)$$

For the mile-run data, *i* = 293/15 = 19.53, which we round to 20.

Grouping scores is much easier if we can place scores into groups that have whole numbers for their starting and ending points. But depending on the range, interval size may be either an integer or a decimal value. To simplify the process of grouping, we usually round interval size to the nearest odd integer. An odd interval size always results in a group with a whole number as the midpoint.

The one exception to the odd rounding rule is when interval size is equal or close to 10 or 20 (as is the case in our example). Because we are more familiar with 10 than with any other number, 10 makes a convenient interval size. Even though

TABLE 2.3

Grouped Frequency Distribution: Times in Seconds for Mile Run

X	f
580-599	3
560-579	9
540-559	13
520-539	15
500-519	17
480-499	21
460-479	19
440-459	25
420-439	23
400-419	18
380-399	15
360-379	12
340-359	9
320-339	5
300-319	2
	$N = 206$

the midpoint will not be a whole number, interval size values close to 10 or 20 are usually rounded to 10 or 20 for convenience in creating groups and tallying raw data into the groups. The decision to round to an odd integer, or to 10 or 20 when interval size is near these numbers, is another arbitrary decision. These choices are based on the nature of the data and on the need for simplicity in tallying the scores into groups.

After interval size is determined, we create the groups. When interval size is a multiple of 5 (or 0.5 if the data are recorded in tenths), the lowest group should include the lowest score in the data, but the lower limit of the lowest group should be a multiple of 5. When interval size is not a multiple of 5, the lower limit of the lowest group should be equal to the lowest score in the data. Because the determination of interval size is an arbitrary decision, the rounding-off procedure may be modified to create groups with which it is easy to work.

Because $i = 20$, we want 20 scores per group, and the first group should start with a multiple of 5. The lowest score is 302, so the first group will include all scores between 300 and 319 (see table 2.3). Note that the upper limit of the first group is 319, not 320. The group starts at 300, and the 20th score up from 300 is 319. In a similar manner, the remaining groups are created, with 20 scores in each group. The group creation process is continued up the data scale until a group is created (580–599) that includes the highest score (595).

After the groups are created, tally the scores that fall into each group. The frequency of each group is determined by adding the tally marks. Usually tallying is done on scratch paper; the final result shows only the groups and the frequency, as is shown in table 2.3. Grouping the scores makes it easier to visually interpret the data. In table 2.3, most values are in the middle groups, and the average is about 450 seconds. Because group frequency tables are used when N is large, a computer can be a great help when dealing with large databases that require frequency distribution tables. Computer software is available that will allow the user to enter data and organize it in various ways.

Some information is lost during grouping. For example, once the scores are grouped, it is impossible to tell exactly which score a given person received. We know that 25 subjects scored between 440 and 459 seconds, but we do not know how the scores are distributed within the group. Did all 25 subjects score 450? Did anyone score 440, 445, or 455? If we assume that the scores in a group are equally distributed within that group, then the score that best represents the whole group is the score that falls at the midpoint of the group. Using whole numbers for the midpoints makes it easier to work with the data.

The process just described for grouping data works well for discrete data in which decimal values are not involved. However, a problem arises when we use continuous data and measure scores in fractions of whole numbers. Consider the grouped frequency distribution representing time in seconds to swim 100 meters shown in table 2.4.

TABLE 2.4

Grouped Frequency Distribution: Times in Seconds for 100-Meter Swim

X	f
115-119	1
110-114	2
105-109	3
100-104	7
95-99	10
90-94	15
85-89	17
80-84	11
75-79	5
70-74	4
65-69	2
60-64	2
	$N = 79$

Swim times are usually measured in tenths, or even hundredths, of a second. The grouped frequency distribution is set up by integers, but the raw data include decimals. A student who swam the distance in 86.4 seconds would obviously be included in the 85-to-89 group, but into which group would we place the student who scored 84.7 seconds? Because time scores can be recorded to any degree of accuracy, we must make some provision to accommodate these in-between scores. We accomplish this by establishing the real limits of the groups.

The **apparent limits** of each group are the integer values listed for each group in the grouped frequency distribution. A one-point gap exists between the apparent end limit of one group and the beginning of the next. The **real limits** of each group are the assumed upper and lower score values in each group, created to the degree of accuracy required by the data; the real limits define the true upper and lower limits of the group. To establish the real limits, the gap between the apparent limits is equally divided; values in the upper half belong to the group above, and values in the lower half belong to the group below.

The lower real limit of the 80-to-84 group is 79.5, and the upper real limit is 84.49999 . . . . Usually the upper real limit is carried out to one decimal place more than the data require (i.e., if the data are in tenths, set the limit at hundredths).

Thus, to establish the real limits of the 100-meter swim groups, we raise the highest score in the group by 0.49 and lower the lowest score in the group by 0.5. Once this is done, it is easy to determine that the swimmer whose time was 84.7 seconds should be included in the 85-to-89 group because the lower real limit of that group is 84.5. The real limits are usually not listed in the grouped frequency distribution, so we need to keep those values in mind when we classify the data into groups according to the real, not the apparent, limits.

Displaying Data

Data can be displayed in several ways. The following sections detail the various display options.

Tables and Spreadsheets

Tables present data in a row-and-column format. In computer terminology, a table is sometimes called a **spreadsheet.** Spreadsheets are useful for organizing and displaying data.

Spreadsheets are especially helpful when presenting **multivariate** data (data with more than one variable measured on each subject). Traditionally, spreadsheets present data with the subjects listed in a column vertically on the left and the variables listed horizontally across the top row. In this arrangement, you can see all the scores for a given subject by scanning across the rows, or all the scores on a given variable by scanning down the columns. Table 2.5 is an example of a spreadsheet representing five subjects on three variables: height, weight, and body mass index (BMI).

TABLE 2.5

Sample of Spreadsheet

Subject no.	Height (in)	Weight (lb)	Body mass index
1	60	150	25
2	70	165	21
3	62	160	25
4	65	130	19
5	67	200	27

Microsoft Excel is an example of a computer spreadsheet. **Excel** can easily summate columns or rows, calculate **descriptive statistics,** and perform other types of statistical tests on uni- or multivariate data. It can also create new variables by manipulating existing variables. For example, the BMI column in table 2.5 was created using a formula to predict BMI from height and weight.

Graphs

Graphs are another way to present information about data. Graphs are visual representations of data that provide us with insight into the data that may not be easily observed in a spreadsheet or table. A **graph** is a figurative, or visual, representation of data. Graphs are helpful for comparing two or more sets of data or for showing trends. All graphs should include at least three characteristics: (a) a title, (b) a label on the X-axis (the abscissa), and (c) a label on the Y-axis (the ordinate). Other explanatory notes and information may be included on the graph, such as numerical values on both the X- and Y-axes (see figure 2.1).

It is a general practice on statistical graphs to place the frequency on the ordinate (the vertical or Y-coordinate) and the scores on the abscissa (the horizontal or X-coordinate), but this practice is not mandatory. Reversing this process would rotate the graph 90°, but the data would not be changed. On the X-axis, scores move from left to right as the numeric values move from low to high. But remember, good scores are not always high scores—for example, faster runners have lower time scores. On the Y-axis, frequencies move from bottom to top as the numeric values move from low to high. Following are examples of various kinds of graphs.

Histogram

Probably the most common graph is the **histogram,** or bar graph. It is usually constructed from a grouped frequency distribution, but it may also originate from a simple frequency distribution. The bars on the X-axis represent the groups, and the height of each bar is determined by the frequency of that group as plotted on

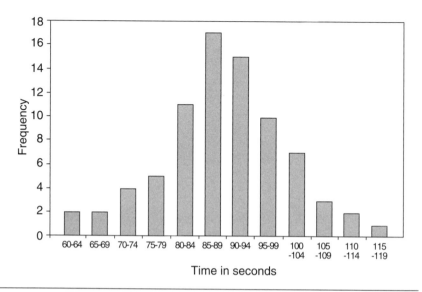

Figure 2.1   Histogram of 100-meter swim.

the Y-axis. The data from table 2.4 (100-meter swim times) are shown in figure 2.1 in a histogram. The apparent limits of each group are plotted on the X-axis.

Another graph that may be constructed from frequency distribution data, called a **frequency polygon,** is a line graph of the scores plotted against the frequency. It is usually formed from simple frequency distribution data by plotting the ordered scores on the X-axis against frequency on the Y-axis. A histogram may be converted to a frequency polygon by plotting the midpoints of the bars. A frequency polygon of the data from table 2.2 on pull-up scores is shown in figure 2.2.

Cumulative Frequency Graph

A third graph often used by statisticians is called a **cumulative frequency graph,** or cumulative frequency curve. It is a line graph of ordered scores on the X-axis plotted against the number of subjects who scored at or below a given score on the Y-axis. An exercise physiologist, studying the effects of resistance exercises, measured upper body strength on a group of subjects and counted the number of arm dips subjects could perform on the parallel bars. The scientist arranged the resulting data into a grouped frequency distribution and constructed a grouped frequency table that included a cumulative frequency column (see table 2.6).

The cumulative frequency column tells us the status of each subject compared with the total group. It is calculated by determining the number of subjects whose score is at or below that of each group. To accomplish this, we add the frequency of a given group to the frequencies of all the groups below it. To determine the cumulative frequency of the 10-to-12 group, we add $6 + 10 + 15 + 18 = 49$.

The cumulative frequency graph is plotted with scores on the X-axis and cumulative frequency on the Y-axis. Because the data in table 2.6 are discrete, the upper

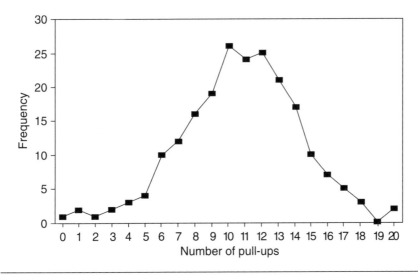

Figure 2.2 Frequency polygon for pull-up scores.

apparent limit of each group is plotted because that value represents all subjects who scored at or below that group. For continuous data, the upper real limits would be plotted to represent the maximum score that could be obtained by anyone in that group or in the groups below. Figure 2.3 is an example of a cumulative frequency graph for the parallel bar dip test data from table 2.6.

TABLE 2.6

Upper Body Strength as Measured by Parallel Bar Dip Test

X	f	Cum. f
34-36	2	130
31-33	4	128
28-30	5	124
25-27	7	119
22-24	9	112
19-21	15	103
16-18	19	88
13-15	20	69
10-12	18	49
7-9	15	31
4-6	10	16
1-3	6	6
	N = 130	

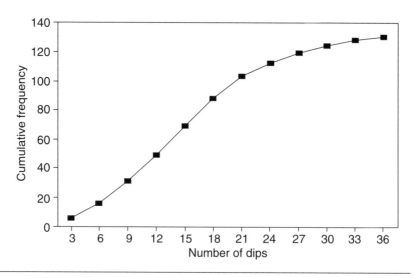

Figure 2.3   Cumulative frequency of dips on parallel bars.

Curves

A **curve** is the line that results when scores (*X*) are plotted against frequency (*Y*) on a graph. The shape of the curve depends on the distribution of the data. A curve presents a visual picture that permits us to see trends in the data that we may not easily observe when looking at a table. The following curves are commonly used in statistics.

Normal Curve

The most widely known curve in statistics is the **normal curve.** This uniquely shaped curve, which was first described by mathematician Karl Gauss (1777–1855), is sometimes referred to as a Gaussian curve, or a *bell-shaped curve.*

> Nature generally behaves according to rule. Karl Friedrich Gauss discovered that fact and formulated his discovery in a mathematical expression of normal distribution . . . and this curve has ever since become the sine qua non of the statistician. (Leedy, 1980, p. 25)

A normal curve is characterized by symmetrical distribution of data about the center of the curve in a special manner. The mean (average), the median (50th percentile), and the mode (score with the highest frequency) are all located at the middle of the curve. The frequency of scores declines in a predictable manner as the scores deviate farther and farther from the center of the curve.

All normal curves are bilaterally symmetrical and are usually shaped like a bell, but not all symmetrical curves are normal. When data are identified as normal, the special characteristics of the normal curve may be used to make statements about the distribution of the scores. Many of the variables measured in kinesiology are normally distributed, so kinesiology researchers need to understand the normal

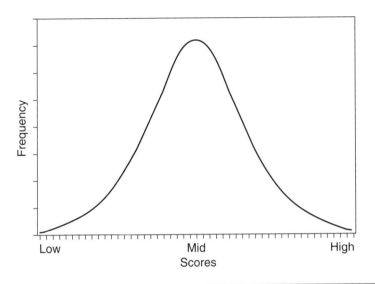

Figure 2.4 Normal curve.

curve and how it is used. Chapter 6 discusses the special characteristics of the normal curve in detail.

A typical normal curve is presented in figure 2.4. Notice that the two ends, or tails, of the curve are symmetrical and that they represent the scores at the low and high extremes of the scale. When scores approach the extreme values on the data scale, the frequency declines. This is demonstrated by the data represented in table 2.4. Most subjects score in the middle range.

We extensively use the normal distribution in statistics, which suggests that the distribution turns up a lot in nature. Of course, we could be using the normal distribution when we shouldn't; that is, maybe things that we think are Gaussian really aren't. Nonetheless, the commonness of the normal distribution in statistics is based on the **central limit theorem.** The central limit theorem can be defined in a variety of ways. However, explained simply, a *sum* of random numbers becomes normally distributed as more and more of the random numbers are added together (Smith, 1997). To see how this works, imagine that a random number generator spit out 20 numbers and you calculated and saved the mean value of those 20 numbers. Then you repeated this process over and over so you had a lot of those means sitting in a pot. If you then made a frequency distribution of those means, it would be about normal (Gaussian). The ubiquity of the normal distribution stems from the idea that randomness leads to Gaussian distributions. The central limit theorem is a key underpinning of our use of statistics.

Mesokurtic, Leptokurtic, and Platykurtic Curves

When most scores fall in the midrange and the frequency of the scores tapers off symmetrically toward the tails, the familiar bell-shaped curve occurs. This is referred to as a **mesokurtic** (*meso* meaning middle and *kurtic* meaning curve) curve. But if

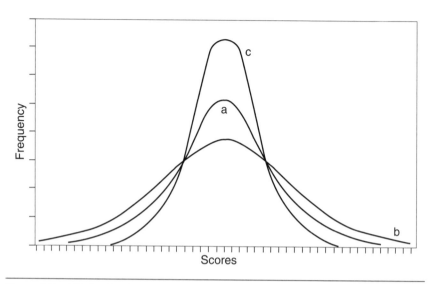

Figure 2.5 (a) Mesokurtic (bell-shaped), (b) platykurtic, and (c) leptokurtic curves.

the results have a very wide range of scores with low frequencies in the midrange, the curve is called **platykurtic,** or flat. The opposite of the platykurtic curve is a **leptokurtic** curve, which results when the range of the group is limited and many scores are close to the middle. The differences among mesokurtic, platykurtic, and leptokurtic curves are shown in figure 2.5.

Bimodal Curves

The **mode** is the score with the highest frequency. On the normal curve, a single mode is always in the middle, but some distributions of data have two or more modes and hence are called **bimodal,** or multimodal. If one mode is higher than the other, the modes are referred to as the major and minor modes. When such data distributions are plotted, they have two or more humps representing the clustering of scores. Bimodal curves are not normal curves. A bimodal curve is shown in figure 2.6.

Skewed Curves

Sometimes the data result in a curve that is not normal; that is, the tails of the curve are not symmetrical. When a disproportionate number of the subjects score toward one end of the scale, the curve is **skewed.** The data in table 2.6 for parallel bar dips show that a larger number of the subjects scored at the bottom of the scale than at the top. A few stronger subjects raised the average by performing 30 or more dips. When the data from table 2.6 are plotted, as in figure 2.7, the hump or mode of the curve is pushed to the left, and the tail on the right is longer than the tail on the left. The curve has a positive skew because the long tail points in a positive direction on the abscissa.

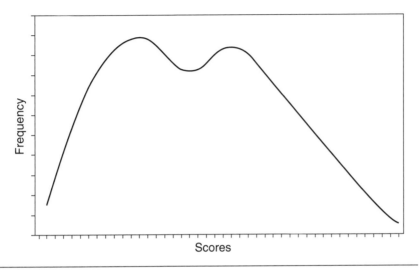

Figure 2.6 Bimodal curve.

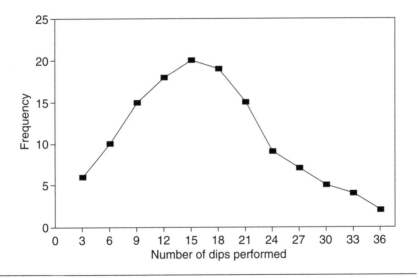

Figure 2.7 Positively skewed curve for dips on parallel bars.

If most subjects score high on a test and only a few do poorly, then the mode is to the right, and the left tail is longer than the right tail, as shown in figure 2.8. This is called a negative skew because the long tail points in the negative direction on the abscissa. Chapter 6 presents a method for calculating the amount of skewness in a data set. It is fairly common to see skewed distributions when the data exhibit ceiling or floor effects. A **ceiling effect** occurs when there is a limit on how high scores can be and scores tend to bunch around that limit. Similarly, a **floor effect**

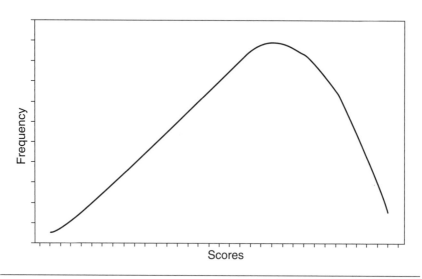

Figure 2.8 Curve with a negative skew.

can occur when there is a limit on how low scores can go and scores tend to bunch around that limit. For example, grades in a graduate-level statistics course cannot be higher than 100% (assuming no extra credit), and we would expect most students to score above 80% on a typical exam, but (hopefully) only a small number of students would score below 70%. Here we might anticipate ceiling effects as scores tend to bunch above 80% and produce a negative skew to the data.

Other Shapes

Other curve shapes are possible, but most of them are relatively rare. **Rectangular curves** occur when the frequency for each of the scores in the middle of the data set is the same. A **U-shaped curve** is the result of a high frequency of values at the extremes of the scale and a low frequency in the middle.

A **J-curve** results when frequency is high at one end of the scale, decreases rapidly, and then flattens and reduces to almost zero at the other end of the scale. This curve is different from a straight line, in which frequency decreases uniformly from one end of the scale to the other. J-curves can be positive or negative in direction, depending on the orientation of the lower tail of the curve. If the tail points to the positive end of the X-axis, the curve is positive. J-curves may also be inverted.

Summary

The purpose of organizing and displaying data is to produce order and parsimony from raw data. After you understand the principles explained in this chapter, you should use a computer to organize and display your data. Organizing and displaying data is the first step in a statistical analysis. Organizing data requires that we present them as a distribution of scores. A rank order distribution can be used when the sample size is less than or equal to 20. When the range is limited to 20 or fewer values, but the number of scores is large ($N > 20$), a simple frequency distribution is used. When both the range and the number of scores are large, the data should be arranged into a grouped frequency distribution.

A graph presents a visual picture of data and reveals trends that may not be obvious in a table. The curve of the graphed data is especially meaningful because it shows how the data are distributed. Many of the variables measured in kinesiology are normally distributed. The normal curve has special characteristics that allow us to make statements about the distribution of scores. Chapter 6 discusses the normal curve in more detail.

Problems to Solve

The following scores are from a 4 × 10-yard shuttle-run test for a group of middle school students. The scores are measured in seconds to the nearest tenth of a second.

Raw Data: 12.1 12.9 13.1 13.6 15.4 12.2 12.3 11.9 11.8 10.4 10.8 13.0 9.3 11.4
13.8 11.8 10.9 13.4 12.7 10.8 14.8 15.9 11.0 11.7 14.2 12.9 10.3 9.8 12.0 13.1
14.7 11.8 10.5 11.7 9.6 12.1 14.4 15.3 12.1 10.3 11.2 11.4 10.1 11.1 13.4 12.6
9.1 13.0 12.6 12.7 11.9 10.4 11.6 12.5 13.2 13.3 11.5 10.6 12.3 12.9 13.5 12.8
11.6 12.8 12.5 12.0 12.4 14.6 12.8 12.4 13.6 11.7

1. Determine the highest and lowest scores, and calculate the range.
2. Determine an appropriate interval size (i) for grouping data. *Hint:* Remember that you are working with data accurate to the nearest tenth of a second.
3. Establish the groups and tally the number of scores per group. Develop a frequency column and a cumulative frequency column. Set the lower limit of the lowest group as a multiple of 0.5. What is N?
4. What are the real limits of group 12.0 to 12.4?
5. Draw a histogram from the data.
6. Draw a frequency polygon from the data. Are the data normal or skewed?
7. What is likely to happen to the frequency polygon if more cases are added to the data?
8. Draw a cumulative frequency curve from the data.
9. Using a statistical computer program, enter the above data in a single column and produce a histogram and a frequency polygon.

See appendix C for answers to problems. *Note:* Ask your instructor for help with statistical computer software. Some programs you may consider using are Microsoft Excel, Statistical Package for the Social Sciences (**SPSS**), or the Statistical Analysis

System (**SAS**). Other programs may be available in computer labs on your campus. If you use Excel, make sure the Analysis Toolpak is installed as an add in.

Key Words

apparent limit
bimodal
ceiling effect
central limit theorem
cumulative frequency graph
curve
descriptive statistics
Excel
floor effect
frequency polygon
graph
grouped frequency distribution
histogram
interval size
J-curve
leptokurtic

mesokurtic
mode
multivariate
normal curve
platykurtic
range
rank order distribution
real limit
rectangular curve
SAS
simple frequency distribution
skewed
spreadsheet
SPSS
U-shaped curve

$$H = \left[\frac{12}{18(18+1)} \right] \left[\frac{82.5^2}{6} + \frac{54.5^2}{6} + \frac{34.0^2}{6} \right] - 3(18+1)$$

$$H = 1.035 | 1822.091 - 57$$

$$H = 6.77$$

Percentiles

When basketball announcers report the performances of various players or teams, they usually use percentages rather than raw data. We see the shooting percentages of certain players compared on free throws, two-point shots, and three-point shots. Percentages are used because everyone understands them and they allow immediate comparisons of two or more players.

Raw data would make it difficult to tell who is a better free throw shooter: player A, who made 46 of 62 attempts, or player B, who made 56 of 78. But when the data are converted to percentages (74% for player A, 72% for player B), comparisons among players can easily be made.

Percentiles are probably the most common statistical tool in use today. Educators and scientists find percentiles helpful in interpreting data for the lay public. This chapter presents methods of calculating percentiles by hand. Although researchers use computers to make calculations on more complicated data or on databases with large numbers of scores, it's best to learn the methods using simple data and paper and pencil.

Cent is the Latin root word for hundred. The word *percent* means by the hundred. When we use percents, we are comparing a raw score with a conceptual scale of 100 points. Percents provide a quick reference for any raw score in relation to the rest of the scores. A percentile represents the fraction (in hundredths) of the ordered scores that are equal to or fall below a given raw score. In statistics, a **percentile** is defined as a point or position on a continuous scale of 100 theoretical divisions such that a certain fraction of the population of raw scores lies at or below that point.

A score at the 75th percentile is equal to or surpasses three-fourths of the other scores in the raw data set. The 33rd percentile is equal to or better than about one-third of the scores but is surpassed by two-thirds of the scores. A student's percentile score on a test indicates how the student's score compares with a perfect score of 100%.

Percentiles are standard scores. A **standard score** is a score that is derived from raw data and has a known basis for comparison. In percents, the center is 50% and the range is 0 to 100%. A middle-aged male may be able to consume 40 milliliters of oxygen per kilogram of body weight per minute. This is the raw score. Without more information, the score is difficult to evaluate. But if we compare this value with other values in the population of all males of like age and calculate a percentile score of 65, then we know that the man's oxygen consumption is equal to or better than that of 65% of people in that population. Raw scores are measured values. Standard scores, which are derived from raw scores, provide more information than do raw scores.

Standard scores allow us to (a) evaluate raw scores and (b) compare two sets of data that are based on different units of measurement. Which is the better score, 150 feet on the softball throw or 30 sit-ups in a minute? With only this information, it is impossible to tell. We must ask, What was the range? What was the average?

But when we compare a percentile score of 57 on the softball throw with a percentile score of 35 on the sit-up test, the answer is clear. Not only are the relative values of the scores apparent, but we also know the middle score (50) and the range (0–100). The student who received these scores is better on the softball throw than on sit-ups when compared with the other students in the class.

The conversion of raw scores to standard scores, of which percentiles are just one example, is a common technique in statistics. It is most useful in evaluating data. (Chapter 6 discusses other standard scores.)

Percentiles may present a problem of interpretation when we consider the extreme ends of the scale for data sets with large numbers of scores. The problem results from imposing a percentile scale on interval or ratio data.

For example, figure 3.1 shows a graph of the number of curl-ups performed in 1 minute by 1,000 seventh-grade boys. Both the number of curl-ups (raw scores) and the percentile divisions have been plotted. Note that a score of 30 curl-ups is equal to the 50th percentile.

If a boy scores 30 curl-ups on the first test and increases his raw score to 35 on a second trial, his percentile score would increase from 50 to about 75, or 25 percentage points. However, if he was already at the upper end of the scale and made the same 5 curl-up improvement from 45 to 50, his percentile would increase only about three points, from 97 to near 100. This phenomenon is an example of the ceiling effect.

Teachers and coaches must account for this when they grade on improvement. It is much easier to reduce the time it takes to run the 100-meter dash from 12.0 seconds to 11.5 seconds than it is to reduce it from 10.5 seconds to 10.0 seconds. An improvement of 0.5 seconds by a runner at the higher level of performance (i.e., lower-time scores) represents more effort and achievement than does the same time improvement in a runner at the middle of the scale.

Learning curves typically start slow, accelerate in the middle, and plateau as we approach high-level performance. The ceiling effect is demonstrated by the plateau

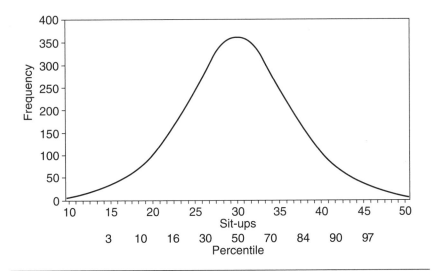

Figure 3.1 Curl-ups per minute ($N = 1,000$).

in the curve. It is more difficult to improve at the top of the learning curve than at the beginning or in the middle. This difficulty in interpretation should not deter us from using percentiles, but we must recognize the ceiling effect when considering scores that represent high levels of performance, particularly if improvement is the basis for evaluation.

Common Percentile Divisions

All measurements contain some error. When the error is relatively large, the subject's true score is only approximated. In these cases, it would be inappropriate to report an exact percentile rank because the raw score is only an estimate of the true score. Reporting a range of scores or a range of percentiles within which the subject's true score probably lies is a common practice.

The two most common ranges of percentile scores are the **quartile** range and the **decile** range. Occasionally, scores are reported in **quintiles.** In the quartile range, the percentile scale is divided into four equal parts, or quartiles. The first quartile extends from zero to the 25th percentile (Q_1). The second quartile ranges from Q_1 to the 50th percentile (Q_2). The third quartile extends from Q_2 to the 75th percentile (Q_3). The fourth, or highest, quartile reaches from Q_3 to the 100th percentile (Q_4). In the quintile range, five divisions of the percentile scale are made.

Deciles follow the same format except the scale is divided into 10 parts, each of which has a range of 10 points. The 1st decile ranges from 0 to D_1, the 2nd from D_1 to D_2, and so forth; the 10th decile extends from D_9 to D_{10} (the 100th percentile). Figure 3.2 shows the quartile and decile divisions of the percentile scale.

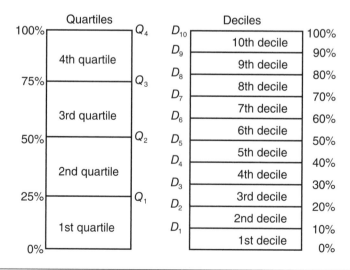

Figure 3.2 Relationships among percentile, quartiles, and deciles.

Sometimes test scores are reported in quartile or decile ranks. If a student ranks in the 3rd quartile, his or her score lies somewhere between the 50th and the 75th percentiles. Likewise, a decile rank of 4 means the score lies between the 30th and the 40th percentiles. This type of reporting is used if the teacher or scientist is aware of errors in testing or if the test is such that only an estimate of the true score can be obtained. Test batteries that report scores on several aspects of a student's ability, so as to form a profile, or total picture, of the student's talents, are commonly reported in quartile or decile ranks.

Calculations Using Percentiles

Converting raw scores into percentiles or determining the raw score that represents a given percentile can be performed using rank order data. We also need to be able to apply these processes to data that have been organized into simple and grouped frequency distributions. In this section we first discuss how to do these processes with rank order distributions. Then we apply the concepts to simple and grouped frequency distributions.

Rank Order Distributions

A basketball coach had 15 players on a high school team complete a free throw test. Each player took 10 shots from the free throw line, and the scores were recorded. Table 3.1 presents the data in a rank order distribution. With scores such as these in a group this small, percentile scores may not be meaningful. But the example is presented here to demonstrate the basic concept of percentile calculation.

Determining the Percentile From the Score

What is the percentile rank for a raw score of 6 baskets? Remember, a percentile is a point on a continuous scale of 100 theoretical divisions such that a certain fraction of the population of raw scores lies at or below that point. To calculate the percentile for a score of 6 baskets, we need to determine what fraction of scores falls at or below 6 baskets.

Percentiles are not based on the value of the individual scores but on the order of the scores. In table 3.1, the value of the bottom score could be changed to 2 without affecting the percentile divisions because it would still be in the same order in the group. The question we must ask is this: How many scores fall at or below 6 in the ordered list of scores?

Counting from the score of 6 down, we note that nine scores fall at or below 6. There are 15 scores, so the person who scored 6 baskets is 9/15 of the way from the bottom of the scale to the top. This fraction is first converted into a decimal (9/15 = .60), and then the decimal is multiplied by 100 to convert it into a percent (.60 × 100 = 60%). Thus, a score of 6 on this test falls 60% of the way from the bottom of the scale to the top and is classified as the 60th percentile.

TABLE 3.1

Rank Order Distribution: Basketball Free Throws Completed in 10 Attempts

X
10
9
9
7
7
7
6
5
5
5
4
4
3
3
1

$N = 15$

Any single score could be converted in a similar manner, but several scores were obtained by more than one person. For example, three persons each received a score of 5. What is their percentile rank? Using the method described previously, we could calculate that the first score of 5, the sixth score from the bottom, represents the 40th percentile ($6/15 = .40$, $.40 \times 100 = 40\%$). But the top score of 5 is eight scores from the bottom and represents approximately the 53rd percentile ($8/15 \times 100 = 53.3\%$).

Do we then conclude that the persons who each made five baskets scored between the 40th and the 53rd percentile ranks? No. Because we define percentile as a fraction at or below a given score, we conclude that all are equal to or below a score of 5. And because they all performed equally, they should all receive the same percentile as a fraction of scores at or below a given score. Therefore, all three have a percentile score of 53.3.

Determining the Score From the Percentile

The most common computation in working with percentiles is calculating the percentile from the raw score, but sometimes the opposite calculation is required. If a coach determines that the 60th percentile is the cutoff point on a test for selecting athletes for a varsity team, the coach needs to know what raw score is represented by the 60th percentile.

Frequently, grades or class divisions are made on the basis of percentiles. A class may be divided into thirds, with the top third receiving one practice schedule, the middle third another, and the bottom third still another. To determine which students should be included in each group, we must establish the raw scores equivalent to the percentile points. The technique of determining the raw score that matches a given percentile point is the opposite of that for finding the percentile from the raw score.

For example, the basketball coach who collected the free throw data (see table 3.1) can take the top two-thirds of the team to away games. If the coach uses the free throw score as the criterion for determining who makes the traveling squad, how many baskets must a player make to qualify?

The coach must ask, "Which score defines the bottom third of the team?" All players with scores above that point are in the top two-thirds and qualify for the traveling team. This is a simple problem with 15 players because it is easy to determine that the top 10 players represent the upper two-thirds of the team. But by thinking through the process with simple data we learn the concepts that may be applied to more difficult data.

The percentile equivalent to 1/3 is found by converting the fraction to a decimal (1/3 = .333) and multiplying this decimal equivalent by the total number of subjects in the group (.333 × 15 = 5). This value (5) is the number of scores from the bottom, not the raw score value of 5 baskets made.

To determine the raw score that is equivalent to a given percentile (P), convert the percentile to a decimal, multiply the decimal equivalent by the number of scores in the group (N), and count that many scores from the bottom up. Counting up five scores from the bottom, we find that a score of 4 free throws represents a percentile score of 33.3. Any player who made four or fewer free throws is included in the bottom third of the group, and any player scoring more than four free throws is included in the top two-thirds of the group. So five free throws is the criterion for making the traveling squad.

With small values of N and discrete data, it may be necessary to find the score closest to a given percentile if none of the raw scores falls exactly at that point. If the product of $P \times N$ is not an integer, round off to the nearest integer to determine the count from the bottom. In table 3.1, the 50th percentile is 7.5 scores from the bottom (0.50 × 15 = 7.5), but there are no half scores. So we round 7.5 to 8 and count 8 scores from the bottom. The eighth score up is 5. Therefore, 5 is the closest score to the 50th percentile.

Simple Frequency Distributions

When more scores are available but the range is still small ($N \geq 20$ and $R \leq 20$), the scores are usually grouped into a simple frequency distribution. Suppose the basketball coach in the previous example gave the free throw test to 60 students in a physical education class. The data are presented in a simple frequency distribution in table 3.2.

Determining the Percentile From the Score

What is the percentile rank of a student who made seven baskets? To answer this question, we need to compute the fraction of scores that are equal to or less

TABLE 3.2

Simple Frequency Distribution:
Basketball Free Throws Completed in 10 Attempts

X	f	Cum. f
10	2	60
9	6	58
8	9	52
7	12	43
6	15	31
5	8	16
4	4	8
3	2	4
2	1	2
1	1	1
	N = 60	

than 7. We do so by adding the numbers in the frequency column (*f*) from the score (*X*) of 7 down to the bottom.

If several percentile calculations are to be performed, it is helpful to create a cumulative frequency column. The cumulative frequency column in table 3.2 indicates that 43 people made seven or fewer baskets. Sixty persons took the test, so 43/60 of the students made seven or fewer baskets. Converted to decimals, 43/60 = .716, and .716 × 100 = 71.6. Therefore, a person who scored seven baskets ranks equal to or better than about 72% of those who took the test.

Determining the Score From the Percentile

In table 3.2, what score is equivalent to the 75th percentile? The concept of determining the score equivalent to a given percentile from a simple frequency distribution is the same as it was for a rank order distribution. First we establish the number of free throws at or below which 75% of the group scored. This value is determined as follows: .75 × 60 = 45 students.

The cumulative frequency column indicates that 43 students scored seven or fewer, and 52 students scored eight or fewer. The nine students who each scored eight baskets represent the 44th through the 52nd students in the ordered distribution. Therefore, the 45th student is one of these nine. We cannot separate the nine students (they are all equal in ability), so the 75th percentile is determined to be eight baskets. Because nine students tied at eight baskets, the bottom three-fourths of the group is approximated to be those who scored eight or fewer, and the top fourth of the group is approximated to be those who scored nine or more.

If the product of the percentile (*P*) in decimal value times *N* is not an integer, we would round the product to the nearest integer and find that value in the cumulative frequency column. For example, if the 14th percentile is desired ($.14 \times 60 = 8.4$), we would find where 8 falls in the cumulative frequency column, then read 4 as the number of baskets closest to the 14th percentile.

Grouped Frequency Distributions

Data accumulated from a test in which a wide range of scores are available ($R > 20$) and many persons participated ($N > 20$) may be presented in a grouped frequency distribution. As an example, we consider the results of a test of softball throw for distance given to 115 students. Table 3.3 presents the raw data in a grouped frequency distribution.

Determining the Percentile From the Score

What is the percentile rank of a student who threw the softball 146 feet? This problem is approached in the same manner as the simple frequency distribution problem except that, because the data are grouped, we do not know where within an interval a given score lies.

Table 3.3 indicates that seven people threw the ball between 140 and 149 feet. We do not know how many of the seven people threw farther than 146 feet and how many threw less, so we must assume that all seven people are equally distributed

TABLE 3.3

Grouped Frequency Distribution: Softball Throw for Distance, in Feet

X	f	Cum. f
220–229	2	115
210–219	5	113
200–209	7	108
190–199	7	101
180–189	9	94
170–179	12	85
160–169	10	73
150–159	15	63
140–149	7	48
130–139	9	41
120–129	10	32
110–119	7	22
100–109	6	15
90–99	5	9
80-89	4	4
	$N = 115$	

between the real limits of the interval (139.50–149.49). Real limits are used so that percentiles for both discrete and continuous data may be calculated.

This problem is now easily solved by applying the following equation:

$$P = \frac{\left(\dfrac{X-L}{i}\right)f + C}{N},$$
(3.01)

where P is percentile, X is raw score, L is lower real limit of interval in which raw score falls, i is size of the interval, f is frequency of the interval in which the raw score falls, C is cumulative frequency of the interval immediately below the one in which the raw score falls, and N is total number of cases.

When equation 3.01 is applied to the raw score of 146, the following results are obtained:

$$P = \frac{\left(\dfrac{146 - 139.5}{10}\right)7 + 41}{115} = .396 \ or \ 39.6\%.$$

Therefore, a throw of 146 feet is approximately equal to the 40th percentile.

Determining the Score From the Percentile

Using the data in table 3.3, the teacher may desire to determine the middle score, or 50th percentile. How far would a student have to throw the softball to be considered in the top half of the class? To solve this problem, we must first multiply $P \times N$ to determine the interval in which the score falls. With 115 students in the class, the score representing the 50th percentile falls exactly halfway up the cumulative frequency column (.50 × 115 = 57.5). This places the middle score somewhere in the interval of 150 to 159.

Equation 3.01 may be solved algebraically for X, to obtain the following equation:

$$X = \left(\frac{PN - C}{f}\right)i + L.$$
(3.02)

The symbolic notation is the same as in equation 3.01. The problem previously discussed is now easily solved:

$$X = \left(\frac{(.5 \times 115) - 48}{15}\right)10 + 149.5 = 155.83 \ \text{feet}$$

The 50th percentile is approximated by a throw of 156 feet.

Summary

Converting raw scores to standard scores is a common technique in statistics. Standard scores let us evaluate raw scores and compare two sets of data that are based on different units of measurement. Because of this, standard scores provide information about data that is not available from raw scores alone. The most common standard scores, percentile scores, are frequently used in kinesiology. Percentile scores can be determined from rank order data, simple frequency data, and grouped data. Percentiles can also easily be converted into raw scores.

Problems to Solve

A test of tennis serve accuracy resulted in the following data: *Note:* Higher scores indicate better performance.

X
15
12
12
10
9
8
8
7
5
4
4

1. What is the percentile rank of a person who scored (a) 7; (b) 10?

2. What is the nearest score to (a) the 70th percentile; (b) the 45th percentile?

The same test of tennis serve accuracy was administered to three college classes. The scores were arranged into a simple frequency distribution:

X	f	Cum. f
18	1	70
17	2	69
16	3	67
15	3	64
14	5	61
13	5	56
12	6	51
11	8	45
10	10	37
9	7	27

(continued)

X	f	Cum. f
8	8	20
7	5	12
6	3	7
5	3	4
4	1	1
	$N = 70$	

3. What is the percentile of a person who scored (a) 16; (b) 5?

4. Which score most closely represents the (a) 47th percentile; (b) the 80th percentile?

A teacher tested her students to find out how many sit-ups they could do in 1 minute. She arranged the scores into a grouped frequency distribution:

X	f	Cum. f
70–74	2	83
65–69	4	81
60–64	7	77
55–59	9	70
50–54	10	61
45–49	12	51
40–44	13	39
35–39	11	26
30–34	7	15
25–29	5	8
20–24	3	3
	$N = 83$	

5. What are the percentile ranks of a student who scored (a) 32; (b) 39; (c) 55?

6. What scores best represent (a) Q_1; (b) Q_2; (c) Q_3?

7. Using the tennis serve data with $N = 70$ and statistical computer software of your choice, calculate the answers to problems 3 and 4 using a computer. How does the computer printout compare with your hand-calculated answers? *Hint:* Enter the data into the computer in a single column. Enter each data point as many times as is indicated by the frequency column. For example, enter 18 once, 17 twice, 16 three times, and so on.

See appendix C for answers to problems.

Key Words

decile
percentile
quartile

quintile
standard score

$$H = \left[\frac{12}{18(18+1)} \right]\left[\frac{82.5^2}{6} + \frac{54.5^2}{6} + \frac{34.0^2}{6} \right] - 3(18+1)$$

$$H = [.035][(1822.09] - 57$$

$$H = 6.77$$

Measures of Central Tendency

In the Olympics, the public is typically interested only in the scores of the athletes who placed first, second, and third in each event. However, scientists who study athletic performance want to (1) know the relative value of scores for athletes who do not place in the top three, (2) discover trends in the data over the years, and (3) compare modern with ancient performances. These scientists need to know the mean (average) score, the median (50th percentile or most typical score), and the mode (most common score) for all athletes in all events. When these three measures of central tendency are known, it is easier to interpret the value of a single score by comparing it to the mean, median, or mode.

Measures of **central tendency** are values that describe the middle, or central, characteristics of a set of data. Some terms that may be used to describe the center of a group of scores are the most common score (mode), the typical score (median), and the average score (mean). These measures provide important information that allows us to calculate the relationship of a given score to the middle scores of a data set.

Mode

The mode is the score that occurs most frequently. There is no formula to calculate it; it is found by inspection or by a computer program. In a rank order listing of scores, the mode can be determined by scanning the list of all the scores to determine which is the most frequent. In a simple or grouped frequency distribution, the score with the highest value in the frequency column is the mode.

In a simple frequency distribution, the mode is the score with the highest frequency, but in a grouped frequency distribution, the mode is considered to be the midpoint of the group with the greatest frequency. A distribution can, of course, have more than one mode. If two or more scores or groups have the same frequency, then the variable is said to be bi- or multimodal.

The ease with which it can be determined is one advantage of the mode. It gives a quick estimate of the center of the group, and when the distribution is normal or nearly normal, this estimate is a fair description of the central tendency of the data.

The mode also has some disadvantages:

- It is unstable; it may change depending on the methods used for grouping.
- It is a **terminal statistic;** that is, it does not give information that can be used for further calculations.
- It completely disregards the extreme scores—it does not reflect how many there are, their values, or how far they are from the center of the group.

Median

The **median** is the score associated with the 50th percentile. It can be determined by using the methods of calculating percentiles that were presented in chapter 3. In this chapter, we discuss the median's importance as a measure of central tendency.

The median is the middle score in that it occurs in the middle of the list of scores; it divides the data set in half. The median is also the typical score because it is the single score that best represents the majority of the other values.

In a rank order distribution, when N is odd, the median is the middle score in the range. When N is even, the median falls between two scores. Consider these two examples:

A	B
9	19
8	18
6 ← median	17
4	← median
1	16
	13
	12

In example A, the median is 6, the middle score. In example B, no single score represents the exact middle. When N is even, and the median falls between two scores, the higher score is usually designated as the median. However, computer programs will often calculate the average of the two middle scores as the median. In example B, choosing the higher score would yield a median of 17, and taking the average of the two middle scores gives a median of 16.5.

The calculation of the median does not take into consideration the value of any of the scores. It is based only on the number of scores and their rank order. For this reason, the median is appropriately used on ordinal data and on data that are highly skewed.

An extreme score that is radically different from the other scores in the data set does not affect the median. Consider the following example. Nine people each perform 40 sit-ups, and one does 100. The median score for the group is 40, but the average (mean) is 46. The median would still be 40 even if the highest score was 200. It is easy to see why the median is called typical of the group. The median is more representative of the majority of scores than is the average when radically extreme scores exist.

That the median is not affected by the size of extreme scores is both an advantage and a disadvantage. Because the median does not consider the size of scores but only how many there are, it neglects some important information provided by the data—namely, the value of the extreme scores.

Mean

The **mean** is the statistical name for the arithmetic average. It is represented by a point on a continuous scale and may be expressed as an integer or a decimal value. The mean is the most commonly used measure of central tendency. Its calculation considers both the number of scores and their values.

Because of this feature, the mean is the most sensitive of all the central measures. Slight changes in any of the scores in the group may not affect the mode or the median, but they always change the mean. The chief advantage of the mean is that it considers all information about the data and provides a basis for many additional calculations that will yield still more information.

These characteristics make it the most appropriate measure of central tendency for ratio and interval data. Many argue that it is not proper to calculate the mean on ordinal data because distances between scores are not known in ordinal data. (The median would be more fitting in this case). Calculation of the mean requires more information than is provided by ordinal data, whereas the median considers exactly the type of information given by ordinal data—order of scores, but not relative distance between scores.)

The mean's sensitivity may also be a disadvantage. When one or more scores, called **outliers,** are considerably higher or lower than the other scores, the mean is pulled toward that extreme. The previous example of the average number of sit-ups by 10 subjects illustrates this phenomenon.

The precision of the mean should not exceed the precision of the data by more than one decimal place. For example, if distance is measured in meters, it is appropriate to calculate the mean to the nearest tenth of a meter, or decimeter, but not to the nearest thousandth of a meter, or millimeter. Convention permits the precision of the mean to be one significant figure beyond the accuracy of the data.

Calculating the Mean

The mean is computed simply by summing all the scores and dividing the sum by the number of scores (N). In statistics, summing scores is represented by the Greek uppercase sigma (Σ). This symbol is read as "the sum of." For example, if a variable is labeled X, then ΣX should be read "the sum of X."

To denote the mean of a variable, we place a line over the symbol for the variable. In equation 4.01, X is the symbol for the variable, so the mean of X is represented by $\overline{X}$, often read as X-bar. If Y represents the variable, $\overline{Y}$, or Y-bar, represents the mean of Y.

The formula for the mean of the variable X is

$$\overline{X} = \frac{\Sigma X}{N}. \tag{4.01}$$

From a Rank Order Distribution

Computing the mean or average of a small set of scores is easy: add all the scores and divide the sum by the number of scores.

X
33.9
31.2
29.6
27.0
24.7
25.0
23.0
$\Sigma X = 194.4.$

$$\bar{X} = \frac{\Sigma X}{N} = 194.4 / 7 = 27.77.$$

The value the calculator returns for 194.4/7 is 27.77142857. But the data are significant only to the nearest tenth, so the mean value is calculated to the nearest hundredth. Rounding the mean to one significant figure beyond the data provides users of the data with the information that the mean is closer to 27.8 than to 27.7, so we can say that $\bar{X} = \sim 27.77$ (the mean of X is approximately 27.77).

Although a researcher should know the techniques for calculating the mean by hand, the process is often tedious and time consuming. Computers perform these functions much more rapidly and with greater accuracy. Computers can also remember large amounts of information, so it is not necessary for a computer to group the data. Because computers are so commonly used to calculate the mean, this chapter will not demonstrate the hand calculation methods for simple and grouped frequency distributions. When one knows how to do the calculations by hand on simple numbers for small values of N, and understands the meaning of the answers, then the computer can add speed, accuracy, and convenience.

Relationships Among
the Mode, Median, and Mean

When the data are distributed normally, the three measures of central tendency all fall at or near the same value. But when the data are skewed, these measures are no longer identical. Figure 4.1 demonstrates the relationships among the three measures on a positively skewed distribution.

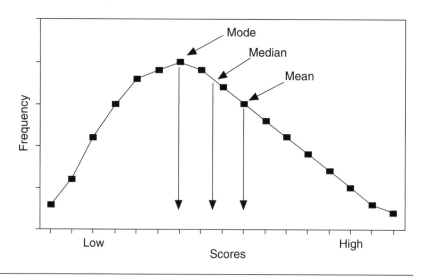

Figure 4.1 Relationships of measures of central tendency on a positively skewed curve.

Note that the highest scores are farther from the mode than are the lowest scores. This characteristic of positively skewed curves shifts the median and the mean to the right of the mode. On a positively skewed curve the three measures of central tendency read from left to right in the following order: mode, median, mean. On a negatively skewed curve the order is reversed: mean, median, mode.

When deciding which of the three measures to use, consider the following:

- Use the mode if only a rough estimate of the central tendency is needed and the data are normal or nearly so.

- Use the median if (a) the data are on an ordinal scale, (b) the middle score of the group is needed, (c) the most typical score is needed, or (d) the curve is badly skewed by extreme scores.

- Use the mean if (a) the curve is near normal and the data are of the interval or ratio type, (b) all available information from the data is to be considered (i.e., the order of the scores as well as their relative values), or (c) further calculations, such as standard deviations or standard scores, are to be made.

Summary

Measures of central tendency are values that describe the central characteristics of a set of data. It is essential to know the central tendency of a set of scores in order to evaluate any of the raw scores in the set. The three measures of central tendency are the mode, the median, and the mean. The mode provides a quick estimate of central tendency for all types of data. The median can be used on ordinal data or when the data are badly skewed. The mean is the preferred measure of central tendency for interval or ratio data that are normally or almost normally distributed. When the data are normally distributed, the three measures of central tendency all fall at or near the same value.

Problems to Solve

1. Calculate the mode, median, and mean for each of the following sets of data.

a.	X		b.	X	f
	17			17	2
	15			16	4
	13			15	5
	12			14	7
	12			13	7
	11			12	8
	11			11	10
	9			10	9
	8			9	5
	8			8	3
	8				
	5				
	4				
	2				
	2				

2. Enter the data from problems 1a and 1b into a statistical software program and calculate the mode, median, and mean. *Hint:* For problem 1b, enter the data into a single column. Enter each data point as many times as is indicated by the frequency (*f*) column. For example, enter 17 twice, 16 four times, 15 five times, and so on.

See appendix C for answers to the problems.

Key Words

central tendency

mean

median

outlier

terminal statistic

$$H = \left[\frac{12}{18(18+1)} \right] \left[\frac{82.5^2}{6} + \frac{54.5^2}{6} + \frac{34.0^2}{6} \right] - 3(18+1)$$

$H = (.035)(1622.09) - 57$

$H = 6.77$

Measures of Variability

onsistency in performance is critical in many sports. For example, in golf, it is important to know how far you hit with each wood or iron so that you can determine which club to use when approaching the green. If the distance you hit with a given iron varies each time you hit, you cannot tell which club to use.

Consider the hypothetical scores of two golfers, one a beginner and one a professional. Each hits 25 balls toward the pin with a 9-iron. The pro is very consistent; all of her balls lie close to and equally spread out around the pin. The beginner is inconsistent; some of his shots land close to the pin, and others land far to either side or far beyond or short of the pin. If we measured the distance of every ball from the pin (consider shots that fell short or left to be negative values and those that went long or right to be positive), then the average, or mean, of each set of values could be near zero, because negative left balances positive right, and negative short balances positive long. If only the average distance of the balls from the pin for both players were evaluated, we might erroneously conclude that both players were equal in ability because each average distance from the pin would be near zero. But when the variability of their performances is considered, the pro is clearly a much more consistent (and better) golfer than the beginner. She has less variability in her performance.

Variability is a measure of the spread, or dispersion, of a set of data. Once the mean, median, and mode have been calculated, the central tendencies of the raw data are known. However, data can be either compact around the central measure or spread out. Leptokurtic curves (see chapter 2) typically represent limited variability, whereas platykurtic curves represent large variability. Consequently, it is important to know the variability, or spread, of the group as well as its central tendency.

When both central tendency and variability are known, two or more sets of data can be completely compared. It is often interesting to compare the means of two or more sets of data to determine which set has a higher average, but it may be more important to compare the variability of the data because the variabilities may differ while the means can be similar.

Most statisticians recognize four main measures of variability: range, interquartile range, variance, and standard deviation. Of these four, the variance and the standard deviation are the most important because they lead to additional calculations that may answer further questions about the data. The applicability of the range and interquartile range is limited, but they may serve as useful, easy-to-calculate rough estimates of variability.

Range

Recall from chapter 2 that the range is simply the difference between the highest and lowest scores in the data set. This is the best quick estimate we can make of the variability of data. It is easily calculated, but it is only a very rough measure. The range is somewhat analogous to the mode in that it is only an estimate of variability and is very unstable; it can change radically if extreme scores are introduced to the data set.

Interquartile Range

Another measure of variability often used when data are ordinal, or when ratio or interval data are highly skewed, is called the **interquartile range** (*IQR*). This range is the difference between the raw scores at the 75th (Q_3) and the 25th (Q_1) percentile points. It is calculated as follows:

$$IQR = Q_3 - Q_1. \tag{5.01}$$

The interquartile range is a useful measure of the spread of data if the investigator is more interested in the middle scores than in the extremes. Because the interquartile range considers only 50% of the data (the scores that fall in the middle half of the data set), it is not affected by highly divergent scores at the extremes. This is why it is useful for skewed data. Like the median, it presents a typical picture, but it does not consider all information available about the data.

An alternative measure of variability when using percentiles is the **semi-interquartile range** (*SIQR*), which is half of the interquartile range. The formula is $SIQR = (Q_3 - Q_1)/2$. This measure is used if a smaller indicator of the variability is desired.

Variance

Both the range and the interquartile range consider only two points of data in determining variability. They do not include the values of the scores between the high and low, or between the Q_1 and Q_3 data points. Another way to assess variability that does consider the values of each data point is to determine the distance of each raw score from the mean of the data. This distance is called deviation (*d*). The computation of deviation scores for two sets of data, each of which has a mean of 25, is demonstrated in table 5.1.

The sum of the deviations around the mean will always equal zero. This is true regardless of the size of the scores. This is one way to verify the accuracy of the mean. If the deviations do not sum to zero, the mean is incorrect. In table 5.1 the sum of all deviations around the mean equals zero in both the X and the Y examples; therefore, we can be assured that our calculations of the means are correct.

In the second example (Y), the deviation scores are larger and the range is larger ($R_X = 27 - 23 = 4$ and $R_Y = 35 - 15 = 20$). In fact, as a comparison of the ranges

TABLE 5.1

Deviations From the Mean

X	$(X - \bar{X})$	$(X - \bar{X})_{ABS}$	Y	$(Y - \bar{Y})$	$(Y - \bar{Y})_{ABS}$
27	+2	2	35	+10	10
26	+1	1	30	+5	5
25	0	0	25	0	0
24	−1	1	20	−5	5
23	−2	2	15	−10	10
$\Sigma = 125$	$\Sigma = 0$	$\Sigma = 6$	$\Sigma = 125$	$\Sigma = 0$	$\Sigma = 30$

$$\bar{X} = 125/5 = 25$$
$$\bar{Y} = 125/5 = 25$$

shows, the Y data are five times more variable than the X data. The interquartile range comparisons confirm this conclusion: $IQR_X = 26 - 24 = 2$ and $IQR_Y = 30 - 20 = 10$. The relationship between the range and the interquartile range is always consistent when the data from the two sets are normally distributed.

If we sum the absolute values of the deviations (ignoring the direction of the deviation) from the mean, we find that the total deviation of X is 6 and the total deviation of Y is 30, or five times more variable than X. However, using the sum of the absolute values to quantify variability is problematic in a couple ways. First, absolute values do not lend themselves to the proper algebraic manipulations that are required in subsequent chapters. Second, in these examples the signs of the deviations have meaning; they indicate whether the raw score is above or below the mean. Because we need to know this, we cannot ignore the signs without losing information about the data.

However, negative signs can be eliminated in another way. If we simply square the deviations, the squared values are all positive. Then we can calculate the average of the squared deviations. This process forms the basis for how variability is typically quantified statistically and is more useful than the range, interquartile range, and sum of absolute deviations. The **variance** is the average of the squared deviations from the mean. The symbol V is used for variance in this text; other texts may use the symbol S^2. Variance is represented in algebraic terms as follows:

$$V = \frac{\Sigma\left(X - \bar{X}\right)^2}{N}. \tag{5.02}$$

Table 5.2 shows how the variance is determined for the previous examples of X and Y. Note that this is a **sums of squares** type of calculation. That is, as part of the calculation, we add up squared values. This is common in many statistical calculations, and we revisit it in subsequent chapters.

TABLE 5.2

Calculation of Variance

X	$(X - \bar{X})$	$(X - \bar{X})^2$	Y	$(Y - \bar{Y})$	$(Y - \bar{Y})^2$
27	+2	4	35	+10	100
26	+1	1	30	+5	25
25	0	0	25	0	0
24	−1	1	20	−5	25
23	−2	4	15	−10	100
$\Sigma = 125$	$\Sigma = 0$	$\Sigma = 10$	$\Sigma = 125$	$\Sigma = 0$	$\Sigma = 250$

$$V_X = 10/5 = 2$$
$$V_Y = 250/5 = 50.$$

Standard Deviation

The calculation of variance shown in table 5.2 suggests that Y is 25 times more variable than X ($50/2 = 25$), whereas we previously concluded from the range and interquartile range that Y was only 5 times more variable than X. This discrepancy is the result of squaring the deviation scores. To bring the value for the variance in line with other measures of variability (and with the unit values of the original raw data), we compute the square root of the variance.

The resulting value is called the standard deviation because it is standardized with the unit values of the original raw data. The **standard deviation** is the square root of the average of the squared deviations from the mean (i.e., it is the square root of the variance). This definition applies to a population of scores; the standard deviation of a sample is discussed later in this chapter.

Definition Method of Hand Calculations

The standard deviation of a population is symbolized by the Greek lowercase sigma (σ) and is represented algebraically as follows:

$$\sigma = \sqrt{\frac{\Sigma \left(X - \bar{X} \right)^2}{N}}. \tag{5.03}$$

In the example from table 5.2, the standard deviation of X is

$$\sigma_X = \sqrt{\frac{10}{5}} = \sqrt{2} = 1.4.$$

and the standard deviation of Y is

$$\sigma_Y = \sqrt{\frac{250}{5}} = \sqrt{50} = 7.1.$$

These values are now consistent with the range and interquartile range because the standard deviation of Y is five times as large as the standard deviation of X ($1.4 \times 5 = \sim7.1$). This statistic gives an accurate and mathematically correct description of the variability of the group while considering each data point and its deviation from the mean. The standard deviation is very useful because many advanced statistical techniques are based on a comparison of the mean as the measure of central tendency and the standard deviation as the measure of variability.

The method just described for calculating the standard deviation is called the definition method; it is derived from the verbal definition of standard deviation. But this is a cumbersome and lengthy procedure, especially when N is large and the mean is not a whole number. Under these conditions a high probability of mathematical error exists during the calculation. When N is large, the use of a computer is necessary to save time and eliminate arithmetic errors.

Calculating Standard Deviation for a Sample

You will recall from chapter 1 that most research is performed on samples taken from populations. The sample is always smaller than the population and should be selected randomly so that the statistics calculated from the sample are representative of the corresponding parameters in the population. The researcher assumes that conclusions based on the sample are applicable to the population from which the sample was drawn.

Samples rarely contain the extreme values that are found in the population. For example, if we randomly sampled the weights of 100 men in a university with 15,000 male students, it is not likely that anyone in the sample would weigh 350 pounds (~160 kilograms) or 100 pounds (~45 kilograms), although some students in the population might weigh these amounts. The variability of the sample is never as large as the variability of the population.

When standard deviation is calculated from a sample and then used to estimate the standard deviation of the population, a correction factor must be applied to the equation so that the estimate of the population is not biased by a small sample. Without this correction factor, an estimate of the standard deviation of a population based on a sample would be erroneously small.

The equations presented previously in this chapter for calculating standard deviation are based on the assumption that an entire population has been measured. If these equations were applied to samples, an error when generalizing from a sample to a population would occur. The correction to the equations is based on the concept of degrees of freedom. The **degrees of freedom** are the number of values in a data set that are free to vary. But what does that mean?

If no restrictions are placed on the data, then all values are free to vary; that is, they may take on any value, limited only by the precision of the measuring instrument and the actual values in the population. But when we take a sample and make the assumption that the sample represents the population, a restriction is placed on the data in the sample.

When the sample mean is assumed to be identical to the population mean (i.e., the sample mean is established), the sum of the sample data is also set because the mean equals the sum divided by N. This limits the numerical value of the last data point in the sample to a number that will create a sample mean theoretically equal to the population mean (even though the population value is unknown).

Assume that the mean of four values must be 5.0. Therefore, the sum of the four numbers must equal 20. Let the values 2, 3, and 7 be the first three numbers. What must the fourth number equal to bring the sum to 20? It must be 8. The last number is not free to vary; it is limited to only one value, that which will create a sum of 20. Therefore, this example has 3 degrees of freedom. Three of the four numbers can assume any value, but the last number must be whatever it takes to make the sum equal 20. The degrees of freedom for a single data set that is a sample representing a population are always $N - 1$. In this example, degrees of freedom $= N - 1 = 3$.

This correction factor, when applied to the definition formula for standard deviation (equation 5.03) reduces the denominator by one. The standard deviation (*SD*) for a sample is

$$SD = \sqrt{\frac{\Sigma\left(X - \overline{X}\right)^2}{N - 1}}. \tag{5.04}$$

Note that the symbol for standard deviation in equation 5.04 is *SD* rather than σ. In this book, we use *SD* to represent the standard deviation for a sample and σ to represent the standard deviation for a population. In some statistics books, standard deviation may be represented by S (pop) or s (sample).

When we apply equation 5.04 to the X data in table 5.2, we get $SD = \sqrt{\frac{10}{4}} = 1.6$. The standard deviation calculated by equation 5.04 yields a slightly larger answer than does that calculated by equation 5.03 ($\sigma = 1.4$). The difference represents the correction for degrees of freedom. Similarly, the sample standard deviation for the Y data (compared with 7.1 from equation 5.03) is

$$\sqrt{\frac{250}{4}} = 7.9.$$

As N increases, the differences between the standard deviations obtained with the sample and the population formulas decrease. When $N \geq 50$, the differences are usually trivial. This demonstrates that samples with large N are more accurate in estimating population parameters than are samples with small N.

Coefficient of Variation

Sometimes it is helpful to compare the standard deviations from different variables. For example, we might ask if the variability in our one-repetition-maximum squat scores is similar to that in our vertical jump scores. However, the two measures are in different units (kilograms for the squat and centimeters for the vertical jump). We can normalize our standard deviations by dividing the standard deviation by the mean and then multiplying by 100 to convert it to a percentage. This is called the coefficient of variation (*CoV*). The equation for the coefficient of variation is

$$CoV = \frac{SD_X}{\overline{X}}(100). \qquad (5.05)$$

For the data in table 5.2, the coefficient of variation for X (using the sample standard deviation formula) is $(1.6/25) \times 100 = 6.4\%$. Therefore, the standard deviation for the X data is equal to about 6.4% of the mean. For the Y data, $CoV = (7.9/25) \times 100 = 31.6\%$. From these normalized scores, we can see that the X data is relatively less variable than the Y data. See page 65 for calculation of SD for X and Y.

Standard Deviation and Normal Distribution

When N is large and the distribution is close to normal, there are usually 5 or 6 standard deviations within the range of a data set. That is, there are about 3 standard deviations from the mean to the largest score and from the mean to the smallest score (see figure 5.1).

If N is small or the data are skewed, this rule does not hold. When N is large and the range contains either fewer or more than about 5 or 6 standard deviation, the researcher would be wise to check the calculations for error. In the next chapter, we examine the relationship between the standard deviation and the normal distribution. In particular, we will see that we can quantify the area of specific regions of a normal distribution using standard deviations.

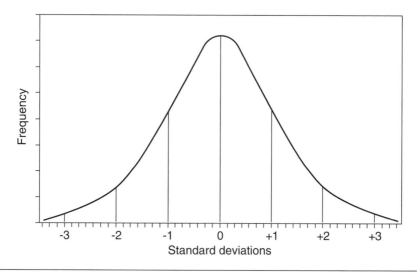

Figure 5.1 Distribution of standard deviations on a normal data set with large N.

Summary

A variety of indices are used to quantify the variability in a set of scores. The variance and the standard deviation are the most useful for data analysts. The standard deviation is the most common method of measuring the variability, or dispersion, of a data set. It considers the deviation of each data point and its distance from the mean. When the standard deviation is small, the group is compact. When it is large, there is more diversity, or spread, among the scores in the group. When we compare several different sets of data using the standard deviations of each, we can determine the relative spread of the different sets using the coefficient of variation.

Because standard deviation and variance are commonly used concepts in statistics, you should be familiar with both their calculation and their meaning. In later chapters, important techniques for determining relationships among variables (correlation) and for comparing differences between group means (t tests and analysis of variance) are discussed. Variance and standard deviation are critical factors in these techniques.

Problems to Solve

In an exercise physiology laboratory, a student collected data on the sum of skin-fold thicknesses in millimeters at the triceps and subscapular sites. The following measurements were obtained: 17, 19, 12, 24, 26, 18, 15, 14, 20.

1. What is the standard deviation if the data represent a population?
2. What is the standard deviation if the data represent a sample?
3. What is the variance if the data represent a population?
4. Assume the following set of data on basketball shots completed in 1 minute represents a sample from a larger population. Calculate N and the mean, median, mode, and standard deviation (SD) using a computer.

X	f
29	3
27	2
26	4
25	10
24	13
23	7
22	7
21	2
20	1

Remember to enter the data into the computer in a single column (e.g., 29, 29, 29, 27, 27, 26, 26, 26, 26, etc.).

See appendix C for answers to problems.

Key Words

degrees of freedom

interquartile range

semi-interquartile range

standard deviation

sum of squares

variability

variance

$$H = \left[\frac{12}{18(18+1)} \right] \left[\frac{82.5^2}{6} + \frac{54.5^2}{6} + \frac{34.0^2}{6} \right] - 3(18+1)$$

X_4

X_1

$H = [.035][1822.09] = 57$

$H = 6.77$

X_3

X_2

The Normal Curve

n the Olympics, all performances (even the last-place finishers) are superior to those of non-Olympic athletes. If we graded scores in a high school physical education class by comparing them with Olympic performances, all the students would fail. How good is a 4-meter long jump in a high school class? Without additional information, we can't be sure. If we know that the average of all long jumpers in the class is 5 meters, then we know that this score is below the mean, but how far below? To evaluate a single raw score, we must compare it with a scale that has known central tendency and variability. Scores from such scales are called *standard scores*. Standard scores provide information that helps us evaluate a given raw score.

When studying the concept of variability, especially while learning about standard deviation, students often ask, What does standard deviation mean? They know how to calculate it, but they do not understand its meaning or its value. The answer lies in understanding the unique characteristics of the normal curve. Further, the characteristics of the normal curve are central in our study of inferential statistics in subsequent chapters.

Observed over a sufficient number of cases, many variables assume a normal distribution. This is fortunate for the statistician because the characteristics of the normal curve are well known. If the data are from an interval or a ratio scale, if enough cases have been measured, and if the curve is normal or nearly normal, the characteristics of normality may be applied to the data.

Figure 6.1 shows a frequency polygon of fabricated data simulating the weights of a population ($N = 15,000$) of college-age males. The mean weight is 175 pounds,

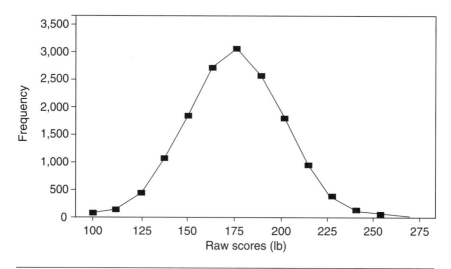

Figure 6.1   Frequency polygon of population distribution of college-age males' weight in pounds.

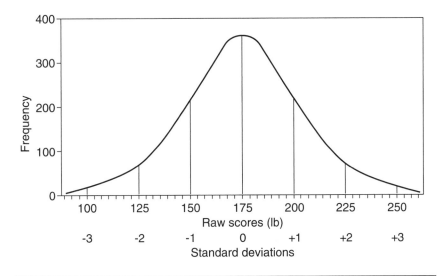

Figure 6.2 Relationship between raw scores in pounds and standard deviations (σ) on a normal distribution for a population of 15,000 college-age males.

and the standard deviation is 25 pounds. The graph shows that most subjects weigh between 150 and 200 pounds ($\overline{X} \pm 1\sigma$) and that the highest frequency is at the mean (175 pounds). A few subjects weigh less than 100 pounds and a few weigh more than 250, but not many.

The weight of most subjects is near the mean. As values progress farther from the mean in either direction, fewer and fewer cases are represented. The standard deviation units are distributed equally above and below the mean, and the majority of the cases fall between the mean and ± 1 σ, or between 150 and 200 pounds. Figure 6.2 shows an idealized normal distribution of that same population; the corresponding standard deviations are noted on the *X*-axis.

Z Scores

It is common for statisticians and researchers to convert raw scores into scores that have been standardized in terms of standard deviations. That is, a score is given a numeric value that corresponds to how far that score is from the mean. A **Z score** is a raw score expressed in standard deviation units. If the standard deviation of the scores in figure 6.2 is 25, then 1 standard deviation unit is equivalent to 25 pounds on the raw score scale. A score of 200 lies 25 raw score units, or 1 standard deviation unit or 1 Z score, above the mean. The raw score of 200 is equivalent to a Z score of +1. Likewise, a raw score of 150 (25 raw units, or 1 standard deviation unit, below the mean) has a Z score of −1.

When the mean and the standard deviation of any set of normal data are known, Z can be calculated for any raw score (X) by using the following formula:

$$Z = \frac{X - \overline{X}}{\sigma}. \tag{6.01}$$

Note: If the data represent a sample, use *SD* in the denominator.

Using this formula, we can calculate that a male student who weighs 200 pounds has a Z score of +1:

$$Z = \frac{200 - 175}{25} = +1.$$

When the raw score is less than the mean, the Z score is negative. If the raw score is 165, then

$$Z = \frac{165 - 175}{25} = -0.4$$

A unique characteristic of the normal curve is that the percentage of area under the curve between the mean and any Z score is known and constant. When the Z score has been calculated, the percentage of the area (which is the same as the percentage of raw scores) between the mean and the Z score can be determined.

In any normal curve, 34.13% of the scores lies between the mean and 1 Z score in either the positive or negative direction. Therefore, when we say that most of the population falls between the mean and ±1 Z score, we are really saying that 68.26% (2 × 34.13 = 68.26), or about two-thirds, of the population falls between these two limits. This is true for any variable on any data provided that the distribution is normal. Figure 6.3 demonstrates this concept.

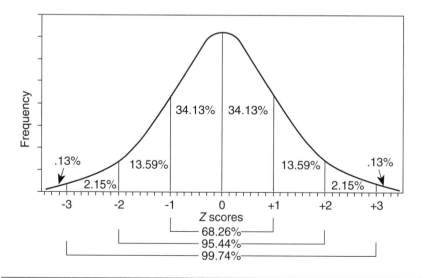

Figure 6.3 Percentage of area under the normal curve for selected Z score values.

Applying this to the body weight data illustrated in figure 6.2, we would expect about 10,239 scores (68.26% of 15,000 scores) to fall between 150 and 200 pounds.

Table A.1 in appendix A may be used to determine the percentage of scores that falls between the mean and any given Z score. The values on the left of each column represent Z scores to the nearest hundredth.

Using table A.1 (p. 314) to determine the percentage of scores that falls between the mean and ±1.00 Z score, we proceed down the Z column to the value 1.0 and move across the row to the Area (%) column; the value in that column is 34.13. This value is the percentage of raw scores that falls between the mean and either +1.00 Z or −1.00 Z. For ±2.00 Z the value in table A.1 is 47.72 (which is equal to 34.13 + 13.59, the percentages of the area from 0 to +1 and from +1 to +2 in figure 6.3). This indicates that 47.72% of the population of scores lies between the mean and +2.00 Z scores or between the mean and −2.00 Z scores.

Doubling that number (47.72 × 2 = 95.44) tells us that slightly more than 95% of all raw scores lies between the mean and ±2 Z. The figure for 3 σ is 99.74% (49.87 × 2 = 99.74). This confirms that most raw scores fall within ±3 σ of the mean (see figure 6.3).

Because the normal curve is bilaterally symmetrical, table A.1 in appendix A provides only the values for one-half of the curve; the values are the same for positive and negative Z scores. The percentage corresponding to a Z score of −1.78 is found by proceeding down the Z column to 1.78, then across to the Area (%) column, where the value 46.25 is read. This indicates that 46.25% of the population of scores lies between the mean and 1.78 Z scores in either direction.

Table A.1 may also be used for the opposite procedure: to find the Z score that corresponds to a given percentage of the population. If we want to know the Z value that represents 30% of the area under the curve, we look for the figure of 30.00 in the body of the table. This exact number cannot be found, so we find the value closest to 30.00 (29.95). This value corresponds to a Z score of ±0.84. So, to be 30 percentage points above or below the mean, a person must have a raw score equal to about ±0.84 Z scores.

Once we know the Z score for a raw score, we can determine the percentile value of that score by looking in table A.1 in appendix A. Because the mean, or a Z score of 0.00, represents the 50th percentile, any figure read from the table for a positive Z score is added to 50 to determine the percentile of that score. A Z score of +1.24 has a corresponding table value of 39.25. Therefore, the raw score equivalent to a Z score of +1.24 has a percentile value of 89.25 (50 + 39.25 = 89.25). If the Z score is negative (−1.24), then the value from the table is subtracted from 50. This results in a percentile score of 10.75 (50 − 39.25 = 10.75).

Just as any raw score has a corresponding Z score, each Z score has a corresponding percentile score. Table A.1 may be used to convert scores from Z to percentiles or vice versa. A positive Z score must be added to 50, and a negative one must be subtracted from 50, to determine the percentile value of the raw score. The values in table A.1 represent only half of the curve; they should be interpreted as the distance from the middle of the curve toward either end.

Standard Scores

Raw scores are the direct result of measurement, usually in units of distance, time, force, or frequency. Raw scores from more than one variable will have different units of measurement, different mean values, and different variability. A standard score is derived from raw data and has a known central tendency and variability. Raw scores from multivariate data can be directly compared only after they are converted to the same standard score base.

It is not logical to compare 50 sit-ups in 1 minute with a mile-run time of 4:36.3. Which performance is better, 20 feet in the long jump or 10.5 seconds in the 100-meter dash? We cannot answer such questions using raw data alone because the values are based on different units of measurement. To make an appropriate comparison, we must first convert raw scores to one of the four standard scores: percentiles, Z scores, T scores, or stanines.

Percentiles

In chapter 3, we discussed the percentile. This type of standard score has several advantages. Percentiles have 50% as their central tendency and zero to 100% as their range. They also have known quartile and decile divisions.

We can easily compare different types of raw scores when we convert them to percentiles. If a jump of 5 feet, 5 inches represents the 75th percentile and a time of 11.5 seconds in the 100-meter dash represents the 80th percentile, then the sprinting score is better than the jumping score. These scores are now directly comparable because they have both been converted to the same base, or standard.

Z Scores

The calculation of Z scores was explained earlier in this chapter. We now discuss how to use Z scores as standard scores. Z scores have a known central tendency ($\bar{Z}$ = 0) and known variability (σ = 1.0). When raw scores are converted to Z scores, two or more sets of data may be directly compared. For example, which score is better, a Z score of −1.3 on a long jump test or a Z score of −0.50 on a gymnastics floor exercise test? The gymnastics score is better because a Z score of −0.50 is higher (thus better) than a score of −1.3.

We can confirm this by using table A.1 in appendix A to convert both scores to percentiles: −1.3 Z equals a percentile score of 9.68 (50 − 40.32 = 9.68), and −0.5 Z equals a percentile score of 30.85 (50 − 19.15 = 30.85). The student who received these scores is not a whiz at either the long jump or floor exercise, but it

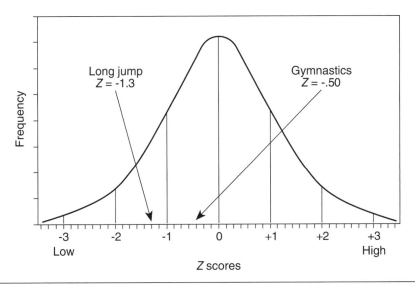

Figure 6.4 Comparison of Z score values on the normal curve.

is safe to say that the student performs better in gymnastics. Figure 6.4 presents a graphic representation of the two scores.

T Scores

A third standard score, called the *T* score, is often used to report norms in educational settings, such as those on national fitness or skill tests. By convention and definition, a **T score** has a mean of 50 and a standard deviation of 10. A *T* score of 60 is 1 standard deviation above the mean, and a *T* score of 30 is 2 standard deviations below the mean. Before *T* can be calculated, the corresponding Z score must be known. The formula for converting Z scores to *T* scores is

$$T = 10Z + 50. \tag{6.02}$$

The *T* equivalent for a Z score of +1.5 is 65.0 [$T = (10 \times 1.5) + 50 = 65.0$]. When the Z score is negative, the *T* score is less than 50 (remember that Z scores have a mean of zero). In the previous example of gymnastics and long jump scores, the *T* score for a Z score of −0.50 in gymnastics is calculated as follows: $T = 10 \times -0.50 + 50 = 45.0$.

The *T* scale was created because the lay public has difficulty understanding Z scores. It is not common to think of scores with zero as the mean. Most people consider zero to be nothing and prefer to have 50 as the middle and a range from zero to 100. Percentiles are widely used because they are easy to understand.

T scores have a mean of 50 and a range from zero to 100, but it is very unlikely that a *T* score would be less than 20 or greater than 80 because these figures represent Z scores of −3 and +3, respectively. As figure 6.2 shows, only $2 \times .13$, or

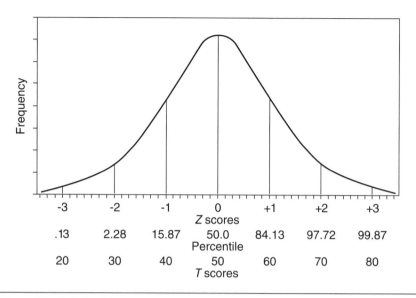

Figure 6.5 Relationships among Z scores, percentiles, and T scores on the normal curve.

0.26%, of the population lies beyond the ±3 limit on the Z scale. Indeed, a T score of 100 would represent a Z score of +5 and a percentile of 99.99999, which is a rather unlikely occurrence. Figure 6.5 shows the relationships among Z scores, percentiles, and T scores.

Stanines

Like the T scale, the **stanine** scale (a derivation of the words standard nine) is commonly used for reporting the results of educational tests. Parents who inquire about their children's scores on standardized tests may find the results presented in stanines. For example, a student may score at the seventh stanine in math, the fourth stanine in reading, and the fifth stanine in verbal skills.

How are stanine scores interpreted? As figure 6.6 shows, in the stanine scale, the standard normal curve is divided into nine sections: 5 as the middle score, 1 as the lowest score, and 9 as the highest score.

To calculate a stanine score, we need to know the raw score, the mean, and the standard deviation. From this information, we calculate the Z score. Once we have the Z score, we can find the stanine score directly from figure 6.6.

Each section on the stanine curve is one-half Z score wide except for stanines 1 and 9. The center section (stanine 5) ranges from −0.25 Z to +0.25 Z. The other sections continue to the left or right in intervals of 0.5 Z.

Scores that fall exactly on a dividing line between two stanines are usually given the higher value. Thus, a Z score of +0.75 is considered to be in the seventh stanine and a Z score of −1.25 is in the third stanine.

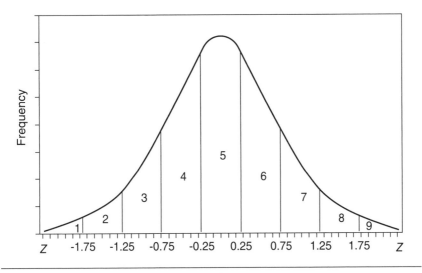

Figure 6.6   Stanine distribution on the normal curve.

Stanine scores do not represent an exact raw score. Rather, they represent a range, or section of the curve, into which the raw or Z score falls. In this way, stanines are similar to quartiles or deciles. When only the stanine is known, it is impossible to tell exactly where a raw score falls within the stanine. Only the section of the curve that best represents the score is known.

A student who scored at the seventh stanine on pull-ups, the fourth stanine on the mile run, and the fifth stanine on the sit-and-reach test of flexibility is considerably above the mean in strength, slightly below the mean in aerobic capacity, and very close to the mean in flexibility.

Probability and Odds

Throughout the remainder of this book, we discuss procedures that employ the concepts of **probability** and **odds.** Probability and odds are similar ideas but they are not equivalent (although we often use them interchangeably in everyday conversation). Probability (p) can be defined as the long-run proportion of a particular outcome (Armitage et al., 2002) and can vary between zero (the outcome is impossible) and 1.0 (the outcome is ensured). For example, when you flip a fair coin the probability of the flip resulting in heads is 1/2 = .50 or 50%. If you flipped that coin many times, you would expect half of those flips to be heads. Similarly, if you rolled a fair die the probability that you would roll a four is 1/6 = ~.167, or about 16.7%. That is, of the six possible outcomes, a four should happen on average one out of every six rolls. A four is not guaranteed when you roll the die six times; rather, 1/6 is the average proportion over a very large number of rolls. Imagine that you rolled a die two times in a row and rolled a four both times. This occurrence is somewhat unlikely (in fact, $p = 1/6 \times 1/6 = .028$, or 2.8%) but not so unlikely that we might suspect something suspicious. What if you rolled a four 10 times in a row? Although not impossible, it is quite unlikely and we might now suspect that the die is rigged. (In subsequent chapters we calculate a p value for statistical tests and ask whether that p value is small enough for us to suspect that something is going on in the data that is not due just to chance.)

Probability values are related to standard scores, percentiles, and the characteristics of the normal distribution. For example, in a normal curve 34.13% of all scores fall between the mean and a Z score of 1 (see figure 6.3). Therefore, if you were to pick one score at random from the distribution, the probability of that score being between the mean and $+1$ σ is .3413. Similarly, .13% of scores in a normal distribution are greater than $Z = +3$. The probability of picking a score at random with a Z score greater than $+3$ is .0013.

Odds are mathematically tied to probability as follows:

$$\text{odds} = p/(1 - p), \qquad (6.03)$$

where p is the probability of the outcome in question. The probability of the flip of a fair coin turning up heads is $1/2 = .5$. Therefore, the odds of a coin flip turning up heads are $.5/(1 - .5) = 1.0$; more commonly, we would say that the odds are 1/1 or one to one. The sum of the odds should equal the total number of possible outcomes. For example, imagine that 10 billiard balls placed in a box are each labeled with a number from one to 10. Without looking, you remove one of the balls from the box. What are the odds of that ball being a six? Ten possible outcomes exist and only one possibility is a six, so nine possible outcomes are not a six. The odds of picking a six are 1/9 or one to nine.

Calculating Skewness and Kurtosis

The process of statistical inference is addressed in detail in chapter 7. A major assumption of the process of statistical inference is that the characteristics of the normal curve can be applied. Consequently, it is critical that we know whether the data deviate from normality. **Skewness** is a measure of the bilateral symmetry of the data, and **kurtosis** is a measure of the relative peakedness of the curve of the data.

By observing a graph of the data and identifying the three measures of central tendency, we can get a general idea of the skewness of the data; however, this method is not exact (see figure 4.1 on p. 56). Using Z scores, we can obtain a numerical value that indicates the amount of skewness or kurtosis in any set of data.

Because Z scores are a standardized measure of the deviation of each raw score from the mean, we can use Z scores to determine whether the raw scores are equally distributed around the mean. When the data are completely normal, or bilaterally symmetrical, the sum of the Z scores above the mean is equal but opposite in sign to the sum of the Z scores below the mean. The positive and negative values cancel each other out, and the grand sum of the Z scores is zero.

If we take the third moment (the cube of the Z scores, or Z^3), we can accentuate the extreme values of Z, but the signs of the Z values remain the same. This places greater weight on the extreme scores and permits a numeric evaluation of the amount of skewness. Computing the average of the Z^3 scores produces a raw score value for skewness. The formula for calculating the raw value for skewness is

$$\text{skewness} = \frac{\Sigma Z^3}{N}. \tag{6.04}$$

When the Z^3 mean is zero, the data are normal. When the Z^3 mean is positive, the data are skewed positive, and when the Z^3 mean is negative, the data are skewed negative. This effect can be seen by examining the data presented in table 6.1. Notice that the data are skewed negative. When these data are graphed (see figure 6.7), the skewness is easily observed.

Kurtosis may also be calculated from Z scores. By taking the fourth moment Z^4 of the Z scores, the extreme Z values are again accentuated but the signs are all converted to positive. When the average of the Z^4 value is 3.0, the curve is normal. To make the units equal for both skewness and kurtosis, the mean of Z^4 is typically reduced by 3.0. The formula for calculating the raw value for kurtosis is (AndersonBell, 1989, p. 173; Spiegel, 1961, p. 91)

$$\text{kurtosis} = \left(\frac{\Sigma Z^4}{N} \right) - 3.0. \tag{6.05}$$

A score of 0 indicates complete normal kurtosis, or a mesokurtic curve, just as a score of 0 for skewness indicates complete bilateral symmetry. When the raw score for kurtosis is greater than 0.0, the curve is leptokurtic (more peaked than normal), and when the raw score is less than 0.0, the curve is platykurtic (more flat than normal).

Raw skewness and kurtosis scores are not easily interpreted because a raw score alone does not indicate a position on a known scale. But when raw scores are converted to Z scores, they are easy to interpret. To convert the raw scores for skewness (equation 6.04) or kurtosis (equation 6.05) to Z scores for skewness or kurtosis, we divide the raw scores by a factor called the standard error. Standard error is a type of standard deviation. (We explore standard error in more detail in subsequent chapters.) Therefore, when we divide the raw skewness and kurtosis scores by the appropriate standard error, the result is a Z score for skewness and kurtosis. According to Dixon (1990, p. 137), the standard error (*SE*) for skewness is

$$SE_{\text{skew}} = \sqrt{\frac{6}{N}}$$

and the standard error for kurtosis is

$$SE_{\text{kurt}} = \sqrt{\frac{24}{N}}.$$

TABLE 6.1

Calculation of Skewness and Kurtosis

X	Z	Z^3	Z^4
5	1.08	1.26	1.36
5	1.08	1.26	1.36
5	1.08	1.26	1.36
4	0.27	0.02	0.01
4	0.27	0.02	0.01
4	0.27	0.02	0.01
4	0.27	0.02	0.01
4	0.27	0.02	0.01
3	−0.54	−0.16	0.09
3	−0.54	−0.16	0.09
2	−1.35	−2.46	3.32
1	−2.17	−10.22	22.17
		$\Sigma Z^3 = -9.21$	$\Sigma Z^4 = 29.80$

Mean = 3.67; SD = 1.23; N = 12.

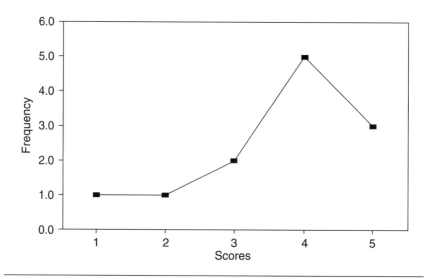

Figure 6.7   Negative skew.

If we divide the raw scores for skewness or kurtosis by the appropriate standard error, we obtain a Z_{skew} or Z_{kurt} value as follows:

$$Z_{skew} = \frac{\Sigma Z^3 / N}{\sqrt{6 / N}}, \tag{6.06}$$

$$Z_{kurt} = \frac{(\Sigma Z^4 / N) - 3.0}{\sqrt{24 / N}}. \tag{6.07}$$

These values may be interpreted as Z scores (i.e., values greater than 1.96 or less than -1.96 exceed $p = .05$, and values greater than 2.58 or less than -2.58 exceed $p = .01$). Typically, data are considered to be within acceptable limits of skewness or kurtosis if the Z values do not exceed ± 2.0.

Using the data from table 6.2, we can find Z_{skew} in the following manner:

$$skewness = \frac{-9.21}{12} = -0.77,$$

$$SE_{skew} = \sqrt{\frac{6}{12}} = 0.71,$$

$$Z_{skew} = \frac{-0.77}{0.71} = -1.08.$$

Z_{kurt} can be found as follows:

$$kurtosis = \frac{29.80}{12} - 3.0 = -0.52,$$

$$SE_{kurt} = \sqrt{\frac{24}{12}} = 1.41,$$

$$Z_{kurt} = \frac{-0.52}{1.41} = -0.37.$$

From these values ($Z_{skew} = -1.08$ and $Z_{kurt} = -0.37$) we can determine that the data in table 6.1 and figure 6.7 are slightly skewed negative and slightly platykurtic; however, neither value approaches significance (± 2.0). Therefore, we may conclude that the data are within acceptable ranges of normality. Data sets with small values of N may appear to be significantly skewed when graphed (see figure 6.7), but the true evaluation of the degree of skewness must by made by Z score analysis.

Summary

Raw scores may be converted to standard scores in the form of Z, percentile, T, or stanine to provide more information about the data and to assist in evaluating raw data. Standard scores are also useful for comparing the results of tests measured using different units of measurement (e.g., comparing time with force or distance). Because standard scores have a common central tendency and variability, data presented in standard score units may be directly compared regardless of the unit of measurement of the raw score. The shape of a distribution can be characterized by skewness, a measure of bilateral symmetry, and kurtosis, a measure of the peakedness of the distribution.

Problems to Solve

1. Find the percentage of values that falls between the mean of a given set of population data and a Z score of +0.35. (*Hint:* Use table A.1 in appendix A.)

2. Given a set of data with a mean of 25.7 and a standard deviation of 5.2, calculate the Z score equivalents of the following raw scores: (a) 21.6, (b) 28.9, and (c) 24.5. What is the interpretation of the Z scores?

3. Use table A.1 in appendix A to determine the equivalent percentile rank of each of the scores in problem 2.

4. Convert the Z scores calculated in problem 2 to T scores.

5. What is the stanine value for each of the scores in problem 2? (*Hint:* Use figure 6.6.)

6. Calculate the odds associated with the following probability (*p*) values: (a) .50, (b) .25, (c) .10, (d) .05, and (e) .01.

See appendix C for answers to problems.

Key Words

kurtosis   stanine
odds   T score
probability   Z score
skewness

$$H = \left[\frac{12}{18(18+1)} \right] \left[\frac{82.5^2}{6} + \frac{54.5^2}{6} + \frac{34.0^2}{6} \right] - 3(18+1)$$

$$H = 0.035 [1822 (9)] - 57$$

$$H = 6.77$$

Fundamentals of Statistical Inference

Aresearcher wishes to characterize the vertical jump performance of Division I female collegiate basketball players. However, there are thousands of players and they are geographically dispersed. It is impractical to test the vertical jump of all the players. Instead, the researcher recruits a random sample of players from several Division I schools and measures their vertical jump performance. The researcher can then use the sample data and the tools of statistical inference to estimate the characteristics of the population.

Predicting Population Parameters Using Statistical Inference

In many sciences, little research is conducted on entire populations. Often the population is so large that it would be impossible to measure each member. In such cases, the researcher takes a sample of the population and assumes that the sample represents the population and that the sample statistics are indicative of the population parameters. For this assumption to be valid, the sample must be randomly selected. For more information on sampling selection, see chapter 1. The process of estimating population parameters based on sample statistics is called inferential statistics.

Earlier we discussed the problem of determining the mean weight of all men at a university. If the population was 100 or fewer men, we could measure all of them, and a sample would not be needed. But if the population of all men in the university was 15,000, it would be too time consuming to measure all of them. So we would take a random sample to estimate the mean of the population.

The sample size is limited by such factors as time constraints, finances, facilities, and equipment. If we wanted to measure height or weight, a large sample could be collected because it is easy to measure these variables. But if we were interested in hydrostatically measured body composition or $\dot{V}O_2max$ on a treadmill or bicycle ergometer, perhaps only a few subjects could be measured.

Let us assume that a sample size of 50 men is desired. If the university administration permitted us access to the student files, a computer could select 50 males from the files in a completely random fashion. These randomly selected subjects could then be invited to participate in the study and could be measured by appointment. (It would be wise to invite more than 50 subjects because a few may not participate.) The mean weight of this sample could then be used to represent the mean weight of all men at the university.

Estimating Sampling Error

Sampling error refers to the amount of error in the estimate of a population parameter that is based on a sample statistic. Even if the sample is randomly drawn, it is unlikely that the mean of the sample will be identical to the mean of the population. Also, the true population mean is never known exactly because all members of the population are never measured. Consequently, we need a way to determine how accurate the sample mean is and what the odds are that it is deviant from the population mean by a given amount.

The **standard error of the mean** (SE_M) is a numeric value that indicates the amount of error that may occur when a random sample mean is used as a predictor of the mean of the population from which the sample was drawn. By accepting this error, we admit that the exact population mean can never be known (unless every member of the population is measured). We can only know the sample mean and how much error it is estimated to have.

The prediction of the population mean is always an educated guess and is accompanied by a probability statement. That is, the population mean is assumed to exist between some set limits, and the chance of this assumption being correct is stated as odds such as 90 to 10 ($p = .10$), 95 to 5 ($p = .05$), or 99 to 1 ($p = .01$).

To explain the theory behind this technique, let us use the example of estimating the mean weight of men at a university. Assume that the mean is about 175 pounds (a value we can never know precisely unless we measure all 15,000 men) and that the standard deviation is 25 pounds. Therefore, in the population about 68% of all weights fall between 150 and 200 pounds and about 95% of all weights fall between 125 and 225 pounds (see figure 6.3).

Suppose a large (theoretically infinite) number of random samples is taken and that each sample comprises 50 subjects. After each sample is taken, the subjects are returned to the population pool so that they have an equal chance of being chosen again in a subsequent sample. Suppose the mean of most samples is between 165 and 185 pounds. The range of the population will be larger than the range of any one sample because it is unlikely that the extremes of a population of 15,000 will be randomly selected in a single sample of 50.

If a true random sample of sufficient size is taken each time, the sample mean will not vary greatly from the actual population mean. It is unlikely that a random sample of $N = 50$ would have a mean value near one of the extremes of the population because, for this to happen, all subjects in the sample would have to be from one extreme of the population. But this unusual occurrence is more likely to happen if the sample size is small (e.g., $N = 5$). To prevent this potential error, we make our samples as large as possible within the limits of our resources. The probability of obtaining a biased sample increases as the sample size decreases. The larger the sample, the smaller the error in predicting the population mean.

After all of the samples have been taken and the mean of each has been calculated, the sample means could be arranged into a graph that would approximate a normal curve. The means of this series of random samples would be normally distributed, even if the population from which they were drawn is not normal. This concept is known as the central limit theorem, which was briefly introduced in chapter 2.

Each sample has its own mean and standard deviation, and the total group of sample means also has a mean (the mean of the means) and a standard deviation (the standard deviation of the means). This value, the standard deviation of the means, is called the standard error of the mean.

Figure 7.1 shows a frequency distribution of a computer simulation of the scenario described previously. In this simulation, the computer generated 100,000 samples; $N = 50$ per sample from a normally distributed population of scores with a mean of 175 pounds and standard deviation of 25 pounds. This frequency distribution is called a **sampling distribution of the mean** because it is a frequency distribution of sample means.

Because the distribution of sample means form a normal distribution, all of the characteristics of normality apply to the curve of the sample means. The mean of the 100,000 sample means is 175 pounds and the standard deviation is 3.52 pounds. Notice that the mean of the sampling distribution is equal to the population mean (175 pounds) from which the 100,000 sample means were drawn; this is expected based on the central limit theorem. The mean of the sample means becomes the best estimate of the true population mean because it represents the results of a large number of randomly drawn samples of 50 subjects each. Because the standard deviation from the distribution of sample means is about 3.52 pounds, we can say that the standard error of the mean is approximately 3.52 pounds.

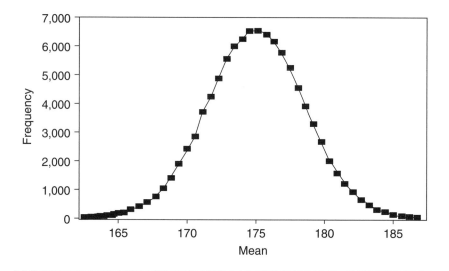

Figure 7.1 Sampling distribution of the mean from a computer simulation of 100,000 samples. $N = 50$ per sample from a population of scores with a mean of 175 pounds and a standard deviation of 25 pounds.

Applying our knowledge about the relationship of percentile points to Z scores on a normal curve (see figure 6.3), and assuming that the mean of the sample means is the best estimate of the population mean, we can state that the population mean probably lies somewhere between $175 \pm (1 \times 3.52)$ pounds (171.48–178.52).

Recall that approximately 68% of the area under any normal curve lies between ±1 standard deviation (see figure 6.3) and that the remaining 32% lies outside these limits (16% on each end). The mean of the means (175) becomes the estimate of the population mean, and the chance that the estimate of $175 \pm (1 \times 3.52)$ pounds is correct is slightly better than 2 to 1 (actually, it is 68 to 32).

If we widen the population estimate (making it less precise but more encompassing) to ±2 standard deviations [$175 \pm (2 \times 3.52)$] and estimate that the population mean lies somewhere between 167.96 and 182.04, we increase the odds of being correct to better than 95 to 5. Note that 95.44% of the area under the normal curve lies between $\pm 2\,\sigma$ of the mean (see figure 6.3). And if we make the estimate accurate to $\pm 3\,\sigma$ [$175 \pm (3 \times 3.52)$] and estimate that the population mean is between 164.44 and 185.56 pounds, then the odds that our estimate is correct are better than 99 to 1.

Using this technique, we can make probability statements about the population mean with various degrees of accuracy. The more precise, or narrow, the estimate, the lower the odds of being correct. As the estimate becomes more general, or broad, the odds of being correct improve.

The process just described is intended to show the theory behind the concept of the standard error of the mean. In practice, it may be as difficult to measure a large number of samples of 50 subjects each as it would be to measure all 15,000 males in the population. Fortunately, an equation has been derived that estimates the standard deviation of a series of theoretical sample means based on only one random sample.

The standard error of the mean estimated from one random sample (SE_M) can be calculated using the following equation:

$$SE_M = \frac{SD}{\sqrt{N}}, \tag{7.01}$$

where SD is the standard deviation of the sample and N is the sample size. (Refer to equation 5.04 for the formula for standard deviation.) Notice that the standard error of the mean varies directly with the sample standard deviation and inversely with the sample size. Therefore, samples that are more homogenous will have smaller standard errors. Similarly, samples with larger sample sizes will have smaller errors.

Using only one randomly drawn sample and equation 7.01, we can estimate where the mean of the population lies, within certain limits. For example, suppose we wish to estimate the population mean of the sit-and-reach test for high school girls. We take one random sample of 50 subjects that has a mean of 35 centimeters and standard deviation of 10 centimeters. Then we can estimate the standard error of the mean as follows:

$$SE_M = \frac{10}{\sqrt{50}} = 1.4 \text{ centimeters}$$

This value of 1.4 centimeters can be interpreted like any other standard deviation on a normal curve; therefore, it is equivalent to a Z score of ± 1.0. We may infer from this calculation that the mean of the population from which this sample was drawn has a 68% chance of being within the limits of $35 \pm 1(1.4)$ centimeters. The process of inference is represented by the following equation:

$$\mu = \overline{X} \pm 1(SE_M),$$

where $\overline{X}$ represents the sample mean and μ (mu in the Greek alphabet) represents the population mean. In our example, this equation would indicate $\mu = 35 \pm 1(1.4)$, or somewhere between 33.6 and 36.4 centimeters. This is sometimes written as $33.6 \leq \mu \leq 36.4$. Notice that if the standard error of the mean was smaller, the range of values about the mean would tighten. Because the researcher cannot control the standard deviation, the most straightforward way of decreasing the size of the standard error of the mean is to increase sample size. A similar formula for calculating the standard error of a proportion is available when data are presented as percentiles (see discussion on the t test for proportions in chapter 10).

Levels of Confidence, Confidence Intervals, and Probability of Error

A **level of confidence** (LOC) is a percentage figure that establishes the probability that a statement is correct. It is based on the characteristics of the normal curve. In the previous example, the estimate of the population mean (μ) is accurate at the 68% LOC because we included 1 SE_M (i.e., 1 Z) above and 1 SE_M below the predicted population mean.

However, if there is a 68% chance of being correct, there is also a 32% chance of being incorrect. This is referred to as the **probability of error** and is written as $p < .32$ (the probability of error is less than .32). The area under the normal curve that represents the probability of error is called **alpha** (α). Alpha is the level of chance occurrence. In statistics, this is sometimes called the error factor (i.e., the probability of being wrong because of chance occurrences that are not controlled). Alpha is directly related to Z because it is the area under the normal curve that extends beyond a given Z value.

Remember that standard error of the mean is a standard deviation on a normal curve. From our previous sit-and-reach example, the standard error was 1.4 centimeters. If we include 2 SE above and below μ [$35 \pm (2 \times 1.4)$, or 32.2–37.8 cm], we increase our level of confidence from about 68% to better than 95% and decrease the error factor from about 32% ($p < .32$) to about 5% ($p < .05$).

To be completely accurate when we use the 95% LOC, or $p < .05$, we should not go quite as far as 2 Z scores away from the mean. In table A.1 in appendix A, the value in the center of the table that represents the 95% confidence interval is 47.50 (95/2 = 47.50, because table A.1 represents only half of the curve).

This corresponds to a Z score of 1.96. The value 1.96 is the number of Z scores above and below the sample mean that accurately represents the 95% LOC, or $p = .05$. The correct estimate of μ at $p = .05$ is $35 \pm (1.96 \times 1.4)$, or 35 ± 2.7 centimeters (32.3–37.7). The range of values associated with a level of confidence is referred to as **confidence interval** (CI). In our example, the 68% confidence interval is about 33.6 to 36.4 centimeters and the 95% CI is about 32.3 to 37.7 centimeters.

A similar calculation could be made for the 99% LOC by using table A.1 to find the value that reads 49.5 (99/2 = 49.5). This exact value is not found in the table. Because 49.5 is halfway between 49.49 and 49.51 in the table, we choose the higher value (49.51), which gives us slightly better odds. The Z score correlate of 49.51 is 2.58. To achieve the 99% LOC, we multiply the standard error of the mean by ±2.58. The estimate of the population mean at the 99% LOC ($p = .01$) is $35 \pm (2.58 \times 1.4)$ or 35 ± 3.6 centimeters. Thus, the 99% CI is 31.4 to 38.6 centimeters. This may be expressed as $31.4 \leq \mu \leq 38.6$, $p = .01$. Likewise, we could establish the 90% LOC by looking up 45% (90/2 = 45) in table A.1. The Z score correlate of 45% is 1.65, so $\mu = 35 \pm 1.65 \times 1.4$, or 35 ± 2.3 centimeters ($p = .10$). Note that percent values from table A.1 are rounded to thousandths.

The level of confidence (chances of being correct) and probability of error (chances of being incorrect) always add to 100%, but by tradition the level of confidence is reported as a percentage and the probability of error (p) is reported as a decimal. The Z values to determine p at the most common levels of confidence are listed in table 7.1. By far, the most common level of confidence used is the 95% CI. Other values may be determined for any level of confidence by referring to table A.1 in appendix A.

The generalized equation for determining the limits of a population mean based on one sample for any level of confidence is as follows:

$$\mu = \bar{X} \pm Z\,(SE_M), \qquad\qquad (7.02)$$

where Z is a Z score that will produce the desired probability of error (i.e., $Z = 1.65$ for $p = .10$, 1.96 for $p = .05$, and 2.58 for $p = .01$).

TABLE 7.1

Corresponding Values for Z, LOC, and p

Z	LOC (%)	p
1.0	68	.32
1.65	90	.10
1.96	95	.05
2.58	99	.01

LOC = level of confidence; p = probability of error for a two-tailed test (sum of both tails of the curve).

An Example Using Statistical Inference

A researcher was interested in the average height of first-grade children in a school district. Eighty-three students randomly selected from throughout the district were measured with the following results: $\bar{X}$ = 125 centimeters and SD = 10 centimeters. The population height was estimated at the 95% LOC, p = .05, as follows:

and
$$SE_M = \frac{10}{\sqrt{83}} = 1.1$$

$$\mu = 125 \pm 1.96\,(1.1) = 125 \pm 2.2;\ p = .05.$$

The researcher concluded with 95% confidence that the mean height of all of the first-grade children in the school district was between 122.8 and 127.2 centimeters ($122.8 \le \mu \le 127.2$). However, a 5% chance (p = .05) exists that this conclusion is incorrect.

In kinesiology, as in behavioral and biomedical sciences, the most common minimum level of confidence used is 95% (p = .05). Some have argued that more liberal level of confidence values, such as the 90% (p = .10) level, may be appropriate in some situations, such as exploratory studies (Franks and Huck, 1986). The researcher decides which level to use, but the reader of the research must be the ultimate judge of what is acceptable. By consulting table A.1 in appendix A, we can determine any level of confidence and its equivalent Z score. The decision of which level to use is based on the consequences of being wrong.

In medical research, if an incorrect conclusion may result in serious injury to or death of the patient, then a very high level of confidence is desired. Even the 99% (p = .01) level may not be sufficient. The reader may wonder, Why not always use p = .01, given that it is the least likely to result in error? Because, although the prediction of the population mean is more accurate at p = .01 than p = .05, it is more broad (wider) and thus less precise. In statistics, if you want less error, you must sacrifice precision. When an incorrect conclusion will not result in bodily harm or excessive financial loss, lower levels may be used. The user of the research must determine the consequences of being wrong in each case and accept or reject the conclusions accordingly. Franks and Huck (1986) provide an excellent review of the history and procedures for selecting a level of confidence.

Statistical Hypothesis Testing

As noted previously, the process of statistical inference is the attempt to estimate population values from sample values. That is, we try to estimate parameters from statistics. If we had access to all the scores in our population of interest, the rest of this book would be unnecessary because our job would be done. We could precisely calculate the population values. But this happens rarely. What we do instead is take a representative sample(s) from the population (e.g., a random sample) and then apply the tools of statistical inference to make our population estimates.

How does this process work in practice? Two general approaches exist, and both are based on the same underlying statistical model. First is the more traditional hypothesis testing approach, which descends from Sir Ronald A. Fisher, whose work was based on studies in agriculture. The second approach de-emphasizes and often dismisses hypothesis testing in favor of estimating effects and calculating confidence intervals. We first cover the hypothesis testing approach and later address estimating effects and use of confidence intervals. Hypothesis testing is addressed first because it is more commonly used. Even if you eschew hypothesis testing, it is important to understand the concept in order to competently read studies that use it.

In hypothesis testing, we must understand the difference between a research hypothesis and a statistical hypothesis. A research hypothesis is a statement of the expected or predicted (or hoped for) outcome of a study. It addresses what you are really interested in. With statistical hypotheses, we create two mutually exclusive and exhaustive mathematical statements about the outcome of the analysis. These are the null hypothesis (symbolized H_0) and the alternate hypothesis (symbolized H_1). These are mutually exclusive in that we set them up so that only one of the two will be true. These are exhaustive in that no third option exists. The H_0 is typically a statement that the independent variable has no effect on the dependent variable (or no relationship exists between the independent variable and the dependent variable). The H_1 is the logical alternative to the H_0 and is usually consistent with the research hypothesis.

The simplest case is one in which we test whether two groups are the same on some dependent variable. Pretend that our research hypothesis is that biomechanists (B) and exercise physiologists (EP) differ with respect to body mass index (BMI). To test this, we might compare the BMI values of biomechanists and exercise physiologists. (Here we ignore substantive questions like whether BMI is a good measure of fatness.) The dependent variable is BMI and the independent variable is the type of movement scientist (B vs. EP). We could set up our statistical hypotheses like this:

$$H_0: \bar{X}_B = \bar{X}_{EP} \text{ and}$$

$$H_1: \bar{X}_B \neq \bar{X}_{EP}.$$

Notice that these are mutually exclusive. If H_0 is true then H_1 is false and vice versa. Notice also that these are exhaustive in that no other logical option exists other than H_0 and H_1. The H_1 is called a nondirectional hypothesis in that we are not interested in whether biomechanists or exercise physiologists have a higher BMI; rather, we are concerned only with whether they are different. We can also set up directional hypotheses. Imagine that we hypothesize that biomechanists are on average fatter than exercise physiologists. We would set up our statistical hypotheses like this:

$$H_0: \bar{X}_B < \bar{X}_{EP} \text{ and}$$

$$H_1: \bar{X}_B > \bar{X}_{EP}.$$

Notice that, as before, the statistical hypotheses are both exhaustive and mutually exclusive.

Whether our research hypotheses are directional or nondirectional, the focus of statistical hypothesis testing is on testing H_0. We will either accept or reject H_0. If we reject H_0, we must then accept H_1. The trick then is to come up with criteria for accepting or rejecting H_0.

To develop these criteria for accepting or rejecting H_0, we must revisit the idea of a sampling distribution. A sampling distribution is a theoretical distribution. A real sampling distribution does not exist; a sampling distribution is more of a thought experiment. Different sampling distributions are possible, but let us first recall the sampling distribution of the mean, which was addressed earlier when we introduced the standard error of the mean. Imagine that you know the BMI of every person in the United States; in other words, you have access to the population. Imagine also that the mean BMI in the population is 25 kilograms per square meter. Then you take a random sample of 50 of these BMI scores from the population, calculate the mean of the 50 scores in the sample, and put those 50 scores back into the pot of all BMI scores. You then repeat this process 10,000 times and have 10,000 mean values. Because each of those 10,000 samples was drawn at random from a population with a mean of 25 kilograms per square meter, we would expect each of those 10,000 means to be representative of the population from which they were drawn. Therefore, each mean should be about 25 kilograms per square meter. But due to sampling error, we would not expect each mean to be exactly equal to 25 kilograms per square meter; some means would be greater and some would be less. Sampling error does not mean you did anything wrong. The sampling error is simply the difference between the sample value and population value from which the sample was drawn. If we attribute differences between parameters and statistics to chance, we are attributing these differences to sampling error. We can now create a frequency distribution of the 10,000 sample means. This frequency distribution is a sampling distribution, and it has a mean and a standard deviation. The mean of the sampling distribution of the mean should be equal to the population mean, in this case 25 kilograms per square meter. It also has a standard deviation, which as you recall was called the standard error of the mean.

Now consider a slightly more complicated situation. Think back to our study of whether biomechanists or exercise physiologists are fatter. Imagine you have the population of BMI scores for all exercise physiologists and biomechanists in the United States. Assume for a moment that the population of exercise physiologists has the same mean BMI as the population of biomechanists; that is, assume that H_0 is true. If we take a random sample from the biomechanists and a random sample from the exercise physiologists, we could calculate the difference in sample means for BMI between the biomechanists and the exercise physiologists. Because we have assumed that the population means between the two types of movement scientists is the same, then we should expect the mean difference score between biomechanists and exercise physiologists to be about zero. Due to sampling error, we would not expect the mean difference between sample values to be exactly zero, but it should be close if H_0 is true. Now repeat this process over and over just like for the sampling distribution of the mean. Only now, the sampling distribution is called a **sampling distribution of mean differences.** Because we started with the premise that the population mean values for BMI were the same in biomechanists and exercise physiologists, the mean of the sampling distribution of mean differences will equal zero. The standard deviation of this distribution is called the standard error of mean differences.

Now we know that the mean of the sampling distribution of the mean equals the population mean and that the mean of the sampling distribution of mean differences equals zero if H_0 is true (i.e., the mean values are equal in the two populations). What should we expect the standard deviations of these two sampling distributions to be? (Because the standard deviations of these sampling distributions are called standard errors, we refer to them as such from now on.) The size of the standard errors is affected by the standard deviations of the population distributions from which they were drawn and, more importantly, is affected by the size of the samples that were drawn. The smaller the standard deviations of the population distributions, the smaller the standard errors, and the larger the N of the samples, the smaller the standard errors. The ability to detect effects using inferential statistics is increased by decreasing the size of the standard errors. Similarly, all else being equal, smaller standard errors result in narrower (tighter) level of confidence. Therefore, researchers try to shrink standard errors, and the most straightforward way of doing that is to increase sample size.

How is this used in practice? When you conduct a study, you don't have access to the population values and you don't run 10,000 iterations of your study. You have a sample, and you do one analysis. In a hypothesis testing model, you can estimate the probability (p) that you could have obtained the data you did, assuming that H_0 is true. In testing whether the BMI of exercise physiologists and biomechanists are different, we take a sample of exercise physiologists and a sample of biomechanists. We then measure the BMI of all subjects in both samples, calculate the means and standard deviations, and calculate the p value. If the p value is small, we get suspicious about the truth of the null hypothesis. If the p value is small enough, we get so suspicious that we reject the null hypothesis and accept the alternate hypothesis.

Type I and Type II Error

Traditional statistical hypothesis testing sets up a binary decision process. We either accept or reject the null hypothesis. When we perform our calculations, we get a p value; this value tells us the probability that we could have obtained the data that we did if H_0 is true. We then reject H_0 if we get a p value that is too small. How small is too small? That is defined a priori by your alpha level.

Because we are making our decision regarding H_0 based upon probabilities, we know that we may be wrong. Table 7.2 is the Orwellian-sounding truth table. For any statistical decision, either we make a correct decision or we make an error. If

TABLE 7.2

Truth Table for Hypothesis Testing

	H_0 true	H_0 false
Fail to reject H_0	Correct decision	Type II error (β)
Reject H_0	Type I error (α)	Correct decision

we reject H_0, we have either correctly rejected H_0 or we have committed what is called a **type I error.** If we commit a type I error, we have said that the independent variable affects the dependent variable, or that a relationship between the independent variable and the dependent variable exists, when in reality no such effect or relationship exists. This is a false positive. In contrast, our statistical decision may have been to fail to reject H_0. This may be correct, or you may have committed what is called a **type II error**. A type II error occurs when you accept H_0 when H_0 is false. This is a false negative. You miss an effect or relationship when one exists.

Keep in mind a couple of points when thinking about type I and type II errors. First, you can commit only one type of statistical error for any statistical decision. If your statistical decision is to reject H_0, either you have made a correct decision or you have committed a type I error. You cannot commit a type II error if you have rejected H_0. Similarly, if your statistical decision is to fail to reject H_0, then either you have made a correct decision or you have made a type II error. You cannot commit a type I error if your statistical decision was to fail to reject H_0. Second, you will not know whether you have committed an error. All you will know is that you either accepted or rejected H_0. If you accepted H_0, you know that either you were right or you made a type II error. If you rejected H_0, you know that either you were right or you committed a type I error.

Notice in table 7.2 that type I error is associated with the term α and type II error is associated with the term β. These are probabilities. α was introduced earlier this chapter as a probability of error. More specifically, it is the probability, or risk, of committing a type I error, if the null hypothesis is true. **Beta** (β) is the probability, or risk, of committing a type II error, if the null hypothesis is false. Remember that you set alpha, which is typically set at .05. When you set alpha at .05, you indicate that you are willing to take a 5% chance of committing a type I error (if H_0 is true). You can also set beta; we will come back to this later.

Statistical **power,** related to β, is the probability of rejecting H_0 when H_0 is false. Power is the probability of finding an effect or relationship if one exists. It is mathematically defined as $1 - \beta$. Power is a good thing. You can do two things to increase power in a study. First, you can control the noise in the data; we talk about noise and ways to minimize it in subsequent chapters. Second, you can increase sample size. The larger the N, the more statistical power you will have.

Here is an overview of how the process works. You set α (typically at .05, but no rule says that it has to be .05). You run your statistical analysis and calculate a p value. The p value is the probability that you would have obtained the data that you did if the null hypothesis is true. If your p value is less than or equal to α, your statistical decision is to reject H_0. In statistical jargon, when H_0 is rejected we often say that the result is statistically **significant,** or sometimes just say that it is significant. That is, the probability of obtaining the data that you did if H_0 is true is less than or equal to the risk you were willing to make of committing a type I error. In contrast, if your p value is greater than α, you accept H_0. Here we might say that the result is not statistically significant. As is discussed in subsequent chapters, just because a result is statistically significant does not necessarily mean that the effect is of practical significance. An effect may be real in the sense that is it greater than zero, but it might still be trivially small in terms of usefulness.

Neophyte researchers are sometimes accused of making type I errors in their zeal to find significant differences. But failure to find a difference does not render the research worthless. It is just as important to know that differences do not exist as it is to know that differences do exist. The experienced and competent researcher is honestly seeking the truth, and a correct conclusion from a quality research project is valuable even if the research hypothesis is rejected.

Table 7.3 demonstrates the conditions under which type I and type II errors may be made. The dilemma facing the researcher is that one can never absolutely know which, if either, type of error is being made. The experimental design provides the means to determine the odds of each type of error, but complete assurance is never possible.

The researcher must decide which type of error is the most costly and then protect against it. If concluding that a difference exists when it does not (a type I error) is likely to risk human life or commit large amounts of resources to a false conclusion, then this is an expensive error and should be avoided. But if differences do exist, and the study fails to find them (a type II error), consumers of the research will never be able to take advantage of knowledge that may be helpful in solving problems.

TABLE 7.3

Possible Causes of Error

Type I	Type II
1. Measurement error	1. Measurement error
2. Lack of random sample	2. Lack of sufficient power (N too small)
3. α value too liberal ($\alpha = .10$)	3. α value too conservative ($\alpha = .01$)
4. Investigator bias	4. Treatment effect not properly applied
5. Improper use of one-tailed test	

Setting an appropriate α level to protect against either of the possible errors is critical. When using the null hypothesis, if we set an α level too low (e.g., $\alpha = .10$ rather than $\alpha = .05$) and found the result to be significant at just barely $p = .10$, then we would reject the null hypothesis and conclude that a real difference exists. If no difference exists between the means, and the result just happens to be that 1-in-10 event that occurs by chance alone, we have committed a type I error.

It is also possible to err in the other direction. For example, if we test the null hypothesis with $\alpha = .01$ and we find that p is greater than α (perhaps due to a small N or to measurement or other experimental errors), then we accept the null hypothesis. But if in reality the means do differ, we have committed a type II error.

We can never know absolutely when we have made either of these kinds of errors. Statistical techniques permit us only to make probability statements about the truth. To reduce the probability of type I errors, use a more conservative α value ($\alpha = .01$ instead of $\alpha = .05$). To guard against type II errors, set a more liberal α value. Researchers must make a trade-off decision. Protecting against one type of error increases the probability of committing the other type.

The critical factor in this decision is the consequence of being wrong. The confidence level should be set to protect against the most costly error. Is it worse to accept the null hypothesis when it is really false or to reject it when it is really true? The answer to this question depends on the context of the study and the subjective opinion of the researchers. However, by far the most common decision is to set the α level at 0.05.

Figure 7.2 is a sampling distribution of mean differences under conditions in which H_0 is true and, hence the mean of the distribution is zero. For simplicity, the standard error has been set at 1.0. Notice the shaded region to the right, which starts at $SE = +1.96$. This is called the **region of rejection** and is defined by the α level.

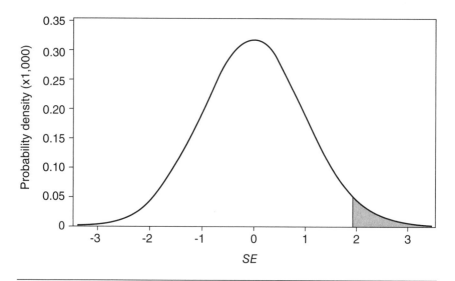

Figure 7.2 Sampling distribution of mean differences when the null hypothesis is true.

The area beyond a Z score of 1.96 equals .025. If we set α at the typical value of .05 and we have a nondirectional hypothesis, then we place half of the region of rejection in each tail of the sampling distribution. For clarity, for now we focus on the region to the right of the sampling distribution. To reject H_0, we would need to show a mean difference between biomechanists and exercise physiologists that is greater than or equal to 1.96 SE from the mean. Because in our example the standard error is 1.0, then a mean difference that is greater than or equal to 1.96 kilograms per square meter in a sample of biomechanists and a sample of exercise physiologists would be of sufficient size to allow us to reject H_0.

Now consider figure 7.3, which illustrates two possible versions of reality. Both curves reflect sampling distributions of mean differences under conditions in which $SE = 1.0$ (for simplicity) so that each integer unit on the X-axis equals 1 Z score from the mean. The curve to the left (solid curve) reflects the sampling distribution of mean differences under the condition that H_0 is true (biomechanists and exercise physiologists have the same mean BMI) and is the same distribution as shown in figure 7.2. The curve to the right (dotted curve) reflects one of an infinite number of possible sampling distribution of mean differences where H_0 is false. In this particular case, this is the sampling distribution of mean differences in which the BMI scores of biomechanists are on average 3.0 kilograms per square meter higher than those of exercise physiologists.

Assume that H_0 is true for the moment, so focus on the solid curve to the left. The region of rejection is positioned starting at where the X-axis is 1.96 on the H_0 is true curve so that the area to the right of the line equals $\alpha/2$. Imagine that now you conduct a study in which you take a sample of biomechanists and a sample of exercise physi-

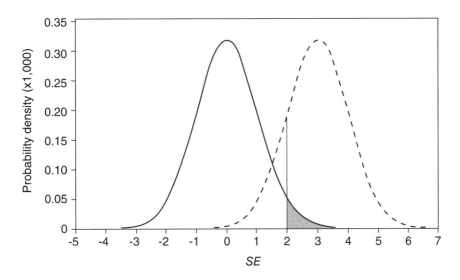

Figure 7.3 Sampling distribution of mean differences when the null hypothesis is true (left) and when the null hypothesis is false (right).

ologists and find that the mean difference in BMI between the two groups equals 2.5 kilograms per square meter. This mean difference is large enough to fall into the region of rejection because 2.5 is greater than 1.96; therefore, we reject H_0.

Now focus for the moment on the dotted curve to the right. The area to the left of the line in this curve reflects β, the probability of committing a type II error. Imagine that the mean difference in your sample between biomechanists and exercise physiologists is 1.0, but in the H_0 is false curve the population difference is 3.0. If you assume that H_0 is true, it is not particularly unlikely that a mean difference of 1.0 could occur. More specifically, a mean difference of 1.0 falls outside the region of rejection defined by α, so you do not reject H_0. But because the true mean difference is 3.0, by failing to reject H_0 you will have committed a type II error. Your score falls in the region defined by beta. You will have not detected a difference that exists.

Degrees of Freedom

Remember from chapter 5 that the calculation of the variance of a sample was sort of an average of the squared deviation scores. We say sort of because we divided the sum of the squared deviation scores not by the number of scores (N) but rather by the number of scores minus one ($N - 1$). The use of $N - 1$ rather than N makes the sample variance calculation an unbiased estimator. The use of $N - 1$ in the denominator of the variance calculation is one example of the use of degrees of freedom. As we will see in subsequent chapters, degrees of freedom are used for all inferential statistical calculations.

What do degrees of freedom have to do with hypothesis testing? The degrees of freedom in a statistical calculation define the shape of the sampling distribution. In chapter 10 we introduce tests that allow us to assess whether groups means are likely to differ. For example, we use the t test to compare two mean values. The larger the number of subjects, the larger the number of degrees of freedom, the taller and narrower will be the sampling distribution, and the easier it will be to reject H_0 (all else being equal). That is, the larger the degrees of freedom, the more statistical power.

In our discussion so far, we have addressed hypothesis testing and sampling distributions in the context of detecting the difference between two means. This is expanded upon in chapter 10. These principles apply to a variety of statistical assessments. In subsequent chapters, we address statistical indices such as r, b, t, and F. Each of these can be assessed in the context of sampling distributions and we can make inferential statements about these indices.

Living With Uncertainty

The primary untidy aspect of inferential statistics is that you have to live with uncertainty. In traditional hypothesis testing, if you reject H_0, either you are right or you are wrong (in this case you made a type I error). Similarly, if you accept H_0, either you are right or you have made a type II error. The problem is that you don't know whether you are right. You make your statistical decision and interpret the data accordingly, but in the back of your mind you know you may be wrong. And you may be wrong after having done nothing wrong; you followed the rules correctly but chance may just have worked against you this time. That is why we must never trust the results of just one study. Replication is a hallmark of science. Also, it is a good idea to be a little humble when reporting your data.

Two- and One-Tailed Tests

Most research is done because the results of the experiment are not known beforehand. If the researcher can answer the research question through logical reasoning or a review of related literature, the experiment is not necessary. When review of all prior research does not yield an answer, the researcher proposes the null hypothesis, H_0, and conducts an experiment to test that hypothesis.

Two-Tailed Test

One way to state the null hypothesis is to predict that the difference between two population means is zero ($\bar{X}_1 - \bar{X}_2 = 0$) and that small differences in either direction (plus or minus) on the sample means are considered to be chance occurrences. The direction, or sign, of the difference is not important because we do not know before we collect data which mean will be larger. We are simply looking for differences in either direction.

Under these conditions, the null hypothesis is tested with a **two-tailed test** (see figure 7.4). If $\alpha = .05$, the 5% rejection area is divided between the two tails of the curve; each tail includes 2.5% of the area under the curve ($\alpha/2$). Use the null hypothesis and a two-tailed test when prior research or logical reasoning does not clearly indicate that a significant difference between the mean values should be expected.

One-Tailed Test

Sometimes the review of literature or reasoned logic strongly suggest that a difference does exist between two mean values. The researcher is confident that the direction of the difference is well established but is not sure of the size of the difference. In this case, the researcher may test the research hypothesis (H_1), but such a situation is rare. The evidence suggesting the direction of the mean difference must be strong to justify testing H_1. The opinion of the investigator alone is not sufficient.

The researcher predicts that two population means are not equal. By convention, the mean expected to be larger is designated as $\bar{X}_1$. Because the first mean is predicted to be greater than the second, the direction of the difference is established as positive.

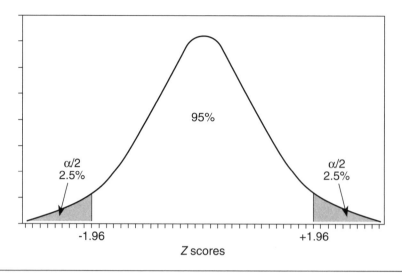

Figure 7.4 Distribution of alpha rejection area for a two-tailed test.

Because the difference to be tested is always positive ($\bar{X}_1 > \bar{X}_2$), we are interested in only the positive side of the normal curve. If an observational comparison of sample means shows $\bar{X}_2$ to be larger than (even by the slightest amount) or equal to $\bar{X}_1$, the hypothesis $\bar{X}_1 > \bar{X}_2$ is rejected.

The **one-tailed test** places the full 5% of the alpha area representing error at one end of the curve (see figure 7.5). The Z score that represents this point (1.65) is lower than the Z score for a two-tailed test (1.96). Therefore, it is easier to find significant differences when a one-tailed test is used. For this reason, the one-tailed

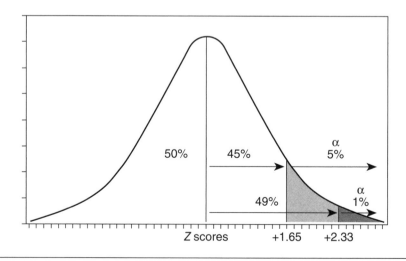

Figure 7.5 Distribution of alpha rejection area for a one-tailed test.

test is more powerful, or more likely to find significant differences, than the two-tailed test if the direction of your hypothesis is correct. If $\bar{X}_1 < \bar{X}_2$, you have zero power to detect that difference.

In practice, journal editors and reviewers generally frown upon one-tailed tests, and most tests are two-tailed. However, most research is not conducted in a theoretical vacuum and researchers typically think in a one-tailed way. Further, sometimes researchers may care only whether one treatment works better than another. Say a therapist is examining a new therapy and comparing it with the current standard of care. The new therapy would be adopted only if it was shown to be superior to the old therapy. An equal or worse performance by the new therapy would not lead to its adoption. Here a one-tailed test seems defensible.

Applying Confidence Intervals

An alternative to the null hypothesis statistical test (NHST) is to use confidence intervals to make inferences about parameters. A confidence interval is an interval that is constructed around a statistic. The approach is based on the same underlying statistical model as the NHST, but instead of making a binary decision about the acceptability of H_0, the analyst simply calculates an interval around which it is estimated that the population value truly exists.

Recall earlier in this chapter when the standard error of the mean (SE_M) was introduced. An example was given in which a sample of 50 subjects resulted in a mean sit-and-reach score of 35 centimeters with a standard deviation of 10 cm. From equation 7.01, $SE_M = \frac{10}{\sqrt{50}} = 1.4$. From the normal curve, we can construct a 95% CI using a Z score of 1.96 such that

$$95\% \text{ CI} = \bar{X} \pm 1.96 \, SE_M.$$

In this example, the 95% CI = 35 ± 1.96 (1.4) = 32.3 to 37.7 centimeters. We estimate then, with 95% confidence, that the true population mean lies somewhere between 32.3 and 37.7 centimeters. This is an inferential calculation because we are estimating a parameter from a statistic.

We can apply the same logic to construct confidence intervals about other statistics. Imagine that the mean difference in BMI between biomechanists and exercise physiologists is 3 kilograms per square meter and that the standard error of mean differences is 2 kilograms per square meter. The 95% CI about the mean difference is 3 ± 1.96 (2) = −0.9 to 6.9 kilograms per square meter. Notice that the 95% CI includes zero, which indicates that we are not confident that the true population mean difference is different than zero. This is tantamount to saying that we fail to reject H_0. Data that results in a 95% CI that includes the null hypothesis value will result in a failure to reject H_0 when the comparable NHST is performed. Similarly, data that result in a 95% CI that excludes the null value will result in a rejection of H_0 when the comparable NHST is performed.

Why worry about confidence intervals? First, confidence intervals are useful even when performing a NHST because they provide more information than just a simple binary decision on the null hypothesis. Second, many statisticians and theorists argue that the NHST is based on flawed logic. When we calculate a p value, the p does not represent the probability that H_0 is false (which is what we would really like for it to mean). Instead, p reflects the probability that we could have obtained data this extreme, or more extreme, if H_0 is false. These are not equivalent statements about p and H_0.

Further, many argue that all null hypotheses are false. That is, all variables are related, even if only in a trivially small way. The researcher's job then is simply to estimate the size of the effect and to construct confidence limits about the estimate. Academic arguments about the validity of the NHST have been going on for many decades. (For example, see the provocatively titled book *The Cult of Statistical Significance* by Zilak and McCloskey, 2008.) We will not enter that fight but rather will simply note that, at a minimum, one needs to understand both the NHST and the use of confidence intervals because both are used in the kinesiology literature and a competent consumer of the literature must be familiar with these ideas.

Summary

The process of statistical inference is to estimate population values based on sample values. Sampling error refers to the differences that will exist between our estimates and the true population values. We use samples that are reflective of the population and employ the tools of probability to make these population estimates. The tools of probability allow us to construct confidence intervals about our estimates and to perform statistical hypothesis testing. Whenever we make hypothesis tests, it is possible our conclusions are in error, but the probability of error can be estimated.

Problems to Solve

1. Using the following sample data, compute the standard error of the mean. What can you say about the value of standard error of the mean in relation to the size of the standard deviation and N?

	SD	N
A.	21.4	36
B.	2.1	106
C.	56.7	19

2. Suppose you wanted to know the mean $\dot{V}O_2max$ in milliliters per kilogram per minute of all females at a university. You randomly select 50 names from

the files of the administration office and invite the subjects to be tested in the lab. Thirty-five accept the invitation and are tested. The mean $\dot{V}O_2max$ of this sample is calculated to be 38.5 with a standard deviation of 4.7. If you want the population mean estimate to be accurate at the 95% LOC, what would be the confidence interval (upper and lower limits) of the predicted $\dot{V}O_2max$ value?

3. Using the best random sample you can get, collect data and estimate the average height in meters of all members of your class. Select the appropriate level of confidence. Look at your computer-calculated values (especially the maximum, minimum, mean, and standard deviation) to see if they make sense. It is always a good idea to eyeball your answers. If time permits, measure the total population (all members of the class). How close did your estimate come to the true population mean?

Key Words

alpha
beta
confidence interval
inferential statistics
level of confidence
one-tailed test
power
probability of error
region of rejection

sampling distribution of mean
   differences
sampling distribution of the mean
sampling error
significant
standard error of the mean
two-tailed test
type I error
type II error

$$H = \left[\frac{12}{18(18+1)} \right] \left[\frac{82.5^2}{6} + \frac{54.5^2}{6} + \frac{34.0^2}{6} \right] - 3(18+1)$$

$$H = [.035][1822.09] - 57$$

$$H = 6.77$$

X_4

X_1

X_3

Correlation and Bivariate Regression

n an exercise physiology class, a student learns that the time it takes to run 1.5 miles is directly related to oxygen consumption as measured in the laboratory by $\dot{V}O_2$max (milliliters of O_2 per kilogram per minute). The instructor explains that subjects with the highest $\dot{V}O_2$max values tend to have the fastest times in the 1.5-mile run. In statistical terms, the instructor is saying these two variables are correlated.

Knowing how fast someone can complete a 1.5-mile run will permit the calculation (using a technique called bivariate linear regression) of an estimate of the person's $\dot{V}O_2$max. This is most useful when one variable (1.5-mile run) is much easier to measure than a related variable ($\dot{V}O_2$max). In this chapter we show how to make this prediction.

Correlation

Correlation is widely used in kinesiology. For example, biomechanists might calculate the correlation between performance in the power clean (a weight-lifting exercise) and performance in the vertical jump (we might expect that athletes who performed well on the power clean would also perform well on the vertical jump). Exercise physiologists may calculate the correlation between skinfold thicknesses and percent body fat. **Correlation** is used to quantify the degree of relationship, or association, between two variables. When we calculate a correlation, we get a number (specifically, a numerical coefficient) that indicates the extent to which two variables are related or associated. Technically speaking, correlation is the extent to which the direction and size of deviations from the mean in one variable are related to the direction and size of deviations from the mean in another variable. This technique is referred to as **Pearson's product moment correlation coefficient;** it is named after Karl Pearson (1857–1936), who developed this concept in 1896 in England (Kotz and Johnson, 1982, p. 199).

The **coefficient,** or number, that represents the correlation will always be between +1.00 and −1.00. A perfect positive correlation (+1.00) would exist if every subject varied an equal distance from the mean in the same direction (measured by a Z score) on two variables. For example, if every subject who was 1 Z score above the mean on variable X was also 1 Z score above the mean on variable Y, and every other subject showed a similar relationship between deviation score on X and deviation score on Y, the resulting correlation would be +1.00.

Similarly, if all subjects who were above or below the mean on variable X were an equal distance in the opposite direction from the mean on variable Y, the resulting correlation would be −1.00. If these interrelations are similar but not perfect, then the correlation coefficient is less than +1.00, such as .90 or .80 in the positive case, and greater than −1.00, such as −.90 or −.80 in the negative case. A correlation coefficient of 0.00 means that no relationship exists between the variables.

Positive correlations result when subjects who receive high numerical scores on one variable also receive high numerical scores on another variable. For example, a positive correlation exists between the number of free throws taken in practice and the percentage of free throws made in games over the season. Players who spend lots of practice time on free throws tend to make a high percentage of free throws in games.

Negative correlations result when scores on one variable tend to be high numbers and scores on a second variable tend to be low numbers. For example, the relationship between distance scores and time scores is almost always negative because greater distance (e.g., a far long jump) is associated with faster sprinters (e.g., low time scores). In adults, as people get older in years, their muscular strength tends to decrease; therefore, a negative relationship exists between age and strength.

Correlation is a useful tool in many types of research. In evaluating testing procedures, correlation is often used to help determine the validity of the measurement instruments by comparing a test of unknown validity with a test of known validity. It is sometimes used to measure reliability by comparing test–retest measures on a group of subjects to determine consistency of performance. However, it is not sensitive to changes in mean values from pre- to posttest. If all subjects improve from pre- to posttest by exactly the same amount, the mean increases but the correlation remains the same. (Procedures for quantifying reliability are addressed in chapter 13.) It may also be used as a tool for prediction. When the correlation coefficient between two variables is known, scores on the second variable can be predicted based on scores from the first variable.

Although a correlation coefficient indicates the amount of relationship between two variables, it does not indicate the cause of that relationship. Just because one variable is related to another does not mean that changes in one will cause changes in the other. Other variables may be acting on one or both of the related variables and affecting them in the same direction.

For example, a positive relationship exists between IQ score and collegiate grade point average, but a high IQ alone will not always result in a high grade point average. Other variables—motivation, study habits, financial support, study time, parental and peer pressure, and instructor skill—affect grades. The IQ score may account for some of the variability in grades (explained variance), but it does not account for all of the variability (unexplained variance). Therefore, although a relationship exists between IQ and grades, it cannot be said that grades are a function of only IQ. Other unmeasured variables may influence the relationship between IQ and grades, and this unexplained variance is not represented by the single correlation coefficient between IQ and grades.

Cause and effect may be present, but correlation does not prove causation. Changes in one variable will not automatically result in changes in a correlated variable, although this may happen in some cases. The length of a person's pants and the length of his or her legs are positively correlated. People with longer legs have longer pants. But increasing one's pant length will not lengthen one's legs!

A visual description of a correlation coefficient may be presented as a **scatter plot,** in which the scores for each subject on two variables are plotted with one variable on the *X*-axis and the other variable on the *Y*-axis. Figure 8.1 shows a scatter plot of the relationship between the long jump and the triple jump (hop, step, and jump). Because both require running speed, and the triple jump contains a long jump, we would expect that the two variables are related—that is, people who are good long jumpers will probably be good triple jumpers, and vice versa.

Each star represents the plot of a person's score on both variables. Subject A long-jumped about 14 feet and triple-jumped about 24 feet. Subject B had scores of about 17 feet and 36 feet, respectively. The other stars represent the scores of additional subjects. In figure 8.1, the correlation between the two variables is about +.80.

Note how all of the stars seem to cluster around the line. This line is known as the **best fit line;** it represents the best linear estimate of the relationship between the two variables. Later in this chapter we explain how to determine this line.

The best fit line for a negative correlation slopes in the opposite direction. Figure 8.2 shows a scatter plot for the relationship between time in the 100-meter dash and distance in the long jump measured in meters. Low scores in the 100-meter dash (good performance) tend to be associated with high scores in the long jump (good performance), and vice versa. This makes sense, because we know that an important factor in long jump ability is the speed at takeoff. Generally speaking, the faster one runs, the farther one jumps.

The value of the correlation coefficient in figure 8.2 is about −.80. The data plots form around the best fit line, which decreases in *Y* value as it increases in *X* value. This is characteristic of a negative correlation.

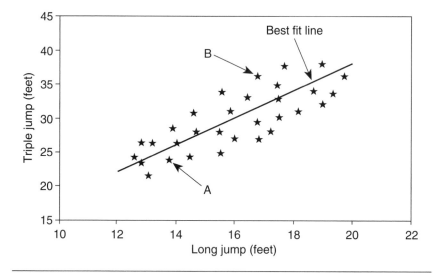

Figure 8.1 Positive correlation.

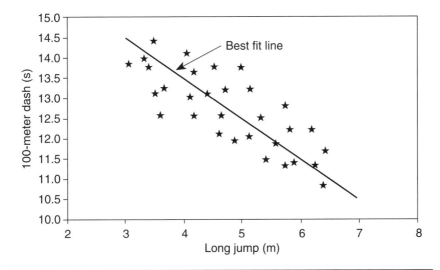

Figure 8.2 Negative correlation.

The positive or negative sign of the coefficient does not indicate the strength or usefulness of the correlation coefficient. The absolute numerical size of the coefficient, not its direction, determines the strength or usefulness of the relationship. As the correlation gets closer to +1.00 or −1.00, the points cluster closer to the line, and the relationship becomes stronger. If the correlation was a perfect +1.00 or −1.00, all of the plotted points would fall exactly on the line. As the correlation approaches 0.00, the data points drift further away from the line until the data points form a random cloud of points. In other words, a 0.00 correlation plot has points all over the graph and they form the shape of the entire graph; the points do not approach the form of a line. With a 0.00 correlation, it is not possible to predict one variable from another.

Figure 8.3 shows a scatter plot of the correlation between long jump distance and grade point average. The plot assumes a rectangular rather than a linear shape. We cannot predict a person's grade point average from his or her distance on the long jump.

Sometimes the best fit line is not straight. Two variables may be related in a **curvilinear** fashion. Pearson's correlation coefficient is a measure of linear relationships. If it is applied to curvilinear data, it may produce a spuriously low value with a corresponding interpretation that the relationship is weak or nonexistent when in fact the relationship is strong but not linear.

Figure 8.4 shows a scatter plot for data that have a curvilinear relationship. The curved line represents the true relationship between the X and Y variables. The straight line represents the relationship assumed by Pearson's coefficient. The true relationship is curvilinear and strong. The spurious linear relationship is weak and incorrect. It is important to examine the scatter plot of the data in addition to calculating the correlation coefficient. The pattern of scores on the scatter plot

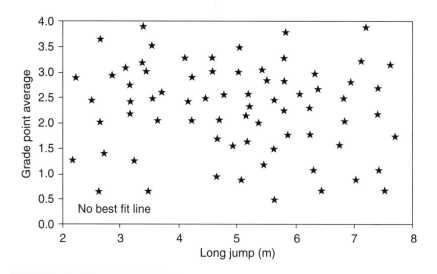

Figure 8.3 Zero correlation.

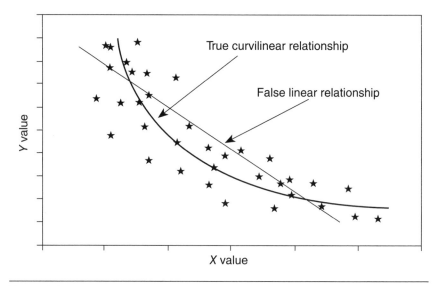

Figure 8.4 Comparison of curvilinear and linear relationships.

can help us decide whether a linear measure is appropriate or whether a nonlinear model might better explain the relationship.

Calculation of the relationship between nonlinear variables is possible but is not discussed here. Determining the shape of the relationship is called trend analysis. This procedure is discussed in more advanced textbooks; a good discussion of trend analysis is found in Keppel (1991, chapter 7).

Calculating the Correlation Coefficient

Several methods can be used to determine the value of Pearson's correlation coefficient (by convention the letter r is used to indicate the Pearson correlation coefficient). We present two equations; if you carefully examine each equation, you can get a sense of what the Pearson r conveys to the data analyst. We first consider what we call the definition formula because it is derived from the definition of correlation stated at the beginning of this chapter. Recall that a correlation coefficient represents the relationship between the Z scores of the subjects on two variables (usually designated X and Y). This can be stated mathematically as the mean of the Z score products for all subjects.

Definition Formula

The definition formula for Pearson's product moment correlation coefficient (r) is

$$r = \frac{\Sigma(Z_X Z_Y)}{N}, \qquad (8.01)$$

where Z_X and Z_Y are the Z scores for each subject on the X and Y variables and N is the number of pairs of scores.

To determine r with this formula, we first compute the mean and standard deviation of each variable and then compute the Z score for each subject on each variable. We demonstrate this method with simple data from table 8.1.

After determining the means and standard deviations, we calculate Z_X and Z_Y for each subject. Then we multiply Z_X by Z_Y (see equation 8.01) for each subject and sum the products. The correlation coefficient r then becomes

$$r = \frac{3.787}{4} = .947.$$

TABLE 8.1

Calculation of r by the Deviation Formula

X	Y	Z_X	Z_Y	$Z_X Z_Y$
2	1	−1.342	−1.342	1.808
3	2	−0.577	−0.447	0.258
5	3	0.962	0.447	0.430
5	4	0.962	1.342	1.291
$\Sigma X = 15$	$\Sigma Y = 10$			$\Sigma Z_X Z_Y = 3.787$

$\bar{X} = 15/4 = 3.75$, $\bar{Y} = 10/4 = 2.5$, $\sigma_X = 1.299$, $\sigma_Y = 1.118$, $r = +3.787/4 = +.947$.

We have carried out the calculation to three decimal places; however, the Pearson r is typically reported to only two decimal places. So for this example, we would round to r to 0.95. Table 8.1 uses equation 5.03 on page 63 (the equation for populations) to calculate standard deviation. The population formula is used in table 8.1 rather than the sample formula because the sample formula is intentionally biased by the placement of $N - 1$ in the denominator to correct for the use of a sample as a predictor of a population. Correlation is not an inferential statistic. It simply states the relationship between the two variables for a given N. Therefore, it would be inappropriate to bias the standard deviation by using $N - 1$ in the denominator. The population formula for standard deviation is necessary for an accurate calculation of r.

The mean and standard deviation values in table 8.1 are carried to three significant places more than the data (which are reported only to the nearest integer value) would seem to permit. This is done to ensure accuracy in the resultant r value. When means are multiplied together, or when standard deviations are squared, the error of rounding is also squared.

The value obtained for r from the data in table 8.1 is positive and close to 1.0. By looking at the data, we can confirm that the values of the X and Y pairs are related. The lowest X (2) is paired with the lowest Y (1), and the highest values of X (two tied at 5) are paired with the highest Y (4) and the second highest Y (3).

Also, the Z score products in table 8.1 are all positive. This is because the values for X that are below the mean of X (and have negative Z scores) are paired with values for Y that are below the mean of Y (and have negative Z scores). The product of two negative numbers is positive. Likewise, the values of X above the mean are paired with values of Y that are above the mean. The products of these pairs are also positive. Therefore, the sum of $Z_X Z_Y$ is positive. If the Z score pairs are not of the same sign, the Z score products will be negative. When low X values are associated with high Y values, and vice versa, the negative Z score products sum to a negative value, producing a negative result for r.

If the Z score products are mixed, some positive and some negative (indicating no special relationship among the X and Y pairs), the positive and negative products tend to cancel each other out, and the sum of $Z_X Z_Y$ is small, approaching zero. This produces a very low value for r.

Understanding the concept of relationships between X and Y pairs is critical to the comprehension and use of correlation. You should study the example in table 8.1 until these relationships become clear.

Because it is laborious to use, equation 8.01 is not applied in most problems. Applying this formula to large values of N, especially when the raw values and means are not whole numbers, takes considerable time and effort. Equation 8.1 is presented here only to provide a theoretical understanding.

Machine Formula

An alternative formula for the calculation of r is

$$r = \frac{\Sigma(x - \overline{X})(y - \overline{Y})}{\sqrt{(\Sigma(x - \overline{X})^2 \Sigma(y - \overline{Y})^2)}}. \tag{8.02}$$

This is called the machine formula because this formula is coded into many software programs and is calculated by machines (computers). However, the machine formula is also useful for understanding what the Pearson r conveys. The denominator reflects the maximum amount of score deviation that is present in the data. The numerator reflects what is called the **covariance** between X and Y. The covariance is an index of how X and Y vary together. Recall the equation for variance in which the numerator is $\Sigma(X - \overline{X})^2$, which if expanded becomes $\Sigma(X - \overline{X}) \times (X - \overline{X})$. Notice the similarity between the covariance and the numerator of the variance equation; we substitute Y for X in the second half of the expansion. Notice that both are also sums of squares operations. Thus, variance is a type of covariance in which we assess how a variable covaries with itself. The Pearson r assesses how one variable (X) covaries with a different variable (Y). The covariance, and therefore the r value, is large when X and Y deviate from their respective means in a consistent manner. In a positive relationship, if X is above the mean then Y is above the mean and vice versa. In a negative relationship, if X is above the mean then Y is below the mean and vice versa. The use of equation 8.02 is presented in table 8.2, which demonstrates the calculation of the relationship between height $(X$, in inches) and weight $(Y$, in pounds).

Substituting the values from table into equation 8.02, we get

$$\frac{\Sigma(x - \overline{X})(y - \overline{Y})}{\sqrt{(\Sigma(x - \overline{X})^2 \Sigma(y - \overline{Y})^2)}} = \frac{2,115.4}{\sqrt{(242.4)(23,264.9)}} = 0.891,$$

TABLE 8.2

Relationship Between Height and Weight

x (height)	$x - \overline{X}$	$(x - \overline{X})^2$	y (weight)	$y - \overline{Y}$	$(y - \overline{Y})^2$	$(x - \overline{X})(y - \overline{Y})$
72	4.4	19.36	167	19.9	396.01	87.56
64	−3.6	12.96	114	−33.1	1,095.61	119.16
71	3.4	11.56	135	−12.1	146.41	−41.14
66	−1.6	2.56	140	−7.1	50.41	11.36
63	−4.6	21.16	115	−32.1	1,030.41	147.66
62	−5.6	31.36	106	−41.1	1,689.21	230.16
73	5.4	29.16	207	59.9	3,588.01	323.46
61	−6.6	43.56	102	−45.1	2,034.01	297.66
76	8.4	70.56	260	112.9	12,746.4	948.36
68	0.4	0.16	125	−22.1	488.41	−8.84
$\Sigma x = 676$		$\Sigma = 242.4$	$\Sigma y = 1,471$		$\Sigma = 23,264.9$	$\Sigma = 2,115.4$
$\overline{X} = 67.6$			$\overline{Y} = 147.1$			

which we round to $r = .89$. As with equation 8.01, the use of equation 8.02 to calculate r by hand is time consuming and is prone to error due to arithmetic and rounding errors. Nonetheless, it is helpful to work through the calculations to get a feel for how the calculation works. For real data collected in a lab or in the field, use a computer with statistical software to perform these calculations.

Statistical Significance and the Confidence Interval

The usefulness of the coefficient is based on the size and significance of r. If r is reliably different from .00, the r value is statistically significant; it did not result from a chance occurrence. If r is significant, we can conclude, with known odds of being correct, that the relationship between the variables is real—that is, it is caused by a factor or factors other than chance. This implies that if we measured the same variables on another set of similar subjects, we would get a similar r value (i.e., the correlation is reliable).

When N is small (for example, 3 or 4 pairs of scores), it is possible that a spuriously high r value can occur by chance. Suppose the numbers 1, 2, and 3 are written on small pieces of paper and placed in a container. The numbers are blindly drawn one at a time on two occasions. It is possible that the numbers could be drawn in the same order twice. This would produce an r value of +1.00. But this r value would be spurious, because no factors other than chance would be operating on the variables to cause the relationship.

In contrast, the odds of 100 numbers being randomly selected in the same order twice are very low. Therefore, if the r value with $N_{pairs} = 100$ is high, we conclude that chance is not a factor because when N is large, it is rare for a high r to occur by chance. Some factor other than chance must cause the relationship. A large N makes us more confident that a high r value is not due to chance. Then we can hypothesize that a common factor causes the relationship. In the example about height and weight, we might hypothesize that body mass is the common factor. As one grows taller, body mass, and hence weight, increases. The common factor of body mass, not chance, causes the relationship between height and weight.

The opposite relationship is not necessarily true. It is certainly possible to gain weight without getting taller! A significant correlation does not prove causation; it only shows that a nonchance relationship exists. The fact that two variables are related does not mean that a change in one will necessarily produce a corresponding change in the other. Causation may be demonstrated with other research techniques (such as comparing changes in an experimental group with changes in a control group after an independent variable has been allowed to operate on the experimental group) but not by correlation alone. Statistical techniques to analyze experimental data are presented in chapters 10, 11, 12, 14, and 15.

When r is evaluated, the number of pairs of values from which the coefficient was calculated (N) is critical for determining the odds that the relationship is not due to chance factors. If N is small, r must be large to be significant. When N is large, small r values may be significant. As we discussed earlier, low r values may indicate a real, or significant, relationship, but they usually are not useful for predicting individual scores.

To test the statistical significance of r, we can set up the statistical hypotheses as follows:

$$H_0: r = 0.0,$$

$$H_1: r \neq 0.0.$$

Table A.2 in appendix A (p. 317) can be used to determine the significance of a correlation coefficient. First we calculate the degrees of freedom (df) for a correlation coefficient: $df = N_{pairs} - 2$, where N_{pairs} represents the number of pairs of XY scores. One degree of freedom is lost for each of the two variables in the correlation. The degrees of freedom represent the number of values that are free to vary when the sum of the variables is set. See chapter 7 for an explanation of degrees of freedom.

Degrees of freedom compensate for small values of N by requiring higher absolute values of r before the coefficient can be considered significant. Notice in table A.2 that as the degrees of freedom increase, the absolute value of the coefficient needed to reach significance at a given level of confidence decreases. When N is large, the odds that a high r value may occur by chance are less than when N is small.

To read table A.2, find the degrees of freedom (df) in the left column. Then proceed across the degrees of freedom row and compare the obtained r value with the value listed in each column. The heading at the top of each column indicates the probability of a chance occurrence, or the probability of error when declaring r to be significant. In the height and weight example from table 8.2, the degrees of freedom are calculated to be 8 ($N_{pairs} - 2 = 10 - 2 = 8$).

Table A.2 indicates that for $df = 8$, a correlation as high as .549 occurs only 10 in 100 times by chance alone ($p = .10$), an r value of .632 occurs only 5 times in 100 by chance ($p = .05$), and an r value of .765 occurs only 1 time in 100 by chance ($p = .01$). Previously we calculated an r value of .891 for the height and weight example. This value is greater than that reported in table A.2 for $\alpha = .01$, so we can report with a better than 99% LOC that the r value of .891 did not occur by chance. This result may also be written as $r = .891, p < .01$. Use $p < .01$ rather than $p = .01$ because .891 is greater than the critical value from table A.2 of .765.

If the obtained r value is between the values in two columns, the left of the two columns (greater odds for chance) is chosen. If the obtained r does not equal or exceed the value in the column associated with the chosen α, it is said to be not significant. Negative r values are read in the same manner, but the absolute value of r is used. A computer is not limited to .10, .05, or .01 levels of confidence. It will produce an exact p value such as $p = .013$, usually carried out to at least 3 significant figures. In practice, use a computer to calculate r and report its significance with actual p values (e.g., $p = .013$ rather than $p < .05$).

When the r value is found to be significant, the cause of the correlation cannot be determined from the correlation data alone, although logic may point to a probable cause. As indicated earlier, further experimental evaluation is needed to determine direct causation in a correlation coefficient.

We should also report a confidence interval about a correlation coefficient. Although many studies in kinesiology do not routinely report confidence intervals for the Pearson r, the confidence interval provides useful information beyond the test of statistical significance and we encourage its more widespread use. We do not explain how to calculate the confidence interval because it is quite cumbersome to calculate by hand and most statistical software packages will calculate it. However, it is important to understand what the confidence interval tells us. For the data in table 8.2, the 95% CI for our Pearson r of .89 (with 8 df) is .60 to .97. We are 95% confident that the true population correlation lies somewhere between .60 and .97. Collectively, we would report that the correlation between height and weight is r = .89 (95% CI = 0.60–0.97). The 95% CI is the most commonly used; however, other confidence interval levels can be used. These include, for example, the 90% CI (analogous to a .10 α level) and the 99% CI (analogous to a .01 α level).

Evaluating the Size of the Correlation Coefficient

When interpreting the magnitude of a correlation coefficient, it is sometimes help-ful to represent the relationship between two variables with a diagram in which each variable is depicted by a circle. This is called a **Venn diagram.** If the circles do not overlap, no relationship exists. If they overlap completely (one directly on top of the other), the correlation is +1.00. If the circles overlap somewhat, as in figure 8.5, the area of overlap represents the amount of variance in the dependent, or predicted, variable that can be explained by the independent, or predictor, variable.

The area of overlap, called the common variance, is equal to r^2. If two variables are correlated at r = .8, then they have 64% common variance ($.8^2$ = .64). This means

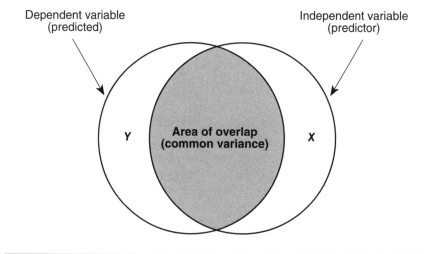

Figure 8.5 Venn diagram for a bivariate relationship.

that 64% of the variability in the Y variable can be explained by (or is shared with) variance in the X variable. The remaining 36% of the variance in Y is unexplained. This unexplained variance is responsible for error when predicting Y from X. For example, if a study reports that strength and muscular movement speed are related at around $-.80$, this indicates that they have 64% common variance. That is, 64% of both strength and speed come from common factors. The remaining 36% is unexplained; that is, it is not accounted for by the correlation coefficient.

The r^2 value is often called the **coefficient of determination** because it reflects the shared variance between X and Y. The variance that is not shared by X and Y is equal to $1 - r^2$ and is called the **coefficient of nondetermination.** Later in this chapter we address the coefficient of determination in more detail. Assume for the moment that the example correlation from above ($r = -.8$) is statistically significant. Is a coefficient of determination of .64 large or important? Guidelines exist (mainly from social sciences) regarding how to characterize r values. For example, an r value of .5 to .7 is considered low, .7 to .8 is moderate, and .9 or above is good for predicting Y values from X. Values lower than .5, if they are statistically significant, can be useful for identifying nonchance relationships among variables, but they are probably not large enough to be useful in predicting individual scores. However, in conjunction with more than one predictor variable (multiple regression), as is addressed in chapter 9, even individually modest predictors may be useful.

Bivariate Regression

When the correlation between two variables is sufficiently high, we can predict how an individual will score on variable Y if we know his or her score on variable X. This is particularly helpful if measurement is easy on variable X but difficult on variable Y. For example, it is easy to measure the time it takes a person to run 1.5 miles and difficult to measure $\dot{V}O_2$max on a treadmill or bicycle ergometer. If the correlation between these two variables is known, we can predict $\dot{V}O_2$max from 1.5-mile-run time. Of course, this prediction is not perfect; it contains some error. But we may be willing to accept the error to avoid the difficult and expensive direct measure of $\dot{V}O_2$max.

The following example illustrates how an individual's score on one variable can be used to predict his or her score on a related variable. Suppose we gave two tests of muscular endurance to a group of 15 high school wrestlers. One (X) tests the number of push-ups they can perform in two minutes. The other (Y) tests the number of seconds they can maintain a hand grip isometric force of 50% of maximum until failure. Table 8.3 shows the data.

The correlation coefficient between the push-up and isometric fatigue scores is $r = .845$ (we use three decimal places to add precision to calculations that use r) and the 95% CI = 0.59 to 0.95. Next, we plot the values for each subject on a scatter plot, as shown in figure 8.6. The push-up scores (independent variable) are plotted on the X-axis and the isometric time to failure scores (dependent variable)

TABLE 8.3

Number of Push-Ups (*X*) and Number of Seconds of Isometric Hand Grip Endurance (*Y*)

X	Y	Y_P	$(x - \bar{X})^2$	$y - Y_P$	$(y - Y_P)^2$	$(y - \bar{Y})^2$	$(Y_P - \bar{Y})^2$
72	68	65.13	386.909	2.87	8.237	293.437	203.348
70	64	63.68	312.229	0.32	0.102	172.397	164.096
69	58	62.955	277.889	−4.955	24.55	50.837	146.047
67	61	61.505	215.209	−0.505	0.255	102.617	113.103
61	62	57.155	75.169	4.845	23.474	123.877	39.501
58	51	54.98	32.149	−3.98	15.84	0.017	16.892
51	55	49.905	1.769	5.095	25.959	17.057	0.931
49	44	48.455	11.089	−4.455	19.847	47.197	5.832
47	43	47.005	28.409	−4.005	16.04	61.937	14.938
43	60	44.105	87.049	15.895	252.651	83.357	45.765
42	46	43.38	106.709	2.62	6.864	23.717	56.100
41	38	42.655	128.369	−4.655	21.669	165.637	67.486
40	36	41.93	152.029	−5.93	35.165	221.117	79.924
40	37	41.93	152.029	−4.93	24.305	192.377	79.924
35	40	38.305	300.329	1.695	2.873	118.157	157.879

$\bar{X} = 52.33$ $\bar{Y} = 50.87$ $\Sigma = 2,267.33$ $\Sigma = 0$ $\Sigma = 477.834$ $\Sigma = 1,673.73$ $\Sigma = 1,191.77$
$SD_X = 12.31$ $SD_Y = 10.56$

are plotted on the *Y*-axis. Note the example of a data point in figure 8.6 for a person who did 43 push-ups and held 50% isometric force for 60 seconds.

As discussed earlier in this chapter, a scatter plot of the data clusters around the best fit line. When the best fit line is known, any value of *X* can be projected vertically to the line, then horizontally to the *Y*-axis, where the corresponding value for *Y* may be read.

The best fit line is graphically created by drawing it in such a way that it balances the data points. That is, the total vertical distance from each point below the line up to the line is balanced by the total vertical distance from each point above the line down to the line. The average distance of the points above the line is the same as the average distance of the points below the line. The vertical distance from any point to the line is called a **residual.** Residuals will be positive and negative and the sum of residuals will be equal to zero (in the same way that the sum of the deviation scores from the mean sum to zero).

Using the best fit line, we can predict the time to failure that would be performed based on the number of push-ups performed. We simply find the push-up score on the *X*-axis—for example, 58 push-ups—and proceed vertically up to the line and then horizontally over to the *Y*-axis to read the estimated time to failure. The estimated time to failure is about halfway between 50 and 60, so we estimate that the wrestler would have reached failure in about 55 seconds. This answer is

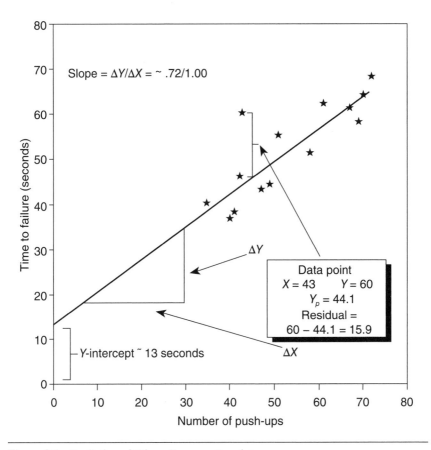

Figure 8.6   Prediction of Y from X on a scatter plot.

not exact because we must visually, or graphically, locate the point on the Y-axis.

The Y-coordinate can be determined more precisely by using the general algebraic formula for a straight line:

$$Y = a + bX, \tag{8.03}$$

where Y is a value on the Y-axis, X is a value on the X-axis, b is the slope of the line, and a is the Y-intercept of the line.

The **Y-intercept** is the point at which the extension of the best fit line intercepts the Y-axis. In figure 8.6, the Y-intercept is about 13.0 seconds.

The slope is the tilt of the best fit line. In figure 8.6, the line advances about 0.72 seconds on the Y-axis for every push-up on the X-axis when it progresses from left to right on the graph. This ratio, the distance traveled on the Y-axis divided by the distance traveled on the X-axis, is called the slope of the line. The slope tells us how much change in the Y-axis we would expect to see, on average, for one unit change in X. The slope is the tangent (opposite over adjacent) of the angle of intercept (the angle formed by the best fit line and a line parallel to the X-axis). In figure 8.6, the slope is approximately +.72. From a table of trigonometric values, we can determine that the tangent of +.72 is about 36°. If the correlation coefficient is negative, the line tilts the other way and the slope is negative.

When the means, standard deviations, and correlation coefficient are known, the slope and the Y-intercept are easily calculated. The formula for the slope (b) of the line is

$$b = \frac{r \; \sigma_Y}{\sigma_X} \tag{8.04}$$

and the formula for the Y-intercept (a) is

$$a = \overline{Y} - \frac{r \; \sigma_Y}{\sigma_X} \overline{X} = \overline{Y} - b\overline{X}. \tag{8.05}$$

The slope and intercept have units. The a units are those on the Y-axis. For the data in table 8.3, a is seconds. The units for slope equal the quotient of Y-axis units over X-axis units. For this example, b is seconds/push-ups.

Using equation 8.04, we solve for slope as follows:

$$b = \frac{0.845 \; (10.56 \; \text{seconds})}{12.31 \; \text{push-ups}} = 0.725 \; \text{seconds per push-up.}$$

We can interpret this as estimating that for every additional push-up the time to failure increases, on average, by 0.725 seconds.

Then we use equation 8.05 to find the Y-intercept as follows:

$$a = 50.87 \; \text{seconds} - 0.725 \; \text{seconds per push-up} \; (52.33 \; \text{push-ups}) =$$
$$12.93 \; \text{seconds.}$$

The generalized equation for Y_p becomes

$$Y_p = 12.93 \; \text{seconds} + 0.725 \; \text{seconds per push-up} \; (X); \; \text{or}$$

$$Y_p = 12.93 + .725(X).$$

If $X = 58$ push-ups, then

$$Y_p = 12.93 + 0.725 \; (58) = 54.98 \; \text{seconds.}$$

This value for Y_p (54.98 seconds) when $X = 58$ push-ups is quite close to the value that we graphically predicted earlier (55), but the algebraic solution is more accurate. Now that the generalized equation $Y_p = 12.93 + 0.725 \; (X)$ is known, we can substitute any value for X and easily predict its concomitant Y value.

This statistical process is called **bivariate regression** analysis (also known as simple linear regression). **Regression** was first observed in 1877 by Sir Francis Galton (1822–1911) when he noted that characteristics of offspring tended to cluster between the characteristics of their parents and the mean of the population (Kotz and Johnson, 1982, p. 275). In other words, the characteristics of the offspring tended to regress toward the mean. By using this technique, scientists could predict the characteristics of the offspring. In application today, the word prediction is probably more descriptive of the process than the word regression. When you hear the term regression, think prediction.

The equation that was defined by the a (12.93 seconds) and b (0.725 seconds per push-up) terms calculated using equations 8.04 and 8.05 allows us to predict Y using X. But as was mentioned earlier, this technique involves some error. How large is the error in the prediction?

Determining Error in Prediction

To demonstrate how to calculate the error in prediction we use the push-up example from figure 8.6. Each data point has an error factor (the distance from the point to the best fit line), called a residual. These distances represent the part left over between each predicted Y value and the actual value. Sometimes they are called residual errors because they represent the error in the prediction of each Y value. One would predict that the wrestler who did 58 push-ups (see figure 8.6) would last about 55 seconds (Y_p). If he actually lasted 51 seconds (Y), the residual error would be −4 seconds, the difference between the actual ($Y = 51$ seconds) and the predicted ($Y_p = 55$ seconds) values.

The best fit line represents the best prediction of Y for any X value. Some residuals are large, and some are quite small; indeed, some points fall right on, or very close to, the line. By using the algebraic solution for the best fit line and the residuals, we can calculate the predicted value for Y and the amount of error in the prediction.

For example, what is the prediction of Y and the error of prediction for a different subject (not one of the original 15) who did 60 push-ups? By substitution into the prediction equation we get $Y_p = 12.93$ seconds + 0.725 seconds per push-up (60 push-ups) = 56.43 seconds. Thus, the predicted Y value is 56.43 seconds. But this prediction has some degree of error. To determine the error in prediction, we could use the generalized equation to predict a Y value for each subject and then compare the predicted Y value (Y_p) with the actual Y value to determine the amount of error in prediction for each subject ($Y - Y_p$). The result of such an analysis is presented in table 8.3. The values for $Y - Y_p$ represent all the errors, or residuals, around the best fit line. Notice that the sum of the residuals is zero. Also, note that the sum of squares of the residuals is 477.834. We can assume that the residuals are randomly distributed around the line of best fit and that they would fall into a normal curve when plotted. In chapter 5, we defined standard deviation as the square root of the average of the squared deviations from the mean. Therefore, the standard deviation of the residuals can be calculated with a modification of equation 5.03 as follows:

$$\sigma_{\text{res}} = \sqrt{\frac{\Sigma\left(y - Y_P\right)^2}{N}}$$

Recall from chapter 5 that when we calculated the standard deviation for a sample, we did not use N in the calculation but rather used $N - 1$, where $N - 1$

was a degrees of freedom term. Here we also divide by a degrees of freedom term, but we lose a degree of freedom for each parameter we estimate. Therefore, in the case of bivariate regression we lose a degree of freedom each for X and Y so we use $N - 2$. (This is the same degrees of freedom value used in testing the statistical significance of the Pearson r.) Therefore, we modify the previous equation so that the denominator under the radical is $N - 2$, and we call the result of this calculation the standard error of estimate (SE_E):

$$SE_E = \sqrt{\frac{\Sigma\left(y - Y_P\right)^2}{N - 2}}, \qquad (8.06)$$

then

$$SE_E = \sqrt{\frac{477.834}{13}} = 6.06 \text{ seconds.}$$

The standard error of the estimate may be interpreted as the standard deviation of all the errors, or residuals, made when predicting Y from X. It is calculated using the sample data that was used to generate the prediction equation, and has the same units as the units of the dependent variable (Y). Because the standard error of the estimate is the standard deviation of a set of scores (the residuals) that are normally distributed, the standard error of the estimate can be interpreted as a Z score of ±1.0. We know that 68% of all errors of prediction will be between ±1 × SE_E, 90% will be between ±1.65 × SE_E, 95% will be between ±1.96 × SE_E, and 99% will fall between ±2.58 × SE_E. (See table 7.1 on page 89 for a review of the relationships between Z, level of confidence, and p.)

Equation 8.06 is generally not used with hand calculations because of the tedious calculations involved. Another formula for the standard error of the estimate is easier to use. This formula requires only the standard deviation of the Y variable and r_{XY}:

$$SE_E = \sigma_Y \sqrt{1 - r^2}. \qquad (8.07)$$

This formula (sometimes denoted by σ_{est}) yields an answer similar to that yielded by equation 8.06:

$$SE_E = 10.93\sqrt{1 - (.845)^2} = 5.84 \text{ seconds.}$$

Equation 8.07 underestimates the standard error of the estimate from formula 8.07 when sample sizes are small (Hinkle et al., 1988), which explains the numerical difference in the standard error of the estimate calculated from equations 8.06 and 8.07. Equation 8.07 is more commonly used in statistics texts, but equation 8.06 is more accurate and reflects the calculation used in most statistical software.

We are now prepared to estimate error for the predicted Y value for a subject who performed 60 push-ups. Remember that we are using sample data to predict population values, so the standard error of the estimate allows us to predict an interval of Y scores where we think Y scores from the population will fall for any given X value. For the data in table 8.3, for a given push-up score (X) we can use the standard error of the estimate to predict an interval of isometric time to failure scores that should occur in the population. To illustrate, the generalized equation predicts the Y value to be

$$Y_p = 12.93 + 0.725\ (60) = 56.3 \text{ seconds.}$$

The standard error of the estimate (from equation 8.06) in this prediction is ±6.06 at the 68% LOC. This means that the interval of 50.17 to 62.23 seconds (56.2 ± 1.0 × 6.06 seconds) should encompass about 68% of all Y scores that occur in the population when $X = 60$. When the Z score is 1.96, the error in the prediction decreases to 5% (1.96 × 6.06 = 11.9). The prediction interval then becomes

$$Y_p = 56.4 \pm 1.96\ (6.06) = 56.4 \pm 11.9 \text{ seconds.}$$

In other words, when a member of the population performs 60 push-ups, our best estimate is that that person would last about 56 seconds on the isometric test, and we are 95% confident that that person will last between 44.5 and 68.3 seconds, or in integer values 45 to 68 seconds. The final form of the generalized regression equation is as follows:

$$Y_p = a + b(X) \pm Z\ (SE_E), \tag{8.08}$$

where Z is a value that produces the desired level of confidence.

Once this formula is established for a given problem, predictions with error estimations can be made from any value of X when the means, standard deviations, and correlation are known. The critical factors are the standard deviation of Y (SD_Y) and r. When SD_Y is small and r is large, the errors in prediction are small. When r decreases below about .7, the errors may become so large that the prediction may not be useful. In the example described previously, if $r = .7$ rather than .845, the error increases:

$$SE_E = 10.93\sqrt{1-(.7)^2} = 7.81 \text{ seconds.}$$

And if r is .5, then

$$SE_E = 10.93\sqrt{1-(.5)^2} = 9.47 \text{ seconds.}$$

Each researcher must determine an acceptable error level for the data being analyzed. That level will depend on the consequences of erroneous predictions.

Another View of the Coefficient of Determination

Recall from earlier this chapter when we noted that r^2 reflects the variance in Y that is accounted for by X. We did not explain that at the time and rather asked you to take our word for it. Consider now the variance in Y. In parametric statistics, the variance in Y, the dependent variable, is partitioned into different pieces. The phrase analysis of variance (ANOVA) is used to describe this partitioning of variance. [ANOVA is traditionally described in the context of determining differences between means (see chapter 11) but applies to all parametric analyses.] Recall also from the discussion of variance in chapter 5 that the calculation of variance is a sums of squares (SS) procedure. First, consider what is called the **total sums of squares** (SS total or SS_T). The SS_T is calculated as

$$SS_T = \Sigma\left(y - \overline{Y}\right)^2, \tag{8.09}$$

where y is an individual Y score in the data set. Equation 8.09 tells us to calculate the difference between each Y score and the mean of Y, square the differences, and then add the squares. This is equal to the numerator of the variance equation (see equation 5.02 on page 62).

Similarly, we can calculate the sums of squares due to how each predicted score (Y_P) varies from the mean. This represents the variance in Y that is due to the regression of Y on X and is called the regression sums of squares (SS regression or SS_R). The SS_R is calculated as

$$SS_R = \Sigma\left(Y_P - \overline{Y}\right)^2. \tag{8.10}$$

Compare equations 8.09 and 8.10 and notice that for each subject, the closer each predicted score (Y_P) is to its corresponding actual score (y), the closer the ratio of the regression sums of squares is to the total sums of squares. Therefore, the better the prediction of Y based on X, the more variance is explained by the regression of Y on X.

Finally, the error sums of squares (sums of squares error or SS_E) reflects the variance in Y that is not explained by X and is calculated as

$$SS_E = \Sigma\left(y - Y_P\right)^2. \tag{8.11}$$

We can summarize these relationships as

$$SS_T = SS_R + SS_E. \tag{8.12}$$

Equation 8.12 reflects the fact that we have partitioned the variance into two pieces: the variance explained by regressing Y on X and the variance that is unexplained. We can then calculate a ratio reflecting how much variance in Y is attributable to the regression of Y on X as follows:

$$r^2 = \frac{SS_R}{SS_T}. \tag{8.13}$$

Using the data in table 8.3, $SS_T = 1{,}673.73$, $SS_R = 1{,}191.77$, and $SS_E = 477.834$. From equation 8.13, $r^2 = 1{,}191.77/1{,}673.73 = \sim .71$. Earlier we noted that the correlation between push-ups and isometric time to failure was $r = .845$. If we square the Pearson r we get $(.845)^2 = \sim .71$. When we square the Pearson r, we are calculating the ratio of the regression sums of squares to total sums of squares.

Statistical Significance and Confidence Intervals for Bivariate Regression

Earlier in this chapter we showed how to test for the statistical significance of the Pearson r and that we can construct a confidence interval about the r value. We can also test whether the regression equation is statistically significant and can construct a confidence interval about the slope coefficient (b) from the regression analysis. As it turns out, for bivariate regression the test of the significance of r and the significance of b is redundant. If r is significant, the regression equation is significant. However, we address the process here because it will help you understand the workings of multiple regression, which is addressed in chapter 9.

To test the statistical significance of b and calculate the confidence interval about b, we use a test statistic called t. The t statistic, which is somewhat analogous to a Z score, is examined in more detail in chapter 10, where techniques to examine differences between mean values are introduced. However, the t statistic is useful for more than just comparing means. The t statistic is based on a family of sampling distributions that are similar in shape to the normal distribution, except the tails are thicker than a normal distribution. When sample sizes are small, the t distributions more accurately reflect true probabilities than does the normal distribution.

When we test the statistical significance of b, we are testing whether the slope coefficient is significantly different from zero. (We can also test whether b differs from some hypothesized value other than zero, but this is rarely done in kinesiology.) The statistical hypotheses for a two-tailed test are as follows:

$$H_0: b = 0$$

$$H_1: b \neq 0.$$

The degrees of freedom for the test are $N - 2$. The test statistic t for this analysis is calculated as (Lewis-Beck, 1980):

$$t = \frac{b}{SE_b}, \tag{8.14}$$

where b is the slope coefficient and SE_b is the standard error of b. We calculate the standard error of b as

$$SE_b = \sqrt{\frac{\left[\dfrac{\Sigma\left(y - Y_P\right)^2}{N - 2}\right]}{\Sigma\left(X_i - \overline{X}\right)^2}}. \tag{8.15}$$

Notice what this equation is telling us. Inside the radical, the numerator is reflecting how much error is in the predictions, because we are adding up the squared differences between the actual Y scores and the predicted Y scores (once again a sums of squares calculation) and dividing it by a degrees of freedom term (a sort of mean square because we are taking a sum and dividing it by a degrees of freedom term). The denominator is the sums of squares for X, the independent variable. The better the prediction of Y based on X, the smaller will be the standard error of b. The smaller the standard error of b, the larger will be the t statistic, all else being equal. The larger the t statistic, the less tenable is the null hypothesis.

Fortunately, most of the work in this calculation has been done already in table 8.3. Substituting the appropriate terms into equation 8.15, we get

$$SE_b = \sqrt{\frac{477.834 / (15 - 2)}{2267.33}} = 0.127$$

To calculate t, we substitute the appropriate values into equation 8.14 so that

$$t = \frac{.725}{.127} = 5.71.$$

To determine the statistical significance of t, we can use table A.3a in the appendix. To be statistically significant at a two-tailed $\alpha = .05$ and with $df = 13$, the calculated t must be greater than or equal to 2.16. Because 5.71 is greater than 2.16, we can reject H_0 and conclude that the number of push-ups is a significant predictor of isometric time to failure. Recall that t is analogous to Z but is used in cases of small sample sizes. Table 7.1 on page 89 shows that a Z score of 1.96 corresponds with a 95% LOC (two-tailed). With 13 df, the 95% LOC (two-tailed) corresponds to a t value of 2.16. The t value associated with the 95% LOC value (2.16) is larger than the Z value associated with the 95% LOC value, which is reflective of the thicker tails of the t distribution relative to the normal distribution. Examine table A.3a for the two-tailed data under the .05 α level. Notice that as the degrees of freedom get larger, the tabled values for t get smaller so that at $df = \infty$, the values for t and Z are equal. That is, as the degrees of freedom increase, the t distribution approaches a normal distribution. (One can consider the normal distribution as just a special case of the t distribution with infinite degrees of freedom.)

We also use the t distribution to calculate a confidence interval about the slope coefficient. The confidence interval is calculated as

$$b \pm t(SE_b). \qquad (8.16)$$

where t is the critical t value from table A.3a for a given α and degrees of freedom. For the push-up data, the t value we need (95% LOC, $df = 13$) is once again 2.16. Previously, we determined that $b = .725$ and $SE_b = 0.127$. Therefore, the 95% CI for the sit-up data is

$$.725 \pm (2.16)(0.127) = 0.44 \text{ to } 0.99.$$

In a journal, we might report this as $b = 0.725$ seconds per push-up, 95% CI = 0.44, 0.99. Thus, we are 95% confident that the population slope lies somewhere from 0.44 and 0.99. Notice that zero is not inside the 95% CI, which means that the slope is statistically significant ($p < .05$).

In practice, these calculations are rarely performed by hand. They are too cumbersome and it is easy to make arithmetic errors. In addition, at each step rounding error will occur. Computers perform these calculations faster and more accurately. Plus, computer software will generate actual p values that are more precise than those generated from looking values up in a table and reporting, for example, $p < .05$. Nonetheless, seeing where the numbers come from is useful in developing an understanding of how these tests work.

Putting It All Together

For the data in table 8.3, we have determined that the correlation (Pearson r) between the number of push-ups performed and the isometric time to failure is $r = .845$ (95% CI = 0.59, 0.95). The regression equation is $Y_p = 12.93$ seconds $+ 0.725$ seconds per push-up (X), $r^2 = .71$, and $SE_E = 6.06$ seconds. The 95% CI for b is 0.44 to 0.99.

Homoscedasticity

Recall from chapter 7 that an assumption of parametric inferential statistics is that the data are normally distributed. A further assumption of regression analysis is called **homoscedasticity.** If data exhibit homoscedasticity, this means that the variance of the residuals does not vary (Lewis-Beck, 1980). To illustrate, figure 8.7 shows a plot of the residuals (Y-axis) from table 8.3 versus the number of push-ups.

An examination of the residuals in figure 8.7 shows no apparent relationship between size of the residuals and the magnitude of the independent variable.

If the variance of the residuals is not constant, the data are said to exhibit **heteroscedasticity.** To illustrate, figure 8.8 shows a simulation where one-repetition-maximum squat performance is regressed against body weight.

Notice that the scatter of the data about the regression line appears to increase as body weight increases. In other words, absolute errors in the prediction of squat strength increase with higher body weights. A residuals plot as shown in figure 8.9 shows this more clearly. The plot of the residuals exhibits a fan pattern.

The effect of heteroscedasticity is to inflate type I error, so tests of statistical significance may be invalid. However, the calculation of the slope and Y-intercept estimates using regression still result in the best linear estimates for these parameters. Both Kachigan (1986) and Tabachnick and Fidell (1996) discuss this condition in more detail. Several computational approaches can be used to test for heteroscedasticity beyond just a visual examination of the residuals. In many cases, heteroscedasticity can be corrected by transforming the data (e.g., converting the scores to logarithms).

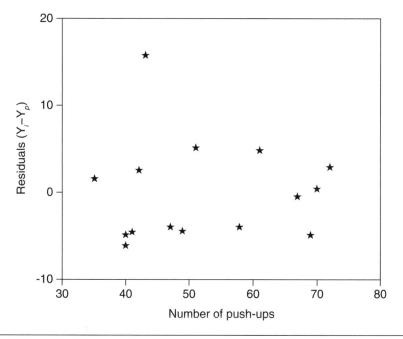

Figure 8.7   Residuals plot.

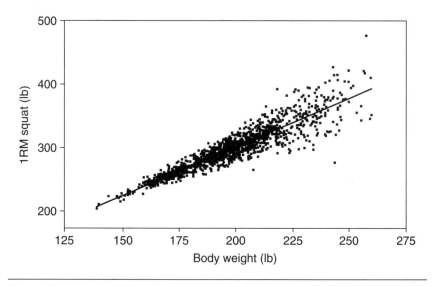

Figure 8.8   Simulated data showing heteroscedasticity.

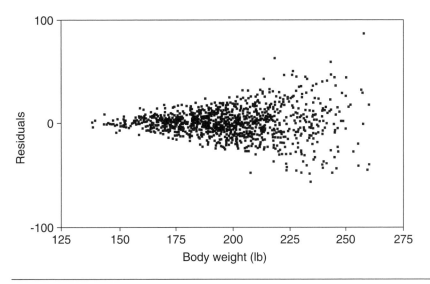

Figure 8.9 Residuals plot showing a fan pattern.

Summary

Correlation is designed to determine the relationships between or among variables. Correlation requires measurements of two variables on the same set of subjects. If the correlation is high and positive, a person is likely to score high on both variables, or low on both variables, or in the middle on both variables. If the correlation is high and negative, a person will tend to score high on one variable and low on the other. The absolute value of the coefficient can be evaluated by using table A.2 in appendix A to determine the probability that the correlation is reliably different from 0.00—that is, the probability that the correlation did not happen by chance. Table A.2 is limited to .10, .05, or .01 values, but a computer can produce the exact p value.

Regression is often used to predict one variable from the value of another. This ability to predict is often useful when it is difficult to measure one of the variables. The prediction always has some error in it, unless the correlation is +1.00 or −1.00. When deciding to use prediction, we must determine whether it is more acceptable to tolerate the error in the prediction or the difficulty of direct measurement.

Problems to Solve

1. Estimate whether the correlation between the following pairs of variables is positive, negative, or near zero:
   a. Height and weight
   b. Upper body strength and distance in the shot put
   c. Arm length and softball throw for distance
   d. $\dot{V}O_2$max and place in a cross-country meet
   e. Standing long jump and vertical jump test scores
   f. Swimming sprint speed and tennis serving accuracy
   g. Golf score and the number of hours spent practicing
   h. School grade point average and hours spent watching television
   i. Daily temperature and the average weight of clothing worn
   j. $\dot{V}O_2$max and 2-mile-run time

2. Use table A.2 in appendix A to determine the p value for the following:

	r	N
a.	.600	13
b.	.600	20
c.	.600	8

3. A physical education department wanted a single test of upper body strength that was easy to administer. Dips on the parallel bars and pull-ups on the horizontal bar were considered good tests. One faculty member thought that both tests were not needed because the correlation between the two was probably high. To evaluate this assumption, 141 students were tested on both criteria. The faculty member let X represent dips on the parallel bars and Y represent pull-ups and calculated the following from the data:

   $\Sigma X = 3,416$, $\Sigma Y = 1,899$, $SD_X = 8.84$, $SD_Y = 4.70$, $\Sigma Z_X Z_Y = 109.416$.

   Calculate the following:
   a. The mean of each variable
   b. The correlation between the two variables
   c. The level of confidence and the p value reached by the coefficient
   d. The predicted number of pull-ups for a student who performed 20 dips
   e. The standard error of the estimate
   f. The predicted range of possible pull-up scores at the 95% LOC for a student who performed 20 dips

4. In a motor learning lab, data on time in milliseconds for a simple reaction time (RT) and movement time (MT) for a linear arm movement were collected on 10 subjects. Does a relationship exist between RT and MT? (Data from California State University Northridge motor learning lab, courtesy of Tami Abourezk.)

RT	MT
271	354
268	435
198	211
345	411
169	288
209	413
322	158
209	333
199	425
216	378

5. A kinesiology major wanted to predict $\dot{V}O_2$max based on the mile run. To develop the regression equation, she obtained $\dot{V}O_2$max values (ml/kg/min) in the exercise physiology laboratory on 18 students. Two days later, she measured the same 18 students on the mile walk–run with scores reported as total time in seconds. The data are as follows.

Subject	Mile walk–run	$\dot{V}O_2$max
1	250	60.3
2	315	57.2
3	420	55.4
4	410	51.4
5	436	52.5
6	511	45.6
7	460	38.4
8	510	41.5
9	530	39.6
10	586	33.2
11	591	37.7
12	600	40.1
13	626	32.0
14	643	35.4
15	650	33.7
16	675	35.9
17	710	27.4
18	720	25.3

a. Using a computer, calculate the means, standard deviation, and correlation coefficient between the two variables.

b. What is the probability that the correlation coefficient happened by chance? Why is the coefficient negative?

c. What is the slope, the Y-intercept, and the standard error of the estimate for the regression line?

d. Estimate $\dot{V}O_2$max for a person who runs the mile in 540 seconds. If you wanted your estimate to be accurate at the 95% LOC, what is the error factor in your estimate?

6. Measure the height in inches and the weight in pounds of about 20 of your classmates.
   a. What is the correlation between the two variables?
   b. Is the correlation significant?
   c. If you concluded that it is significant, what is the probability of error in your conclusion?
   d. Find a classmate who was not one of the original 20 subjects. Develop a regression equation to predict weight from height. Using this person's actual height, make a prediction of their weight with a level of confidence of 95% ($p = .05$). What is the residual for that person?

See appendix C for answers to problems.

Key Words

best fit line
bivariate regression
coefficient
coefficient of determination
coefficient of nondetermination
correlation
covariance
curvilinear
heteroscedasticity
homoscedasticity

Pearson's product moment
   correlation coefficient
regression
residual
scatter plot
standard error of the estimate
total sums of squares
Venn diagram
Y-intercept

$$H = \left[\frac{12}{18(18+1)} \right] \left[\frac{82.5^2}{6} + \frac{54.5^2}{6} + \frac{34.0^2}{6} \right] - 3(18+1)$$

$$H = [.035][1822.09] - 57$$

$$H = 6.77$$

Multiple Correlation and Multiple Regression

Threadhe percentage of fast-twitch muscle fibers can be determined from muscle biopsy, but this is an invasive procedure. An investigator wishes to be able to predict the percentage of fast-twitch muscle fibers in the vastus lateralis noninvasively. Because fast-twitch fibers tend to fatigue rapidly the investigator decides to perform a regression analysis between the percentage of fast-twitch fibers determined from muscle biopsy (dependent variable) and the rate of fatigue from a maximal 30-second cycle test (independent variable). The investigator also knows that fast-twitch fibers are particularly good at generating force rapidly, so it is decided to see if the prediction can be improved by adding another independent variable – the rate of force development from a maximal isometric leg extension task – to the regression analysis. Multiple regression and multiple correlation are tools that can be used to examine the combined relations between multiple predictors and a dependent variable.

In chapter 8, techniques to quantify the degree of relationship between a dependent variable (Y) and an independent variable (X) were presented. The Pearson r and the coefficient of determination (r^2) gave us a sense of the strength of association between X and Y, and the technique of bivariate regression allowed us to develop a prediction equation between X and Y. The standard error of the estimate (and r^2) from the regression analysis provided information about the accuracy of the prediction equation. In this chapter, these ideas are expanded to include multiple independent variables in the prediction of a single dependent variable. The advantage of using multiple independent variables is that, if we choose our variables wisely, we can increase the amount of information available for understanding and predicting the dependent variable. For example, suppose that a researcher in athletic training wishes to predict when an athlete can start running again after a surgical repair of a torn anterior cruciate ligament. Here, the Y variable might be time (in days) from the time of injury to the time that running starts again. Knowing the isokinetic strength of the quadriceps prior to the tear or knee range of motion after surgery may have predictive information. Here, isokinetic strength and knee range of motion would be independent variables. The independent variables together would provide us with more total information about the dependent variable than would any single independent variable alone.

In chapter 8, the Venn diagram was used to illustrate the idea of shared variance as quantified by the coefficient of determination. Figure 9.1 is an expansion of the Venn diagram to illustrate the scenario described previously. The area identified by 1 is the shared variance between X_1 (preinjury isokinetic strength) and Y (time until running after anterior cruciate ligament repair). By adding X_2 (postsurgical range of motion) to the analysis, we have added the explained variance denoted by 2 in the figure. By including a second independent variable, we have accounted for more variance (1 + 2) in Y than by using any one predictor variable alone.

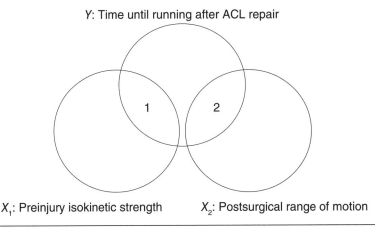

Figure 9.1   Venn diagram illustrating shared variance between one dependent variable and two independent variables.

In this chapter, we do not detail the underlying calculations because multiple regression analysis uses techniques of matrix algebra, which goes beyond the scope of this book. In addition, these procedures are almost always performed by a computer program because the calculations are complicated and tedious. However, we do work through examples to convey the concepts behind the calculations and to facilitate the interpretation of the results of these analyses. A detailed analysis of multiple regression can be found in advanced statistical texts; Tabachnick and Fidell (1996, chapter 5) present a particularly good analysis that includes theoretical, practical, and computer applications.

Multiple Correlation

As in chapter 8, we can calculate indices of correlation between multiple variables. The **multiple correlation** coefficient is symbolized using a capital R to distinguish it from the bivariate Pearson r. Unlike the Pearson r, the multiple correlation coefficient can vary only from 0 to 1.0, where $R = 0$ indicates that the independent variables have no relationship with the dependent variable and where $R = 1.0$ indicates that the X variables have a perfect relationship with Y. In practice, a multiple correlation coefficient is rarely calculated by itself but rather is calculated in conjunction with multiple regression. Furthermore, the squared multiple correlation coefficient (R^2) is typically more useful than the multiple correlation coefficient (R) because we can interpret R^2 values as the multivariate coefficient of determination. For example, if the squared multiple correlation coefficient between Y and the X variables is .50, we can interpret that as 50% of the variance in Y is accounted for (or shared with) the X variables.

Partial Correlation

When we have multiple variables, we can also calculate a type of correlation coefficient that allows us to quantify the relationship between Y and an X variable after removing the effect of another X variable. We do not detail the calculations here because they can be cumbersome and are easily gleaned from statistical software. However, we do address the logic behind and interpretation of the calculation. Let us label one of the variables X_1 and the other X_2. A **partial correlation** coefficient allows us to examine the relationship between Y and X_1 after removing the influence of X_2 from both Y and X_1. We symbolize a partial correlation coefficient between Y and X_1 after removing the effect of X_2 as $r_{YX1.X2}$ where the X_2 after the dot indicates that X_2 has been removed from both Y and X_1. As with the Pearson r we can square the partial r (symbolized $r^2_{YX1.X2}$), but in this case the squared partial correlation coefficient reflects the variance in Y accounted for by X_1 after removing the effects of X_2 from Y and X_1. To illustrate, consider the situation shown in figure 9.2.

In figure 9.2, the variables Y, X_1, and X_2 all share variance with each other. The area identified by A plus the area identified by B reflects the shared variance between Y and X_1. The area identified by C plus the area identified by B reflects the shared variance between Y and X_2. The area identified by B is variance collectively shared by all three variables, whereas the area identified by D is variance in Y that is not shared with either X_1 or X_2. If we calculate $r^2_{YX1.X2}$ we are quantifying the shared variance between Y and X_1 after removing X_2. This is shown in figure 9.3.

In figure 9.3, the variance of X_2 has been removed from both Y and X_1. This is often referred to as partialing out the effect of X_2. The remaining variance in Y after partialing out X_2 is the sum of the variances reflected by A and D (i.e., A +

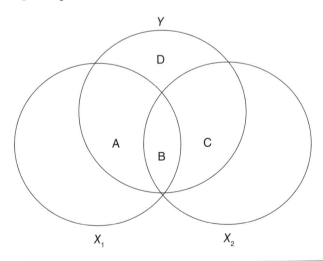

Figure 9.2   Venn diagram illustrating the shared variance between three intercorrelated variables.

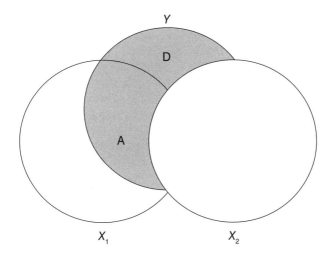

Figure 9.3 Venn diagram illustrating the concept of removing or partialing out the effect of X_2 from Y and X_1.

D reflects what is left over in Y after removing the effect of X_2). The variance in Y accounted for by X_1 after partialing out X_2 is A. Therefore, the percentage of the variance in Y accounted for by X_1 is conceptually A/(A + D) × 100. We say conceptually because Venn diagrams are conceptual models used to convey the idea of shared variance rather than mathematical evidence.

As an example, suppose that a researcher wishes to examine the relationship between muscle strength and age in elementary school children. The researcher knows that as kids get older they simultaneously get stronger and heavier. The researcher may wonder whether the increase in strength with age is simply due to the increase in weight or whether a maturational effect leads to increases in strength beyond that that can be attributed to increased weight. Partial correlation can be used to examine this question. That is, the researcher could examine the effect of age on strength after partialing out the effect of weight. In figure 9.3, the Y variable is strength, X_1 is age, and X_2 is weight. A now reflects the variance in strength that is accounted for by age after partialing out the effect of weight. We revisit this example in the discussion of multiple regression that follows.

Multiple Regression

The general formula for **multiple regression** is represented as

$$Y_p = a + b_1X_1 + b_2X_2 + b_3X_3 \ldots + b_kY_k, \tag{9.01}$$

where $b_1, b_2, b_3, \ldots, b_k$ are slope coefficients that give weight to the independent variables according to their relative contributions to the prediction of Y. The number of predictor, or independent, variables is represented by k, and a is a constant that is similar to the Y-intercept. When raw data are used, the b values are in raw score units. Sometimes multiple regression is performed on standardized scores (Z) rather than on raw data. In such cases, all raw scores are converted to the same scale, and the b values are then referred to as **beta (β) weights.** Beta weights perform the same function as b values; that is, they give relative weight to the independent variables in the prediction of Y. In common usage, b values are sometimes called beta weights, but the word beta is only properly used when the equation is in a Z score format.

Multiple regression is used to find the most satisfactory solution to the prediction of Y, which is the solution that produces the lowest standard error of the estimate (SE_E). Each predictor variable is weighted so that the b values maximize the influence of each predictor variable in the overall equation.

Methods of Multiple Regression

Two approaches exist for generating multiple regression equations. In one approach, the investigator uses a computer algorithm to generate the equation. Common techniques include procedures called forward selection, backward elimination, and stepwise. In all of these approaches, the investigator records the scores on the independent and dependent variables from a representative sample from the population of interest. For example, investigators have developed regression equations to predict percent fat of high school wrestlers from skinfold thickness of various body sites. These are used to determine acceptable weight classes for the competitors (Thorland et al., 1991). That is, the equations are used to put limits on how much weight the wrestlers can lose based on their estimated body fat percentages. The population here is all high school wrestlers. The dependent variable (Y) is the body fat percentage from hydrostatic weighing and the independent variables are the various skinfold thicknesses. All the data are entered into the computer data file and, using the particular algorithm chosen by the investigator, the computer builds the prediction equation.

In the other approach, the investigator specifically tells the computer how to construct the regression equation. **Hierarchical multiple regression** is the process in which we set up a hierarchical order for inclusion of the independent variables. This may be useful if some of the independent variables are easier to measure than others or if some are more acceptable to use than others. Hierarchical approaches are also used if the investigator is examining a specific model. This model building approach is typically used in situations in which the investigator is not interested in prediction per se, but is building a statistical model to test a substantive hypothesis. In the hierarchical approach, the investigator specifies to the computer the order in which to enter variables into the equation.

Forward Selection

Forward selection is a computer-generated approach in which the algorithm starts with no variables in the model and then X variables are added one at a time into the regression equation. To illustrate, consider the data in table 9.1. The data in the table contain age, body size, and strength (isokinetic knee extension torque) measures in adolescent males. For this example we use a forward selection approach to develop a multiple regression equation that allows us to predict isokinetic strength based on body size and age.

A first step in the analysis is to examine the descriptive statistics, including skewness and kurtosis, and to decide whether any variables need to be transformed. The next step is to create what is called a correlation matrix. A correlation matrix is a table that contains all of the bivariate correlations in the data. Most multiple regression software packages will calculate the matrix for you. The correlation matrix for the data in table 9.1 is presented in table 9.2.

TABLE 9.1

Raw Data for Multiple Regression Example

Subject no.	Age (years)	Fat-free weight (kg)	Height (cm)	Weight (kg)	Body mass imdex (kg/m²)	Fat (%)	Isokinetic torque (Nm)
1	11.42	33.13	146.50	35.70	16.63	7.20	40.68
2	12.75	50.10	161.50	56.80	21.78	11.80	89.50
3	14.00	29.30	148.10	31.30	14.27	6.40	44.75
4	13.33	44.36	165.10	49.40	18.12	10.20	89.50
5	13.92	49.83	153.70	55.00	23.28	9.40	89.50
6	11.08	36.41	147.90	41.10	18.79	11.40	65.09
7	13.25	53.93	166.90	60.50	21.72	10.86	89.50
8	11.50	35.21	140.00	39.20	20.00	10.17	61.02
9	11.17	27.84	136.50	30.60	16.42	9.02	31.19
10	13.58	45.06	162.60	49.30	18.65	8.60	73.22
11	11.08	28.58	134.80	30.70	16.90	6.90	36.61
12	12.83	48.13	156.50	51.48	21.02	6.50	65.09
13	12.67	55.17	163.10	59.19	22.25	6.80	105.77
14	15.92	71.76	177.19	90.04	28.68	20.30	148.11
15	17.33	55.49	176.91	69.63	22.25	8.60	146.75
16	17.67	56.22	148.31	70.53	32.07	9.60	163.06
17	14.67	41.48	156.21	52.05	21.33	5.40	77.45
18	16.75	52.96	175.11	66.45	21.67	6.90	146.75
19	16.67	52.15	175.01	65.43	21.36	8.00	114.14
20	14.83	56.04	175.49	70.31	22.83	11.60	97.83
21	15.83	63.54	185.60	79.72	23.14	10.20	146.75
22	16.42	64.89	175.11	81.42	26.55	15.60	146.75
23	16.42	44.74	175.01	56.13	18.33	9.60	114.14
24	15.58	67.69	175.49	84.94	27.58	21.60	146.75
$\bar{X}$	14.20	48.50	161.61	57.37	21.48	10.11	97.08
SD	2.13	12.38	14.62	17.36	4.16	4.03	40.86
ΣX^2	4,940.25	59,979.75	631,741.48	85,924.15	11,474.74	2,826.85	264,584.89

TABLE 9.2

Correlation Matrix for Multiple Regression Example

	Age	Fat-free weight	Height	Weight	Body mass index	Fat	Isokinetic torque
Age	1.00						
Fat-free weight	.696	1.00					
Height	.731	.823	1.00				
Weight	.775	.984	.840	1.00			
Body mass index	.627	.843	.444	.847	1.00		
Fat	.249	.640	.411	.646	.594	1.00	
Isokinetic torque	.876	.889	.777	.925	.819	.485	1.00

See table 9.1 for data units.

Notice that the upper right half of the matrix is empty. This is because half of the matrix is completely redundant with the other half, so it is ignored. Next, notice that the values on the diagonal are all $r = 1.0$; these are the correlations of each variable with itself. The bottom row of the table displays the bivariate correlations between the dependent variable (isokinetic torque) and the independent variables. The independent variables, with the exception of percent fat, all have strong correlations with isokinetic torque. So far, so good. Now examine the other correlations in the table. These are the correlations of the independent variables with each other. For example, the correlation between height and weight is $r = .840$ and the correlation between weight and fat-free weight is $r = .984$. That these body size variables are correlated with each other should not be surprising. The correlations between age and the other variables range from $r = .249$ (percent fat) to $r = .876$ (isokinetic torque). This is also not surprising because we would expect older boys to be, on average, bigger and stronger than younger boys. However, that the independent variables are correlated with each other presents a problem. This means that the variables are somewhat redundant. For example, most of the information that weight provides in the prediction of isokinetic torque is also provided by fat-free weight because these variables are so highly correlated. We address the issue of intercorrelated variables in more detail later.

When using forward selection, at step 1 the program selects the X variable to be entered into the equation first by picking the variable that has the highest Pearson r with the dependent variable. Examination of table 9.2 shows that the variable selected at step 1 is weight because the correlation ($r = .925$) is the largest single correlation with isokinetic torque. The regression analysis at step 1 shows that the y-intercept (a) has a value of −27.92 Newton meters and the slope coefficient (b) has a value of 2.18 Newton meters per kilogram. Therefore, the prediction equation at step 1 is

$$\text{isokinetic torque (Newton meters)} = -27.92 + 2.18 \text{ (weight)}. \quad (9.02)$$

The $R^2 = .86$ and $SE_E = 15.83$ Newton meters. Because at step 1 the equation has only one X variable, it is a bivariate regression and the techniques of assessing statistical significance and confidence interval construction from chapter 8 apply. The slope $b = 2.18$ Newton meters per kilogram tells us that, on average, for every 1 kilogram of body weight we would expect torque to be 2.18 Newton meters higher. The y-intercept $a = -27.92$ Newton meters tells us that if weight is 0 kilograms we would expect isokinetic torque to be -27.92 Newton meters. Of course, this is a bit nonsensical because negative torque and a body weight of 0 kilograms are not plausible. This reminds us that we should not extrapolate beyond the data. In this case the lowest weight is 30.6 kilograms, so any prediction using weights below that value is suspect. At step 1, we do not yet have a multiple regression equation because the equation has only one X variable. However, with this bivariate regression equation, which has only body weight as a predictor, we have accounted for about 86% of the variance in isokinetic torque, so only about 14% of the variance in isokinetic torque is left to explain.

At step 2, the forward selection algorithm selects the variable that increases R^2 the most. Another way of thinking about this is that the algorithm picks as the next variable the variable that adds the most unique variance to the equation. After removing the effects of a variable (in this case, removing the effects of weight), how much variance in isokinetic strength is accounted for by fat-free weight or age or percent fat? For the data in table 9.1, the variable that increases R^2 the most, after having partialed out the effect of weight, is age. The equation at step 2 is

$$\text{isokinetic torque} = -94.36 + 1.45 \text{ (weight)} + 7.61 \text{ (age)}. \quad (9.03)$$

The $R^2 = .92$ and the $SE_E = 12.13$ Newton meters. By adding age to the prediction, we have increased the R^2 by .06 and decreased the standard error of the estimate by 3.7 Newton meters. The investigator will typically ask whether adding this next variable significantly improves the equation, and this is usually determined by testing whether the increase in R^2 is statistically significant. That is, is the .06 improvement in R^2 larger than could be expected just due to chance? Statistical software tests the increment in R^2 for statistical significance, and in this case the result is statistically significant ($p = .001$). The slope coefficient for weight is in units of Newton meters per kilogram and tells us that if we hold age constant, isokinetic torque should be about 1.45 Newton meters higher for every 1.0-kilogram increase in body weight. The slope coefficient for age tells us that if we hold weight constant, isokinetic torque should be about 7.61 Newton meters higher for every year of increase in age. The first subject in table 9.1 has body weight of 87.8 kilograms and an age of 16.08 years. For this subject, the equation at step 2 predicts an isokinetic torque of -94.36 Newton meters + 1.45 Newton meters per kilogram. (87.8 kg) + 7.61 Newton meters per year (16.08 years) = 155.32 Newton meters. That subject's observed isokinetic torque is 142 Newton meters, so the residual (or error in prediction) for that subject is $142 - 155.32 = -13.32$ Newton meters.

After step 2 the combination of the variables weight and age has accounted for 92% of the variance in isokinetic torque and four X variables are left. However, the forward selection process for the example data in table 9.1 is stopped at step 2 and no other variables are added to the equation because none of the remaining X variables add enough unique information to significantly improve the equation. For example, we could force fat-free weight into the equation at step 3 but the increment in R^2 is less than .01, which is not a significant increase ($p = .59$). Therefore, we stop at step 2 and the equation at step 2 is the final prediction equation based on the data in table 9.1.

It is tempting to interpret the order of variable selection in terms of importance of the X variables. That is, we might think that because weight was entered first, it must be the most important or best predictor. However, notice that the algorithm did not perform any statistical test to differentiate predictor variables. At step 1 it merely picked the X variable that had the highest Pearson r. The X variable fat-free weight also has a large correlation with isokinetic torque ($r = .889$), which is quite close to the correlation for weight ($r = .925$). In fact, a lot of overlap exists in the 95% CI for these correlations: $r = .76$ to .95 for fat-free weight and $r = .83$ to .97 for body weight (see chapter 8 for an explanation of the calculation of confidence intervals). A different sample from the same population could quite easily have resulted in a higher Pearson r for fat-free weight than body weight. Therefore, do not infer the importance of independent variables based merely on which get picked first.

Backward Elimination

A second strategy for computer-generated prediction equations is **backward elimination.** In backward elimination, the algorithm starts by first forcing all of the independent variables into the model. In our example in table 9.1, age, fat-free weight, height, weight, body mass index, and percent fat are all initially forced into the equation. Then the algorithm picks for elimination the variable that, when removed, decreases the R^2 the least. If the decrease in R^2 is not significant, then that variable is removed and the algorithm moves to the next step. At this step, the algorithm selects for elimination the next variable that, when removed, decreases the R^2 the least. If the decrease in R^2 is not significant, the variable is removed and the process is repeated by looking for the third variable to remove. If the decrease in R^2 is significant, that variable is not removed and the process stops. In essence, then, the backward elimination process peels independent variables off one at a time until peeling off the next variable makes the equation significantly worse.

Stepwise

A third strategy is called **stepwise multiple regression.** The stepwise procedure follows the same procedure as the forward selection process, with the addition that at each step the algorithm can remove variables that were previously selected. This can occur if the addition of subsequent variables makes a previously included variable no longer useful. That is, the variance accounted for by the subsequently included variables overlaps so much with the earlier variable that its removal does not significantly decrease the R^2 of the equation.

Multicollinearity and Singularity

A potential problem in multiple regression analysis occurs if the independent variables in the regression equation are correlated with each other. This is referred to as **multicollinearity.** Further, in nonexperimental studies, it is quite common for the independent variables to be intercorrelated. We noted earlier the intercorrelations between the independent variables in table 9.2. For example, fat-free weight and weight are correlated with each other at $r = .984$.

Multicollinearity leads to two related problems. First, high multicollinearity widens the confidence intervals around the slope coefficients. In other words, the ability to detect the statistical significance of an independent variable is compromised. Second, the wide confidence intervals mean that the slope coefficients are unstable. That is, under conditions of multicollinearity the magnitude of the slope coefficients can change a lot when, for example, another independent variable is added to the equation (sometimes even changing the sign of the slope coefficient). To illustrate, we have forced a third variable, fat-free weight, into the equation (even though it does not significantly increase the R^2 of the equation). The resulting equation is

$$\text{isokinetic torque} = -107.02 + 0.85 \text{ (weight)} + 8.34 \text{ (age)} + 0.76 \text{ (fat-free weight)}. \tag{9.04}$$

Compare the slope coefficient for weight in the two-variable model ($b = 1.46$) with the slope coefficient for weight in the three-variable model ($b = 0.85$). The magnitude of the slope coefficient has decreased by about 42%.

How much intercorrelation is too much? Unfortunately, no hard and fast rules exist. However, we can calculate some indices that help us quantify the amount of multicollinearity in the data. One such index is called the **variance inflation factor** (VIF). The idea behind VIF is to calculate the R^2 value between an independent variable and the rest of the independent variables in the equation. That is, we treat the independent variable as a dependent variable in a separate multiple regression analysis. Here, when we regress weight on age and fat-free weight, $R^2 = .98$. The equation for VIF is

$$\text{VIF} = \frac{1}{1 - R^2}, \tag{9.05}$$

where R^2 refers to the squared multiple correlation coefficient from regressing one independent variable against the other independent variables. The larger the VIF, the greater the degree of multicollinearity in the data. Variance inflation factor values that exceed 10 should lead investigators to suspect that multicollinearity may be a problem. For the previous example, the VIF = $1/0.02 = 50$, which clearly indicates problematic multicollinearity. The denominator of the VIF equation is referred to as **tolerance,** so that VIF is the reciprocal of tolerance. Some references and software packages evaluate multicollinearity using tolerance and others use VIF, but they convey the same information.

Singularity means two or more independent variables are perfectly related to each other ($r = 1.00$). This may occur if one variable is created from another by a mathematical manipulation such as squaring, taking the square root, or adding, subtracting, multiplying, or dividing by a constant. Most advanced computer programs screen for multicollinearity and singularity and warn the user in the printout if these relationships are detected by producing the squared multiple correlation values for all variables. See Tabachnick and Fidell (1996, p. 84) for further discussion of this issue.

Confidence Intervals

As with bivariate regression, with multiple regression we can calculate the confidence interval about the slope coefficients in the regression equation. Recall from chapter 8 (equation 8.14, page 125) that the confidence interval for b is

$$b \pm t(SE_b), \tag{9.06}$$

where b is the slope coefficient, t is the critical value from table A.3 for a given α and degrees of freedom, and SE_b is the standard error of the slope coefficient. The same procedure is used to construct confidence intervals about the slope coefficients in multiple regression equations. The SE_b is calculated as

$$SE_{b_K} = \sqrt{\frac{SE_E^2}{SSX_k(\text{tolerance}_k)}}, \tag{9.07}$$

where SE_{bk} is the standard error of the slope coefficient for the kth independent variable, SE_E^2 is the square of the standard error of estimate for the multiple regression equation at that particular step, SSX_k is the sum of the squared values for the kth independent variable, and tolerance$_k$ is the tolerance for the kth variable. In practice, you will not calculate the standard error by hand, but rather will use software to generate the values. However, notice what the components of equation 9.07 tell you. Specifically, as multicollinearity increases and then tolerance decreases (or equivalently as VIF increases), the standard error increases. Thus, multicollinearity widens confidence intervals.

To calculate confidence intervals, again examine the two-variable multiple regression equation of equation 9.03. The slope coefficient for the independent variable age is $b = 7.61$ Newton meters per year. The standard error for age is 1.876 Newton meters per year (calculations not shown but are available from multiple regression software). If we wish to construct a 95% CI, the critical t is derived from table A.3 using the column for $\alpha = .05$ with a two-tailed test. The degrees of freedom are calculated as

$$df = N - k - 1, \tag{9.08}$$

where N = sample size and k = the number of independent variables in the equation. For equation 9.02, $df = 24 - 2 - 1 = 21$. From table A.3, the critical $t = 2.08$. The confidence interval is then calculated as

$$95\% \text{ CI} = 7.61 \pm 2.08 \, (1.876) = 3.71 \text{ to } 11.57.$$

That is, we are 95% confident that the true effect of age on isokinetic torque, while holding weight constant, is somewhere between 3.71 and 11.57 Newton meters per year. Notice that zero is not included within the 95% CI, which is equivalent to saying that the variable age, holding weight constant, is a statistically significant predictor of isokinetic torque.

Some Cautions and Assumptions

When performing a multiple regression analysis, be aware of the following assumptions and how they may affect your results.

Ratio of Subjects to Independent Variables

The ratio of subjects to independent variables should be no less than 5:1, and ideally about 20:1. In stepwise regression, an even greater ratio should be sought (40:1). Reducing this ratio seriously limits the ability to generalize the equation. However, it is also possible to have too many cases. With a large enough N, almost any multiple correlation may be found to be significant. For most studies, ratios between 20:1 and 40:1 are considered reasonable (Tabachnick and Fidell, 1996, p. 133).

Outliers

Outliers, cases with excessively large residual values, can produce greater leverage on the resultant equation than is appropriate. They should be eliminated from the database before analysis. Multivariate outliers are especially devious because they are hard to find. It may not be too unusual for a person to be 4 feet 6 inches tall or to weigh 250 pounds; such values are moderate univariate outliers. But a person who is 4 feet 6 inches tall and weighs 250 pounds is clearly not typical of the population and represents an extreme multivariate outlier.

Outliers may represent real but extreme cases in the population, or they may be the result of measurement errors introduced during data collection or data entry. Outliers caused by errors must be found and corrected. Real, extreme case outliers must be evaluated by the researcher to determine whether they truly represent the population to be studied.

L.S. Fidell (personal communication, 1992) stated that

The problem with outliers is that they influence the solution too much; therefore, the solution does not generalize well to normal populations. An outlier may, in fact, have a "smaller than it should have had" residual because it pulled the solution towards itself, thus biasing the prediction equation.

Advanced computer programs such as SPSS and SAS can test for both univariate and multivariate outliers.

Normality and Homoscedasticity of Residuals

The residuals for each value of X are assumed to be normally distributed around the best fit line, and the variances of each set of residuals (X_1, X_2, X_3, etc.) must be approximately equal. The condition of equal residual variance is called homoscedasticity. Both Kachigan (1986) and Tabachnick and Fidell (1996) discuss this condition in more detail. Statistical software programs can be used to test for this assumption.

Cross Validation and Shrinkage

Sound research design requires that the results be tested for accuracy. In the development of either bivariate or multivariate regression equations, the equation should be developed on one sample of the population and then tested on another equivalent sample. This process of developing the equation on one set of subjects and testing the prediction on another set is called **cross validation.** This may require dividing the sample in half, which would reduce the size of N. Without cross validation on an equivalent sample to test the accuracy of the prediction, the results will be suspect.

When developing a multiple regression equation, the equation that results is optimal for the sample from which it was derived. This means that when the equation is applied to another sample, we would expect the error of prediction to be higher than it was when calculated from the original sample. Another way of thinking about this is that the multiple correlation coefficient from the equation applied to a different sample will be smaller than the multiple correlation coefficient from the original sample. This reduction in multiple correlation coefficient in the cross validation process is referred to as **shrinkage.**

Summary

Multiple correlation, partial correlation, and multiple regression can be used to examine the combined relationship between multiple independent variables (predictors) and a dependent variable. While the computations can be readily accomplished with computer software, investigators need to understand how to properly interpret the results and how to avoid common pitfalls. Issues such as multicollinearity, cross-validation, and outliers are important concerns.

Problems to Solve

The same 24 subjects who performed the tests summarized in table 9.1 performed an additional isokinetic test of isokinetic knee flexion at 300° per second. Their scores are summarized in the following table.

Subject no.	Age (years)	Fat-free weight (kg)	Height (cm)	Isokinetic flexion torque (Nm at 300°/sec)
1	11.42	33.13	146.50	28.53
2	12.75	50.10	161.50	40.76
3	14.00	29.30	148.10	24.46
4	13.33	44.36	165.10	40.76
5	13.92	49.83	153.70	40.76
6	11.08	36.41	147.90	29.89
7	13.25	53.93	166.90	40.76
8	11.50	35.21	140.00	24.46
9	11.17	27.84	136.50	17.76
10	13.58	45.06	162.60	32.61
11	11.08	28.58	134.80	24.46
12	12.83	48.13	156.50	32.61
13	12.67	55.17	163.10	48.92
14	15.92	71.76	177.19	50.28
15	17.33	55.49	176.91	62.50
16	17.67	56.22	148.31	80.17
17	14.67	41.48	156.21	48.92
18	16.75	52.96	175.11	51.63
19	16.67	52.15	175.01	81.53
20	14.83	56.04	175.49	32.61
21	15.83	63.54	185.60	81.53
22	16.42	64.89	175.11	67.94
23	16.42	44.74	175.01	40.76
24	15.58	67.69	175.49	65.22

Use statistical software to perform a hierarchical multiple regression analysis where isokinetic torque is the dependent variable and the independent variables age, fat-free weight, and height are entered in this order.

1. Report results at each step, including slope coefficients (with proper units), squared multiple correlation coefficient (R^2), and standard error of the estimate.
2. Calculate the 95% CI for the slope coefficients.
3. Repeat the analysis while reversing the order of entry between age and fat-free weight. How does reversing the order of entry affect the equation and interpretation of the data?

See appendix C for answers to problems.

Key Terms

backward elimination

beta weights

cross validation

forward selection

hierarchical multiple regression

multicollinearity

multiple correlation

multiple regression

partial correlation

shrinkage

singularity

stepwise multiple regression

tolerance

variance inflation factor

$$H = \left[\frac{12}{18(18+1)}\right]\left[\frac{82.5^2}{6} + \frac{54.5^2}{6} + \frac{34.0^2}{6}\right] - 3(18+1)$$

$$H = [.035][1822.09] - 57$$

$$H = 6.77$$

The *t* Test: Comparing Means From Two Sets of Data

The gymnastics coach at Mountain View Community College wanted to know whether female collegiate gymnasts at Mountain View had greater hip and low back flexibility than male gymnasts. To answer this question, she measured the flexibility of athletes on the men's and women's teams using a sit-and-reach box. At the end of the season, the average distance reached beyond the toes for men was 12 centimeters, whereas women averaged 15 centimeters. At this college, the women clearly were more flexible than the men. No further statistical analysis was needed.

Next, the coach wanted to compare all collegiate gymnasts in the entire state. This required her to measure all the athletes in the state, which was not feasible. Instead, she took random samples of male and female gymnasts from several colleges around the state. She then performed a statistical test to determine whether the means in the two samples accurately reflected the means of the populations from which they came. The test she performed is called a *t* test. In this chapter we learn the techniques for performing this type of analysis.

Recall from chapter 7 that a sample mean may be used as an estimate of a population mean. Remember also that we can determine the odds, or probability, that the population mean lies within certain numerical limits by using the sample mean as a predictor. This same technique can be used in reverse to determine whether a given sample is likely to have been randomly selected from a specific population.

If the population mean is known or assumed to be a certain value, and if the sample mean is not close enough to the population mean to fall within the limits set by a selected level of confidence, then one of the following conclusions must be true: (a) the sample was not randomly drawn from that population, or (b) the sample was drawn from the population, but it has been modified so that it is no longer representative of the population from which it was originally drawn.

Using similar logic, we can make conclusions about two sets of data. If we draw two samples from the same population, and the means of these samples differ by amounts larger than would be expected based on normal distributions, one of the following conclusions must be true: (a) one or both of the samples were not randomly drawn from the population, or (b) some factor has affected one or both samples, causing them to deviate from the population from which they were originally drawn.

The *t* Tests

When a sample is drawn from a population with a known or estimated mean (μ) and standard deviation (σ), the probability (or odds) that the mean of a randomly drawn sample ($\overline{X}$) will lie within certain limits of μ can be determined. To ascertain the probability that a given sample came from a certain population, the value of the **standard error of the mean** must be calculated by one of the following formulas. If the standard deviation of the population (σ) is known, the following formula used is

$$\sigma_M = \frac{\sigma}{\sqrt{N}}, \tag{10.01}$$

where σ_M is the actual standard error of the mean for a population with known mean and standard deviation. If μ is not known, the formula used is

$$SE_M = \frac{SD}{\sqrt{N}}, \tag{10.02}$$

where SE_M is the standard error of the mean estimated from the sample. Note that this is the same as equation 7.01 on page 87.

Using either standard error of the mean we can determine the odds that a sample is representative of the population from which it was drawn by doing a Z test, if the population is known:

$$Z = \frac{\overline{X} - \mu}{\sigma_M}. \tag{10.03}$$

Or we can perform a *t* **test** if the population is estimated:

$$t = \frac{\overline{X} - \mu}{SE_M}. \tag{10.04}$$

Evaluating Z

When σ_M is known, Z indicates the significance of the difference between the sample mean ($\overline{X}$) and μ. Under these conditions, we can determine the significance of Z by comparing it with critical ratios of 1.65 at $\alpha = .10$, 1.96 at $\alpha = .05$, and 2.58 at $\alpha = .01$. If Z exceeds one of these levels, we may conclude at the given level of confidence that the sample was not randomly drawn from the population or that it has been modified in some way so that it no longer represents the population from which it was drawn.

A Z Test Example From Physical Education

The faculty of a physical education department became concerned that student skill was not developing during 5-week volleyball units. During such a unit, students were assigned to various instructors alphabetically based on last name. Typically, the students played during the entire period and received little or no practice on specific skills. For many past years, every student had taken a standardized test of volleyball serving ability (50 points maximum), and the data have been saved. The mean for more than 1,000 students (the population) on this test is 31 points (μ), and the standard deviation is 7.5 (σ).

One teacher decided to try a different approach. In one class (the sample), half of the period was devoted to teaching volleyball skills, especially serving skills, and the students practiced under the teacher's direction for 20 minutes. Games were played only at the end of the period, and a tournament among the class squads was held during the last week of the volleyball block.

The students in this class ($N = 30$) were also given the standardized serving test. Their average score ($\overline{X}$) was 35 points out of the 50 possible, and the standard deviation was 8.3. Was the teacher effective in improving serving skills? In other words, if we assume the class of 30 students to be a random sample of the population of more than 1,000 students, is it likely that the average score for the class ($\overline{X} = 35$) is representative of the population mean ($\mu = 31$)? To test the hypothesis that the sample class represents the population, we calculate the standard error of the mean for the population:

$$\sigma_M = \frac{\sigma}{\sqrt{N}} = \frac{7.5}{\sqrt{30}} = 1.37.$$

Then we conduct a Z test (because μ and σ are known) to determine the odds that the mean of a sample randomly drawn from the population would differ from the population mean by as much as 4 points:

$$Z = \frac{35 - 31}{1.37} = 2.92.$$

What are the odds that a Z score of 2.92 would be found if the sample was drawn from the population and not treated? Because Z is greater than 2.58, the odds that $\overline{X}$ did not come from μ are greater than 99 to 1; hence, $p < .01$.

Two possibilities must be considered:

1. The class was not a random sample of the population and therefore does not represent the population. Perhaps by luck, or by design, these 30 students were better at the beginning of the 5-week block than the typical students assigned to the other classes.

2. The sample was random at the beginning of the 5-week block, but the treatment (instruction and practice in serving skill) changed the students in such a way that they no longer represent the population. In other words, the students changed so that they represent another population, one that has instruction and practice rather than free play.

If random assignment to instructors can be demonstrated so that it is certain that the class was representative at the start of the 5-week block, then only one conclusion is left. A difference of four points between the mean of the sample and the mean of the population would occur less than 1 time in 100 by chance alone. In other words, the odds that the instruction was effective are better than 99 to 1 (LOC = 99%). We reject H_0 and conclude that the instruction was effective at $p < .01$.

Logic of the t Test

The approximation of SE_M is not accurate in small samples ($N < 60$). This was first demonstrated by an English statistician named William Sealy Gossett (1876–1937), who wrote under the pseudonym Student (Kotz and Johnson, 1982). He developed a series of approximations of the normal curve to account for the bias in the estimate of SE_M called Student's *t* distribution. Table A.3a in appendix A lists the values for Student's *t* distribution. If no error existed in the estimation, table A.1 could be used in every case. But when samples are used to estimate population parameters, especially when the samples are small, the *t* distribution (table A.3a) must be used to evaluate the *t* statistic. The values in table A.3a at the given α levels are called **critical ratios.** They represent the *t* ratio that must be reached to reject chance at a given level of confidence.

In this chapter, several versions of the *t* test are presented. All *t* tests have the same basic form. Specifically, a *t* ratio is calculated where the numerator reflects the difference between mean values that are being evaluated. The denominator reflects a version of the standard error of mean differences (see chapter 7). In generic equation form, the *t* ratio is

$$t = \frac{\text{mean difference}}{SE_{\text{mean difference}}}. \tag{10.05}$$

All *t* ratios can be conceptualized as a signal to noise ratio. The difference between the means is the signal we are trying to detect. We are trying to detect that signal in the context of the noise, which is the typical amount of difference between means that could be expected given the characteristics of our data (i.e., number of subjects, variability in the scores). If the difference between means is bigger than we could reasonably expect if H_0 is true, then we reject H_0 and infer that the means are significantly different.

Assumptions for the *t* Test

Several assumptions must be met for the *t* test to be properly applied. If these assumptions are not met, the results may not be valid. When the investigator knows that one or more of these criteria are not met, a more conservative (i.e., α = .01 rather than α = .05) level should be selected to avoid errors. This allows us to be confident of the conclusions and helps to compensate for the fact that all assumptions were not met. The *t* test is quite **robust;** it produces reasonably reliable results, even if the assumptions are not met totally. The *t* test is based on the following assumptions:

- The population from which the samples are drawn is normally distributed. (See chapter 6 for methods of determining the amount of skewness and kurtosis in a data set.)
- The sample or samples are randomly selected from the population. If the samples are not randomly selected, a generalization from the sample to the population cannot be made.
- When two samples are drawn, the samples have approximately equal variance. The variance of one group should not be more than twice as large as the variance of the other. This is called **homogeneity of variance.**
- The data must be parametric. The differences between parametric and nonparametric data are examined in more detail in chapter 16. For now we

focus on analyzing data that are on an interval or ratio measurement scale (see chapter 1).

Types of *t* Tests

We cover three types of *t* tests. First, we address a *t* test in which we compare a single sample mean against the mean from a known population value. Second, we examine the independent *t* test, in which we compare the means from two **independent samples**. Finally, we cover the dependent *t* test, in which two means that come from correlated samples are compared. The most common situation for the dependent *t* test is one in which the same subjects are tested twice (e.g., before and after some intervention) and the two means are compared.

Evaluating *t* From a Single Sample

In the previous example with the Z test, we used σ because we knew its value. But in most research, μ and σ are not known. If σ is not known, we estimate the standard error of the mean using equation 10.02 and the standard deviation of the sample. Then we use the *t* test rather than Z to determine the significance of the difference. The *t* test for a single sample produces the ratio of the **actual mean difference** between the sample and the population to the **expected mean difference** (that amount of difference between $\bar{X}$ and μ that can be expected to occur by chance alone). The expected mean difference is estimated by equation 10.02 and is called the standard error of the mean. To interpret *t* for a single sample, we must first find the degrees of freedom, which can be calculated by the formula $df = N - 1$. The *t* ratio is compared with the values for a two-tailed test (one- and two-tailed tests are explained later in this chapter) from the *t* distribution in table A.3a for the appropriate degrees of freedom.

When *t* exceeds the value in table A.3a for a given α level, we may conclude that $\bar{X}$ was not drawn from the population with a mean of μ. When *t* is less than the critical ratio in table A.3a, the null hypothesis (H_0) is accepted; no reliable difference exists between $\bar{X}$ and μ. When *t* exceeds the critical ratio, H_0 is rejected and H_1 is accepted; we infer that some factor other than chance is operating on the sample mean. Notice that in table A.3a when degrees of freedom are large ($df > 99$), the values for a two-tailed *t* test at a given α value are the same as the values read from table A.1 for a Z test (1.65, 1.96, 2.24, and 2.58).

This technique is useful for determining whether influences introduced by an experiment have an effect on the subjects. If we know or estimate the population parameters and then draw a random sample and treat it in a manner that is expected to alter its mean value, we can determine the odds that the treatment had an effect by using a *t* test. If *t* exceeds the critical ratios in table A.3a, we can conclude that the treatment was effective because the odds are high that the sample is no longer representative of the population from which it was drawn. The treatment has caused the sample to change so that it does not match the characteristics of the parent population.

To calculate the *t* test between the sample and the estimated population mean (using equation 10.02), we use the previous volleyball serving example. Recall that the sample mean is 35 (sample size = 30 and $SD = 8.3$) and the estimated population mean is 31. First we calculate the standard error of the means for one sample (SE_M):

$$SE_M = \frac{8.3}{\sqrt{30}} = 1.52.$$

Then we use SE_M to determine *t*. Because $\bar{X} = 35$, then

$$t = \frac{35 - 31}{1.52} = 2.63.$$

Notice that the answer for *t* is slightly smaller than the *Z* score (2.92). This demonstrates that the power to detect differences between samples and populations is greater when the population is known than when it is estimated. To find the level of confidence, we compare the *t* ratio (2.63) with the critical ratios of a two-tailed test from the *t* distribution in table A.3a for $df = 30 - 1 = 29$. The critical ratio at $df = 29$ in table A.3a for $\alpha = .05$ is 2.045 and for $\alpha = .01$ is 2.756. Our *t* value falls between these values, so we would reject H_0 at $\alpha = .05$ but not at $\alpha = .01$. Note that we have a lower level of confidence (95%) for rejecting H_0 when the population is estimated than when the population parameters are known (99%). Note that although table A.3a is limited to four α levels (.10, .05, .025, and .01), computer software will generate exact *p* values. When using a computer to analyze data, if $p \leq \alpha$, reject H_0. For this example, with $df = 29$ and $t = 2.63$, the computer-generated two-tailed *p* value is $p = .0135$. We conclude that the sample mean (35) is significantly different than the estimated population mean (31), which indicates that the teacher's instruction significantly improved volleyball serving skill.

We can also calculate a confidence interval about a sample mean using the *t* distribution. The formula for a confidence interval about a sample mean is

$$CI = \bar{X} \pm t_{CV} (SE_M), \tag{10.06}$$

where t_{CV} is the critical *t* value (two-tailed) from table A.3a for a desired level of confidence and a given df value. If we desire a 95% CI for our example, from table A.3a the critical value with 29 df is 2.045, so the 95% CI = $35 \pm 2.045 (1.52) =$ 31.89 to 38.11. We can interpret this to mean that we are 95% confident that the true mean value for the population that this sample mean represents is between 31.89 and 38.11. Similarly, the 99% CI = $35 \pm 2.756 (1.52) = 30.81$ to 39.19. Notice that the population value of 31 is outside the 95% CI but is inside the 99% CI. This is consistent with the results of the tests of statistical significance: the difference between 35 (sample mean) and 31 (estimated population mean) was significant at $\alpha = .05$ but not significant at $\alpha = .01$. That is because testing whether a sample mean differs from a population mean at a given α is equivalent to determining whether the population mean is outside the confidence interval for that sample mean.

Comparing Two Independent Samples (a Between Comparison)

The same concepts used to compare one sample mean to a population mean may be applied to compare the means between two samples. In essence, we are testing whether the two samples were drawn from the same population. If the t ratio exceeds the critical ratio from table A.3a, the null hypothesis H_0 is rejected and we infer that the two samples were not drawn from the same population. If the t ratio does not exceed the critical ratio, H_0 is accepted and we infer that the samples were drawn from the same population. The t test in this situation is referred to as an **independent t test.**

As an example, say an athletic trainer wishes to examine whether a new treatment procedure for ankle sprains results in less ankle swelling than standard treatment 24 hours following the injury. Over several months, 30 athletes with ankle sprains are randomly assigned to receive either the standard care or new treatment procedure. Ankle swelling is measured using water displacement and the difference in foot and ankle volume (ml) between the injured and noninjured limbs is calculated. Table 10.1 shows example data from this study. Note that, in order to minimize the influence of rounding error on the calculations, we are reporting $\bar{X}$ and SD to decimal values beyond the level of precision of the original measurements.

TABLE 10.1

Comparison of Standard Treatment and New Treatment for Ankle Sprains

Standard treatment	New treatment
42	39
44	43
38	38
44	40
41	37
40	42
44	40
42	38
37	39
39	38
43	41
45	42
41	37
39	39
44	43
$\bar{X} = 41.53$	$\bar{X} = 39.73$
$SD = 2.503$	$SD = 2.052$
$N = 15$	$N = 15$

Equation 10.02 can be used to calculate the standard error of the mean, or the amount that a single, randomly drawn sample mean can be expected to deviate from a population mean by chance alone. A similar formula permits us to calculate the **standard error of the difference** (SE_D), the amount of difference between two randomly drawn sample means that may be attributed to chance alone.

We can use equation 10.07 to estimate the amount of difference between the two means attributable to chance:

$$SE_D = \sqrt{\frac{SD_1^2}{N_1} + \frac{SD_2^2}{N_1}}, \tag{10.07}$$

where the SD^2 values represent the respective sample variances (the squared standard deviations) and the N values represent the respective sample sizes. For the data in table 10.1, the SE_D is calculated as

$$SE_D = \sqrt{\frac{2.503^2}{15} + \frac{2.052^2}{15}} = 0.836.$$

This formula (based on only two samples) estimates the size of the standard deviation of an infinitely large group of difference scores, each of which has been derived from randomly drawn pairs of sample means. Because it is a standard deviation, it may be interpreted in the same manner as any Z score on a normal curve.

Equation 10.08 presents the t test for independent samples:

$$t = \frac{\overline{X}_1 - \overline{X}_2}{SE_D}. \tag{10.08}$$

This t test is a ratio of the actual difference between two means (the numerator of t) and the difference that would be expected due to chance for an infinite set of sample means (the denominator of t). If the t ratio exceeds the critical ratio at a given α level, then chance, or H_0, is rejected as a probable cause for the difference between the sample means, and it is concluded that another factor or factors caused the difference (H_1 is accepted). We may then say that the mean difference is significant. Notice what will make t a large number (i.e., a large difference between the sample means and a small SE_D). From equation 10.07 we can see that large N and small SD values will make SE_D small and therefore will increase statistical power.

For the data in table 10.1, the t value is calculated as

$$t = \frac{41.53 - 39.73}{.836} = 2.15.$$

To use table A.3a in appendix A for an independent t test, we must first calculate the degrees of freedom:

$$df = (N_1 - 1) + (N_2 - 1). \tag{10.09}$$

One degree of freedom is lost for each group. For the data in table 10.1, $df =$ $(15 - 1) + (15 - 1) = 28$. Then we compare the calculated t ratio with the critical ratios in table A.3a for the appropriate degrees of freedom. From table A.3, a two-tailed test at $\alpha = .05$ with 28 df has a critical value of 2.048. The calculated t (2.15) exceeds the critical t (2.048), so we reject H_0 and state that a statistically significant difference in ankle swelling exists between the standard treatment and new treatment. From any common statistical software we can find that with $t = 2.15$ and $df = 28$, $p = .0403$.

The statistics in the t test do not identify the causative factor. Careful controls and proper experimental design are needed to identify the factor. If the control group is closely monitored to ensure that only chance operates on it, and if only one independent variable is permitted to influence the experimental group, then the independent variable may be identified as the causative factor.

As with the example calculating a confidence interval about a single mean value, we can calculate a confidence about the difference in two mean values. The formula for the confidence interval of mean differences from independent samples is

$$CI = (\bar{X}_1 - \bar{X}_2) \pm t_{CV} (SE_D). \tag{10.10}$$

To calculate the 95% CI for this example, the critical t from table A.3a (two-tailed $\alpha = .05$, $df = 28$) is 2.048. Substituting the relevant data from the example, we find that 95% CI = $(41.53 - 39.73) \pm 2.048 (0.836) = 1.80 \pm 1.71 = 0.09$ to 3.5 milliliters. If the null hypothesis is true, then the true mean difference should be zero. Here we see that the 95% CI excludes zero, so we infer that the two samples represent different populations and we are 95% confident that the true mean difference between those populations is somewhere between 0.09 and 3.5 milliliters. This is equivalent to rejecting H_0 at $\alpha = .05$.

The t Test With Unequal Values of N

The example in table 10.1 involves two groups with equal values of N. In practical research, this is almost never the case. Subjects often drop out of experiments (this phenomenon is interestingly referred to as subject mortality), and the two groups usually do not have equal numbers of subjects. The formula for standard error of the difference must be modified to account for the differences in N.

The standard error of the difference formula (equation 10.07) estimates the standard error of the difference based on the assumption that both samples contribute equally to the standard error of the difference. This is true if $N_1 = N_2$. But if N_1 is twice as large as N_2, N_1 should contribute two-thirds of the total value of standard error of the difference. Yet equation 10.07 permits it to contribute only half. For this reason, the following alternative formula for standard error of the difference is used when values of N are unequal:

$$SE_D = \sqrt{\left[\frac{(N_1 - 1)(SD_1^2) + (N_2 - 1)(SD_2^2)}{N_1 + N_2 - 2}\right]\left[\frac{1}{N_1} + \frac{1}{N_2}\right]}. \tag{10.11}$$

Equation 10.11 produces the same answer as equation 10.07 when $N_1 = N_2$, so it could be used in every case. Computers use this equation so it can compute SE_D in any case.

When the values of N are large and only slightly unequal, the error introduced by using the simpler equation 10.07 to compute the standard error of the difference is probably not critical. But when the values of N_1 and N_2 are small and approach a ratio of 2:1, the error introduced by equation 10.07 is considerable. If any doubt exists about which equation is appropriate, equation 10.11 should be used so that maximum confidence can be placed in the result.

An Example From Biomechanics

The following example applies equation 10.11 to the mean values obtained in a laboratory test comparing hip and low-back flexibility of randomly selected males and females. The following measurements in centimeters were obtained using the sit-and-reach test:

Males	Females
$\bar{X}_1 = 22.5$	$\bar{X}_2 = 25.6$
$SD_1 = 2.5$	$SD_2 = 3.0$
$N_1 = 10$	$N_2 = 8$

Using equation 10.09, we calculate that $SE_D = 1.29$:

$$SE_D = \sqrt{\left[\frac{(10-1)(2.5^2)+(8-1)(3.0^2)}{10+8-2}\right]\left[\frac{1}{10}+\frac{1}{8}\right]} = 1.29.$$

Once the standard error of the difference is known, *t* can be calculated:

$$t = \frac{22.5 - 25.6}{1.29} = -2.40.$$

The degrees of freedom are determined using $df = (10 - 1) + (8 - 1) = 16$. Table A.3a indicates that for $df = 16$, a *t* ratio of 2.40 is significant at $p < .05$. So H_0 is rejected and H_1 is accepted. The researcher concludes with better than 95% confidence that females are more flexible than males in the hip and low-back joints as measured by the sit-and-reach test.

The *t* value is negative because a larger value was subtracted from a smaller value in the numerator. The sign of the *t* ratio is not important in the interpretation of *t* because it may be positive or negative depending on which group is listed first in the numerator. Only the absolute value of *t* is considered when determining significance and it is considered bad form to report negative *t* values (Streiner, 2007).

Repeated Measures Design (a Within Comparison)

The standard formulas for calculating *t* assume no correlation between the groups. Both groups must be randomly selected from the population and independent of each other. However, in situations in which a researcher tests a group of subjects twice, such as in a pre–post comparison, the groups are no longer independent. **Dependent samples** assume that a relationship, or correlation, exists between the scores and that a person's score on the posttest is partially dependent on his or her pretest score. The type of t test used in this situation is called a **dependent t test** (also called a **paired t test**).

This is always the case when the same subjects are measured twice. A group of subjects is given a pretest, subsequently treated in some way, and then given a posttest. The difference between the pretest and posttest means is computed to determine the effects of the treatment. This arrangement is often referred to as a repeated measures design or within comparison.

Because both sets of scores are made up of the same subjects, a relationship, or correlation, exists between the scores of each subject on the pre- and posttests. The differences between the pre- and posttest scores are usually smaller than they would be if we were testing two different groups of people. Two test scores of a single person are more likely to be similar than are the scores of two different people.

If a positive correlation exists between the two sets of scores, a high pretest score is associated with a high posttest score. The same is true of low scores. Consequently, we have removed a source of noise from our data; specifically, differences between subjects. That is, some of the noise (reflected by the denominator of the t ratio) is due to the fact that different people make up the means in an independent t test. In a dependent t test, these between subjects differences are eliminated.

This same argument holds true for studies using matched pairs, pairs of subjects who are intentionally chosen because they have similar characteristics on the variable of interest. These matched pairs—sometimes called research twins—are then divided between two groups so that the means of the groups on the pretest are essentially equal. One group is treated, and the other group acts as control; then the posttest means are compared.

We expect the matched group data to have less noise than if the two groups were not matched on the pretest. In effect, we have forced the groups to be equal on the pretest so that posttest comparisons may be made with more clarity. The matched twins in each group may be considered to be the same person, and the correlation between them on the dependent variable can be calculated.

To accomplish this matching process, all the subjects are given a pretest and then ranked according to score. Using a technique sometimes referred to as the ABBA assignment procedure, the researcher places the first (highest scoring) subject in group A, the second and third subjects in group B, the fourth and fifth in group A, the sixth and seventh in group B, and so forth until all subjects have been assigned. The alternation of subjects into groups ensures that for each pair of subjects (1 and 2, 3 and 4, 5 and 6, and so on) one group does not always get the higher score of the pair.

This technique usually results in a correlation between the groups on the dependent variable and in smaller mean differences on the posttest. But because the two groups start with almost equal means on the pretest, it is easier to identify the independent variable as the cause of posttest differences.

If correlated samples are used (either the same subjects or matched pairs) and if no correction is made, the researcher may falsely conclude that no difference exists between the means, when in fact a real, or significant, difference does exist but is smaller than expected because the pretest and posttest scores are correlated. This is called a Type II error.

Correction for Correlated Samples

To illustrate the correction for correlated samples, note that equation 10.07, the equation for the standard error of the difference for independent samples, can also be expressed as

$$SE_D = \sqrt{SE_{M1}^2 + SE_{M2}^2} \qquad (10.12)$$

where SE_{M1} and SE_{M2} are the respective standard error of the mean values. The correction for correlated samples is made in the formula for standard error of the difference because this value indicates the difference to be expected by chance alone. By adjusting the standard error of the difference formula, we can regulate t to more correctly reflect any real difference that may exist in dependent samples. The correction is made by factoring out, or subtracting, the effects of the correlation between the two samples in the formula for standard error of the difference:

$$SE_D = \sqrt{\left(SE_{M1}\right)^2 + \left(SE_{M2}\right)^2 - 2r\left(SE_{M1}\right)\left(SE_{M2}\right)} \qquad (10.13)$$

where r is the Pearson r between the two sets of scores. A positive value for r reduces the standard error of the difference, increases t, and provides a greater chance of finding significance. When r is negative (which is very unlikely with matched pairs or repeated measures), SE_M becomes larger and t becomes smaller.

When r is zero, the term $2r(SE_{M1})(SE_{M2})$ becomes zero and the formula reverts to its original form. Actually, equation 10.13 is the more generalized form of equation 10.12; however, $2r(SE_{M1})(SE_{M2})$ is usually not included as a component when the groups are independent because r is assumed to be zero.

Note that all subjects must have data points on both variables (usually a pre–post comparison). If any subject is missing data on either variable, they must be eliminated from the study.

Suppose that the following values resulted from a study:

Vertical jump [pretest]	**Vertical jump [posttest]**	$r_{\text{pretest–posttest}}$ **= .60**
$\bar{X}_1 = 15$	$\bar{X}_2 = 17.5$	
$SE_{M1} = 0.9$	$SE_{M2} = 1.5$	

If we do not correct for correlated samples, $SE_D = 1.75$

$$SE_D = \sqrt{(0.9)^2 + (1.5)^2} = 1.75.$$

and $t = -1.43$ (which is not significant at 29 df):

$$t = \frac{15 - 17.5}{1.75} = -1.43.$$

When we apply the correction,

$$SE_D = \sqrt{(.9)^2 + (1.5)^2 - 2(.60)(.9)(1.5)} = 1.20$$

and

$$t = \frac{15 - 17.5}{1.20} = -2.08, \ p < .05.$$

At $df = 29$, the uncorrected t (-1.43) is not significant, but the corrected t (-2.08) is significant at $p < .05$. Because the subjects in both tests are the same people, a serious error would be made without the correction factor in the formula for standard error of the difference. The problem of correcting for unequal N never arises with matched pairs or repeated measure designs because the pairs are matched and the N are always equal.

An Example From Leisure Studies and Recreation

A graduate student in leisure studies wanted to know the short-term effect of a 4-day bicycle tour on the self-esteem of the participants. To measure self-esteem, the student administered the Cooper-Smith self-esteem survey to 45 bicycle riders immediately before and after the 4-day trip. The results are shown in table 10.2.

The correlation between the pre- and posttests is quite high (.92). Because of this high correlation, the standard error of the difference is very small (.37). This low error value permits the researcher to find significant differences between the two mean values ($df = 44$, $p < .01$). Based on this analysis, the graduate student rejected H_0 and concluded that the 4-day bicycle tour did have a positive effect on self-esteem.

The use of equation 10.13 and this example nicely illustrates the statistical advantage of using a dependent t test, namely, the decrease in the denominator of the t ratio, which increases statistical power, all else being equal. Another way to

TABLE 10.2

Effects of a 4-Day Bicycle Tour on Self-Esteem

Variable	Mean	SD	SE_M	SE_D	r	t
Pretest	38.4	6.1	.90	.37	.92	−6.35
Posttest	40.7	6.3	.94			

think about this is that in a dependent *t* test, the scores that make up $\bar{X}_1$ come from the same subjects as those that make up $\bar{X}_2$. In contrast, in the independent *t* test situation different subjects make up the scores for the two means. Therefore, as previously noted, we have removed a source of noise relative to the situation with the independent *t* test.

In practice, use of equation 10.13 is a bit cumbersome and we can instead calculate the dependent *t* test using the following equation:

$$t = \frac{\bar{d}}{\frac{SD_d}{\sqrt{N}}}, \tag{10.14}$$

where $\bar{d}$ is the mean difference score (which will equal the difference between the two mean values) and SD_d is the standard deviation of the difference scores (the difference scores are simply the differences between the two scores for each subject). The denominator in equation 10.14 reflects another way of calculating the standard error of the difference (SE_D):

$$SE_D = \frac{SD_d}{\sqrt{N}}. \tag{10.15}$$

An Example From Exercise Physiology

An investigator wished to examine the efficacy of an exercise program on aerobic fitness. Eight subjects were recruited and given a maximal cycle ergometer test for the determination of maximal oxygen consumption (VO_2max; units = ml/kg/min). After a 10-week training program, the subjects were again tested for VO_2max. (We will ignore that the study design would be strengthened with a control group.) The data are presented in table 10.3.

From equation 10.14,

$$t = \frac{3.3}{\frac{3.7}{\sqrt{8}}} = \frac{3.3}{1.31} = 2.5.$$

With $df = N - 1 = 7$ and $\alpha = .5$ (two tailed), a *t* value of 2.5 is statistically significant (critical $t = 2.365$) and we can state that is unlikely that an increase in VO_2max of 3.3 milliliters per kilogram per minute would occur if H_0 is true. However, note that the mean difference was only 3.3 milliliters per kilogram per minute (43.2–39.9). Although this difference did not happen by chance, is it large enough to be meaningful? The subsequent sections provide methods to help answer this question.

Confidence Interval for Correlated Samples

As with our previous *t* test examples, we can construct a confidence interval about the mean difference for correlated samples to help us quantify the magnitude of the effect. The formula for the confidence interval for correlated samples is

$$CI = \bar{d} \pm t_{CV}(SE_D), \tag{10.16}$$

TABLE 10.3

Maximal Oxygen Consumption Before and After 10-Week Training Program

Pretest	Posttest	Difference (d)
43.5	46.5	3.0
38.7	38.8	0.1
35.2	40.3	5.1
42.5	41.8	−0.7
36.5	43.1	6.6
39	39.9	0.9
46.8	48.2	1.4
<u>36.0</u>	<u>46.9</u>	<u>10</u>
$\Sigma = 319.1$	$\Sigma = 345.5$	$\Sigma = 26.4$
$\bar{X}_{pre} = 39.9$	$\bar{X}_{post} = 43.2$	$\bar{X}_d = 3.3$
$SD_{pre} = 4.0$	$SD_{post} = 3.6$	$SD_d = 3.7$

where $\bar{d}$ is the mean difference and t_{CV} is the critical t value from table A.3a for a given level of confidence. For the example data in table 10.3, the 95% CI = 3.3 ± 2.365 (1.31) = 3.3 ± 3.1 = 0.20 to 6.4 milliliters per kilogram per minute. We can interpret this to indicate that we are 95% confident that the true mean difference in VO$_2$max in the population is somewhere between .2 and 6.4 milliliters per kilogram per minute. Since 0.0 is not included, we can reject H_0 and conclude that the training was successful.

Magnitude of the Difference (Size of Effect)

It is common to report the probability of error (p value) reached by the t ratio. Declaring t to be significant at $p = .05$ or some similar level only indicates the odds that the differences are real and that they did not occur by chance. This is often termed statistical significance. But we must also consider practical significance. If the values of N are large enough, if standard deviations are small enough, and especially if the design is repeated measures, statistically significant differences may be found between means that are quite close together in value. This small but statistically significant difference may not be large enough to be of much use in a practical application. How important is the size of the mean difference?

Thomas and Nelson (2001, p. 139) suggest the use of omega squared (ω^2) to determine the importance, or usefulness, of the mean difference. Omega squared is an estimate of the percentage of the total variance that can be explained by the influence of the independent variable (the treatment). For a t test, the formula for omega squared is

$$\omega^2 = \frac{t^2 - 1}{t^2 + N_1 + N_2 - 1}.$$ (10.17)

Applying equation 10.17 to the data from the earlier problem comparing treatments for ankle swelling (p. 156) yields

$$\omega^2 = \frac{(2.15)^2 - 1}{(2.15)^2 + 15 + 15 - 1} = .11.$$

In this case, about 11% of the variance in ankle swelling can be attributed to the different treatments. Note that this is analogous to the coefficient of determination (r^2) from chapter 8. The remaining 89% of the variance is due to individual differences among subjects, other unidentified factors, and errors of measurement.

How large must omega squared be before it is considered important? The answer to that question is not statistically based. Each investigator or consumer of the research must determine the importance of omega squared. In this example, it is meaningful to know that 11% of the variance can be explained. But other factors, such as severity of injury and individual differences in swelling, also contribute to the variance in ankle swelling.

Another method of determining the importance of the mean difference is the **effect size** (ES), which may be estimated by the ratio of the mean difference over the standard deviation of the control group, or the pooled variance of the treatment groups if no control group exists:

$$ES = \frac{\overline{X}_1 - \overline{X}_2}{SD_{control}}.$$ (10.18)

The control group is normally used as an estimate of the variance because it has not been contaminated by the treatment effect. In the ankle sprain example, the effect size is

$$ES = \frac{41.53 - 39.73}{2.503} = 0.72.$$

Jacob Cohen (1988, p. 21), as quoted in the work of Winer and colleagues (1991, p. 122), proposes that effect size values of 0.2 represent small differences, 0.5 represent moderate differences, and 0.8+ represent large differences. Winer and colleagues (1991) also suggest that effect size may be interpreted as a Z (standardized) score of mean differences. In the ankle swelling example, ES = 0.72 indicates that the effect of the treatment was moderate. Although these guidelines are helpful, do not slavishly follow these standards because the magnitude of effect is best judged in the context of the research question. An effect size of 0.5 may be extremely important in one situation but trivially small in another.

The amount of improvement from the pretest to the posttest in repeated measures designs can be determined by assessing the percent of change. The following formula determines the percent of change (improvement) between two repeated measures:

$$\text{percent improvement} = \left(\frac{\overline{X}_1 - \overline{X}_2}{\overline{X}_1} \right) \times 100, \tag{10.19}$$

where $\overline{X}_1$ and $\overline{X}_2$ represent the pre- and posttest mean values.

In the example of the effects of participation in a 4-day bicycle tour on self-esteem (table 10.2), the pre–post improvement is small—only 6% [(40.7 − 38.4)/38.4 × 100 = 6%]. Although the paired t value (−6.35) is significant at $p < .01$, the improvement analysis indicates that the change was minimal at best.

Omega squared, effect size, and percent improvement are important attributes to report when mean differences are studied. They provide additional information to the consumers of the research to assist them in determining the usefulness of the conclusions. These values may be more meaningful than the p value, especially if p just misses being significant (e.g., $p = .06$).

Determining Power and Sample Size

Power is the ability of a test to detect a real effect in a population based on a sample taken from that population. In other words, power is the probability of correctly rejecting the null hypothesis when it is false. Generally speaking, as sample size increases, power will increase. Using the concepts and formulas presented in this section, it is possible to calculate how large N must be in a sample to reach a given level of power (i.e. 80% power, or 80% probability that a real effect in the population will be detected). See Tran (1997) for an excellent discussion of the importance of calculating power. Because the critical t values are lower for a one-tailed test than for a two-tailed test (see table A.3), the one-tailed test is considered more powerful at a given α value. Also, a value of $\alpha = .05$ is more powerful than $\alpha = .01$ because the t value does not need to be as large to reach the critical ratio.

In figure 10.1, the range of the control group represents all the possible values for the mean of the population from which a random sample was taken. The most likely value is at the center of the curve, and the least likely values are at the extremes. The range of the experimental group represents all possible values of the population from which it was taken after treatment has been applied. The alpha point (Z_α) on the control curve is the point at which a null hypothesis is rejected for a given mean value in the experimental group.

Any value for the mean of the experimental group that lies to the right of Z_α (1 − β area) will be judged significantly different from the mean of control at $p < .05$, (Z = 1.96). If H_0 is really true, this represents a type I error. Conversely, any value for the mean of the experimental group that lies to the left of Z_α (the beta area) will be judged to be not significantly different from the mean of control group. If H_0 is false, this represents a type II error.

It then follows that the area of 1 − β in the experimental group is the area of power, the area where a false null hypothesis will be correctly rejected. This area represents all of the possible values for the mean of the experimental population

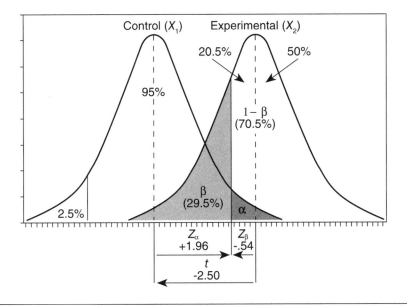

Figure 10.1   Calculation of power for an independent *t* test.

that fall beyond the Z_α level of the control population. Power is calculated by determining Z_β, converting it to a percentile using table A.1 in appendix A, and adding this percent of area to the 50% of the curve to the right of the experimental mean.

In figure 10.1, $1 - \beta$ represents 70.5% of all the possible mean values for the experimental group. A 70.5% chance exists that a false null hypothesis will be rejected; power = 70.5%. Let's consider how power is calculated.

As figure 10.1 shows, power is dependent on four factors:

1. The Z_α level set by the researcher (the level set to protect against type I errors; $\alpha = .05$, $\alpha = .01$, and so on). It is represented by a Z_α score from the normal distribution table (table A.1) [$Z_\alpha (0.10) = 1.65$, $Z_\alpha (0.05) = 1.96$, $Z_\alpha (0.01) = 2.58$].

2. The difference (Δ) between the two mean values being compared ($\Delta = \bar{X}_1 - \bar{X}_2$, where $\bar{X}_1$ is the mean of the control group and $\bar{X}_2$ is the mean of the experimental group).

3. The standard deviations of the two groups, which determine the spread of the curves.

4. The sample size, N, of each of the two groups.

Only N and Z_α are under the control of the researcher, and Z_α usually cannot be radically manipulated because of the need to protect against type I errors. Therefore, the researcher can control power primarily by manipulating the size of N.

Calculating Power

The following process is used to calculate power—that is, to determine the $1 - \beta$ area in figure 10.1.

The researcher sets Z_α based on the α value set to reject the null hypothesis. The values for the means, standard deviations, N for each group, the standard error of the difference, and t are calculated. Figure 10.1 demonstrates that t is the sum of Z_α and Z_β. To determine the power area of the experimental curve, we must find the value of Z_β, which is the percent of the area on the experimental curve between $\bar{X}_2$ and Z_β.

In figure 10.1, Z_α is a positive value; it proceeds to the right of the control mean ($\bar{X}_1$). The t value is negative (-2.50) because $\bar{X}_1 < \bar{X}_2$. The Z_β value (-.54) is also negative; it proceeds to the left of the experimental mean X_2. If the analysis was made with $\bar{X}_1 > \bar{X}_2$, then t and Z_β would be positive values and Z_α would be negative. In order for the following formulas to be applied toward either tail of the curve, the values of t, Z_α, and Z_β will all be considered absolute. Then t is equal to the sum of Z_α and Z_β:

$$t = Z_\alpha + Z_\beta. \tag{10.20}$$

Conversely,

$$Z_\beta = t - Z_\alpha. \tag{10.21}$$

Let us assume the following data apply to figure 10.1:

$\bar{X}_1 = 30, \bar{X}_2 = 32.5$ ($\Delta = 2.5$),

$SD = 5$ for each group,

$N = 50$ for each group,

$SE_M = 0.71$ for each group ($5 / \sqrt{50} = 0.71$),

$SE_D = 1.00$ ($\sqrt{.71^2 + .71^2} = 1.00$),

$t = 2.50$ ($2.5/1.0 = 2.5$), and

$Z_\alpha = 1.96$ ($\alpha = .05$).

Equation 10.21 can now be used to determine Z_β:

$$Z_\beta = 2.50 - 1.96 = 0.54.$$

We convert the Z_β value of 0.54 to a percentile from table A.1 ($Z_\beta = 0.54 = 20.5\%$ of the area under the normal curve) and compute the area of $1 - \beta$ (20.5% + 50% = 70.5%). Therefore 70.5% of all possible values of the experimental population mean lie to the right of Z_α. In other words, a 70.5% chance exists of rejecting the null hypothesis if the values given in the previous data section are true; power = 70.5%.

Calculating Sample Size

The only factor in these equations that is easily manipulated by the researcher is *N*. We could increase our power by increasing *N*, but how large does *N* need to be to produce a given power level?

Equation 10.20 may be solved for *N* as follows. The formula for *t* is

$$t = Z_\alpha + Z_\beta.$$

Because $t = \Delta/SE_D$, we may substitute Δ/SE_D for *t*:

$$\frac{\Delta}{SE_D} = Z_\alpha + Z_\beta.$$

Recall that for $N_1 = N_2$ (equation 10.12),

$$SE_D = \sqrt{SE_{M1}^2 + SE_{M2}^2}$$

and (equation 10.02)

$$SE_M = \frac{SD}{\sqrt{n}}.$$

When $N_1 = N_2$ and $SD_1 = SD_2$,

$$SE_D = \sqrt{\left[\frac{SD_1}{\sqrt{N_1}}\right]^2 + \left[\frac{SD_2}{\sqrt{N_2}}\right]^2} = \sqrt{\frac{2SD^2}{N}}.$$

Substituting this value for the standard error of the difference, we obtain the following:

$$\frac{\Delta}{\sqrt{\frac{2SD^2}{N}}} = Z_\alpha + Z_\beta$$

$$\Delta = \left[\sqrt{\frac{2SD^2}{N}}\right](Z_\alpha + Z_\beta)$$

$$\Delta^2 = \left[\frac{2SD^2}{N}\right](Z_\alpha + Z_\beta)^2$$

$$N\Delta^2 = 2SD^2 (Z_\alpha + Z_\beta)^2.$$

Therefore,

$$N = \frac{2SD^2 (Z_\alpha + Z_\beta)^2}{\Delta^2}. \tag{10.22}$$

With equation 10.22, we can determine the *N* needed for a given power level if we know the other values. Suppose we want power of .80 at two-tailed $\alpha = .05$. To find Z_β from table A.1, we must find the *Z* value where 30% of the area of the normal curve is below Z and 20% is above Z. This will correspond to $\beta = .20$ and power = .80. We look up 30% in table A.1 to find that $Z_\beta = \sim 0.84$. If $\Delta = 5$, and for each group $SD = 6$, and we set Z_α at 1.96, then

$$N = \frac{2(6)^2 \ (1.96 + .84)^2}{5^2} = 22.6.$$

We conclude that to achieve 80% power under these conditions, we must use a sample size of approximately 23 in each group.

The calculation of power is a major factor in experimental design. It is important to know what the odds are that real differences between group means may be detected before we conduct expensive and time-consuming research. Research performed with insufficient power (i.e., N is too small) may result in a type II error (failure to reject a false null hypothesis) or may waste valuable resources on a study that has little chance of rejecting the null.

In a power calculation, the values for the means and standard deviations are not usually known beforehand. To calculate power before the data are collected, these values must be estimated from pilot data or from prior research on similar subjects.

The previous power calculation example is applicable only to a t test of independent means, with equal values of both N and SD. This is the simplest application of the concept of power. Similar calculations may be made for unequal values of N or for dependent tests.

A variety of software programs can be used to perform power calculations for simple research designs. Additional discussions of power and sample size may be found in Kachigan (1986, p. 185) and Thomas, Nelson, & Silverman (2005), p. 116–119.

The t Test for Proportions

The techniques for estimating error in sample means and determining the significance of the difference between two means may be modified to apply to proportions. The following example illustrates this procedure.

An Example from Administration

A teacher surveyed 150 girls in a large school to determine their favorite subject and found that 60% chose physical education. The principal doubted these findings and claimed that the true population value of those favoring physical education couldn't be more than 50%. He challenged the 60% figure and asked the teacher to prove it.

If we assume that the teacher's survey was properly conducted and that subjects were randomly drawn from the population, what are the odds that another random survey from the same population could result in a value of 50%? To answer such a question, we need to know the error that can be expected in a proportion. This error can be estimated with the following formula for the standard error of a proportion (SE_p):

$$SE_P = \sqrt{\frac{pq}{N}}, \tag{10.23}$$

where p is the obtained proportion, $q = 1 - p$, and N is sample size. Applying this formula to the problem at hand yields

$$SE_P = \sqrt{\frac{.60\,(1-.60)}{150}} = .04.$$

The standard error of a proportion may be interpreted as a Z score. Therefore, the odds that the true proportion lies between .56 and .64 (.60 ± .04) are 68 to 32. Multiplying the standard error of a proportion by 1.65 ($p = .10$), 1.96 ($p = .05$), or 2.58 ($p = .01$) will produce the limits of the population at a given level of confidence. For example, at $p = .05$, .60 ± (1.96).04 = .60 ± .078, indicating that the odds that the true population mean lies between .522 and .678 are 95 to 5. Based on this analysis, the principal agreed that the true proportion for the population was probably not 50%.

An Example Comparing Two Proportions

This concept may be applied to a *t* test between two proportions (Bruning and Kintz, 1977, p. 222). The formula for a *t* test between proportions, t_p, is

$$t_P = \frac{P_1 - P_2}{\sqrt{\frac{p(1-p)}{N_1} + \frac{p(1-p)}{N_2}}}, \tag{10.24}$$

where P_1 and P_2 are the proportions to be compared, and p under the radical is

$$p = \frac{N_1 P_1 + N_2 P_2}{N_1 + N_2}. \tag{10.25}$$

The *t* test for proportions should not be used when either p or q times N is less than five. Under these conditions, use nonparametric statistics (see Witte, 1985, p. 155).

If 60% of 150 girls and 70% of 125 boys chose physical education as their favorite subject, does a significant difference exist between the girls and the boys at $\alpha = .05$? To answer this question, we calculate

$$p = \frac{(150)(.60) + (125)(.70)}{150 + 125} = .65$$

and

$$t_P = \frac{.60 - .70}{\sqrt{\frac{.65(1-.65)}{150} + \frac{.65(1-.65)}{125}}} = 1.74.$$

Then we look in table A.3 to interpret *tp* for $df = N_1 + N_2 - 2$. The *t* value of 1.74 is not large enough to reject chance at $\alpha = .05$ because it does not reach the critical ratio of 1.96 for $df = 273$ (∞). No significant difference exists between the proportion of girls who chose physical education as their favorite subject and the proportion of boys who did the same.

Summary

It is very unlikely that means of two random samples from the same population will be identical. Differences will almost always be observed. This is not unexpected; people do not always perform exactly the same, and even if they did, the measurement of their performance is not perfect. Because of these random errors, we always expect mean values to differ. The question is, how much can they differ before we suspect that the difference is caused by something other than chance?

As we discussed earlier in this chapter, the purpose of a t test is to determine whether the difference between two mean values is large enough to reject chance as a probable cause. Using the concepts of the normal curve, we can determine the amount of difference between any two sample means that can be attributed to chance alone. If the observed difference is larger than this estimated difference, then we reject chance as a cause and look for another reason to explain the mean difference.

The t test is the technique by which we perform this analysis. The t value is simply the ratio of the observed difference (the numerator) to the expected difference (the denominator). If the observed difference is larger than the expected difference, the t ratio can be compared to the critical ratios in table A.3a for two-tailed tests or A.3b for one-tailed tests to determine the probability that the observed difference occurred by chance. See chapter 7 for a discussion of one- and two-tailed tests. The table values, or critical ratios, are the values of t for selected sample sizes, or degrees of freedom, that would be expected to occur by pure chance. When our obtained t exceeds these values, we reject the null hypothesis (H_0), accept H_1, and declare the differences to be significant (i.e., not caused by chance).

The t test is useful for conducting experimental research. If we want to know the effect of some treatment, we compare a group that has had the treatment with one that has not. If the treatment is ineffective, we expect only chance differences between the groups. If the treatment is effective, the differences will exceed the expected difference. The t test may be modified to make comparisons between observed proportions.

Following is a list of essential steps that need to be completed to properly determine the significance of the difference between two population means using randomly selected sample groups.

1. Define the population of interest.
2. State the problem.
3. Review the literature. Determine if the problem is solved by prior research.
4. If the problem is not solved, state a hypothesis (H_0 = null, H_1 = directional).
5. Select a level of confidence (consider consequences of error).

6. Select a power level and determine appropriate sample size.

7. Randomly select two samples from the population. Treat one sample with the independent variable. Provide appropriate controls for the other sample.

8. Compute mean, standard deviation ($N - 1$), and standard error of the mean for each sample.

9. Compute standard error of the difference and *t*.

10. Determine degrees of freedom.

11. Compare obtained *t* to critical values in a table of *t* or read *p* value from a computer.

12. Make a conclusion by accepting or rejecting the hypothesis at the given level of confidence, or by interpreting the confidence interval at the given level of confidence.

13. Determine the practical importance of the conclusion by calculating the size of the effect.

Problems to Solve

1. Calculate the standard error of the difference for each of the following sets of independent data.

	$\overline{X}_1$	$\overline{X}_2$	N_1	N_2	SD_1	SD_2
A.	172	175	50	50	20	18
B.	9.7	7.0	10	15	2.5	3.1

2. What are the *t* values for problems 1A and 1B? Is either significant? At what level of confidence?

3. Define the standard error of the difference.

4. In a study of absolute errors in active versus passive arm positioning, an investigator collected data (in centimeters) on 20 college-age subjects (from the motor learning laboratory, California State University Northridge, courtesy of Tami Abourezk). Does a significant difference exist in the errors in arm positioning made by the active group ($N_1 = 10$) compared with those made by the passive group ($N_2 = 10$)?

Subject	Active	Subject	Passive
1	2.65	11	3.30
2	2.42	12	2.00
3	3.30	13	0.09
4	0.19	14	0.04
5	1.25	15	4.56
6	2.00	16	3.33

(continued)

Subject	Active	Subject	Passive
7	3.34	17	1.02
8	4.08	18	0.89
9	0.70	19	2.78
10	2.89	20	1.65

   a. Compute the t value and the p value by hand, then accept or reject the null hypothesis.
   b. Confirm your results on a computer. How does the computer output compare with your hand-calculated results?

Hint: In the computer database, create two columns. In column 1 (the grouping variable) enter the number 1 in rows 1 to 10 and the number 2 in rows 11 to 20. In column 2 (the score column), enter the absolute error values for subjects 1 to 10 in rows 1 to 10 and the values for subjects 11 to 20 in rows 11 to 20.

5. A graduate student in biomechanics was interested in stride length of cross-country skiers. Stride length is an important factor in the development of speed for racing. Data were collected on 20 athletes from the cross-country ski team (the experimental group). The data were compared with those of a second group of 17 students (the control group) who were not varsity athletes but did participate in recreational skiing. The researcher assumed that no significant differences would be found. Does a significant difference in stride length exist between the athletes and the nonathletes? What is the level of confidence? What is ω^2, and what is the ES (effect size)? (Adapted from Duoos, 1984; data fabricated.)

Athletes	Nonathletes
$\bar{X}_1 = 0.90$ meters	$\bar{X}_2 = 0.70$ meters
$N_1 = 20$	$N_2 = 17$
$SD_1 = 0.17$	$SD_2 = 0.23$

6. Find the t value for the following data on dominant versus nondominant grip strength. Twenty subjects were measured twice, once with the dominant hand and once again with the nondominant hand. Which hypothesis, H_0 or H_1, might be appropriate in this study? Is the difference significant, and if so, at what level of confidence?

Dominant		Nondominant
$\bar{X}_1 = 40$ pounds		$\bar{X}_2 = 35$ pounds
$SD_1 = 12.7$	$r = .83$	$SD_2 = 14.2$

7. A researcher in exercise physiology wanted to know whether body composition differs among prepubescent males and females. To test the null hypothesis, she measured skinfolds (in millimeters) on 5 males and 6 females and obtained the following results.

Males	Females
21	22
25	19
19	18
17	24
18	21
	23

a. Is the difference in the means significant? If so, at what level of confidence?
b. Do you accept or reject the null hypothesis?
c. Check your hand calculations on a computer. See the hint in problem 4.

8. A sport psychologist wondered whether motivation could improve aerobic capacity. To answer the question, he measured 10 subjects on $\dot{V}O_2$max (ml/kg/min) on a treadmill. Students were verbally instructed to do their very best. One week later, the subjects were measured again and were told they would receive $100 if their $\dot{V}O_2$max was higher than it was the first time. Following are the data.

First	Second
45	54
33	50
59	58
32	38
30	42
27	35
29	38
59	66
44	48
40	49

a. Did the money cause them to significantly increase their $\dot{V}O_2$max values?
b. If so, what is the probability of error in your conclusion? Should you accept or reject the null hypothesis?
c. Check your hand calculations on a computer. *Hint:* In the computer database, enter values for each subject on the same row. Use column 1 for the first test and column 2 for the second test.

9. An aerobic dance teacher wanted to know whether two workouts per week of 30 minutes each were enough to increase $\dot{V}O_2$max in sedentary middle-aged women. The teacher proposed to compare $\dot{V}O_2$max of some women who had been in the aerobic dance classes for 6 months with $\dot{V}O_2$max of a group of sedentary women. Based on related literature, the teacher estimated that

$\dot{V}O_2$max would be about 7 milliliters per kilogram per minute higher in the aerobic dancers, with a standard deviation of 4.5. If this is a fair estimate of the data to be expected, how many women must she test in each group to produce a power coefficient of .90 at $\alpha = .05$?

See appendix C for answers to problems.

Key Words

actual mean difference
critical ratio
dependent sample
dependent *t* test
effect size
expected mean difference
homogeneity of variance

independent sample
independent *t* test
paired *t* test
robust
standard error of the difference
standard error of the mean
t test

$$H = \left[\frac{12}{18(18+1)}\right]\left[\frac{82.5^2}{6} + \frac{54.5^2}{6} + \frac{34.0^2}{6}\right] - 3(18+1)$$

$$H = [0.035][1822.09] - 57$$

$$H = 6.77$$

Simple Analysis of Variance: Comparing the Means Among Three or More Sets of Data

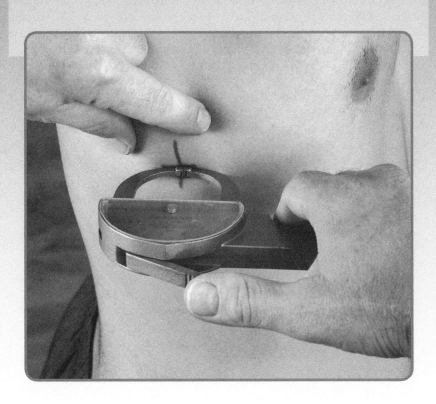

team of researchers (exercise scientists and nutritionists) were interested in the relative effects of dietary carbohydrate on body composition. They conducted a study in which subjects were randomly assigned to one of four groups with different carbohydrate compositions: (1) a very-low-carbohydrate group (~10% of kilocalories from carbohydrate, 45% fat, 45% protein), (2) a low-carbohydrate group (~30% carbohydrate, 35% fat, 35% protein), (3) a moderate-carbohydrate group (~50% carbohydrate, 25% fat, 25% protein), and (4) a high-carbohydrate group (~70% carbohydrate, 15% fat, 15% protein). Percent body fat of the subjects in all four groups was measured at the end of the 6-week program.

If the researchers wanted to analyze the data using *t* tests, six tests would be required to make all possible group comparisons. This procedure is both cumbersome and invites a type I error. Is there a better way? In this chapter we discuss the disadvantages of multiple *t* tests and we learn how to compare three or more means with a single *F* (analysis of variance) test.

Analysis of variance (ANOVA) is a parametric statistical technique used to determine whether significant differences exist among means from three or more sets of sample data. In a *t* test, the differences between the means of two groups are compared with the difference expected by chance alone. Analysis of variance examines differences between the means of three or more groups by comparing the variability between the group means (the between-group variability) with the variability of scores within the groups (the within-group variability). This produces a ratio value called *F* (*F* = average variance between groups divided by average variance within groups). The symbol for ANOVA (*F*) is named after the English mathematician Ronald Aylmer Fisher (1890–1962), who first described it (Kotz and Johnson, 1982, p. 103).

If the between-group variability exceeds the within-group variability by more than would be expected by chance alone, it may be concluded that at least one of the group means differs significantly from another group mean. The null hypothesis (H_0) for an *F* test is designated as

$$\mu_1 = \mu_2 = \mu_3 = \ldots \mu_k.$$

The null hypothesis assumes that the means of all the samples are equal. Another way to say this is that a single population is the parent of the several random untreated samples drawn from it. Therefore, untreated means of samples randomly drawn from the same population(s) should not differ by more than chance. When at least one sample mean is significantly different from any other, *F* is significant and we reject the null hypothesis for one or more of the samples.

Like the *t* test, the theoretical concepts of ANOVA are based on random samples drawn from a parent population and the characteristics of the normal curve. If a large number of samples are randomly drawn from a population, the variance among the scores of all subjects in all groups is the best estimate of the variance of the population. When the variance among all the scores in a data set is known, it may be used to determine the probability that a deviant score is not randomly drawn from the same population. This argument may be expanded to infer that if randomly drawn scores are randomly divided into subgroups, the variance among the subgroup means may be expected to be of the same relative magnitude as the variance among all of the individual scores that comprise the several groups.

With untreated data, when only random factors are functioning between the group means and within the scores of the groups, the between-group and within-group variances should be approximately equal. The **F ratio**, the ratio of the average between-group variance divided by the average within-group variance, is expected to be about 1.00. When the value of the F ratio exceeds 1.00 by more than would be expected by chance alone, the variance between the means is judged to be significant (i.e., the difference was caused by a factor other than chance) and H_0 is rejected.

The F ratio is analogous to the t ratio in that both compare the actual, or observed, mean differences with differences expected by chance. When this ratio exceeds the limits of chance at a given level of confidence, chance as a cause of the differences is rejected. In the F test, the actual differences are the variances between the group means, and the expected differences are the variances within the individual scores that make up the groups. The t test is actually a special case of ANOVA with two groups. Because t uses standard deviations and ANOVA uses variance (V) to evaluate mean differences, and $SD^2 = V$, when there are only two groups in ANOVA, $t^2 = F$.

Analysis of variance is one of the most commonly used statistical techniques in research. But students often ask, Why is this new technique needed when a t test could be used between each of the groups? For example, in a four-group study, why not conduct six t tests—between groups A and B, A and C, A and D, B and C, B and D, and C and D? There are three reasons why multiple t tests are not appropriate.

1. A greater probability of making a type I error (rejecting the null hypothesis when it is really true) exists when one conducts multiple t tests on samples taken from the same population. When a single t test is performed, the findings are compared with the probability of chance. If a confidence level of 95% is set, we are willing to refute chance if the odds are 19 to 1 against it. When multiple t tests are conducted on samples randomly drawn from the same population, the odds of finding the one deviant conclusion that is expected by chance alone increase.

When 20 t tests are performed at $\alpha = .05$ on completely random data, it is expected that one of the tests will be found significant by chance alone (1-to-19 odds). Therefore, if we perform 20 t tests on treated data, and one of the 20 tests produces a significant difference, we cannot tell whether it represents a true difference due to treatment or whether it represents the one deviant score out of 20 that is expected by chance alone. If it is due to chance but falsely declared to be significant, a type I error has been made. This dilemma is sometimes referred to as the **familywise error rate** (the error rate when making a family of comparisons).

Keppel (1991, p. 164) reports that the relationship between the single comparison error rate (α) and the familywise error rate (FW_α) is

$$FW_\alpha = 1 - (1 - \alpha)^c, \tag{11.01}$$

where C is the number of comparisons to be made. If we conduct six t tests, comparing all possible combinations of four groups (A, B, C, and D), at $\alpha = .05$, then

$$FW_\alpha = 1 - (1 - .05)^6 = .26.$$

In this example, conducting multiple t tests raises the probability of a type I error from .05 to .26. Keppel suggests that familywise error rate may be roughly estimated by the product of the number of comparisons to be made and alpha (C and α). In this example, $FW_\alpha = \sim 6 \times .05 = .30$. This method will always overestimate familywise error rate but is fairly close for small values of the number of comparisons to be made and alpha. Analysis of variance eliminates the problem of familywise errors by making all possible comparisons among the means in a single test.

Multiple t tests on samples from the same population may be required on occasion. When this is the case, a commonly used modification of the alpha level called a **Bonferroni adjustment** is recommended. To perform the adjustment, divide the single-test alpha level by the number of tests to be performed. If five tests are to be made at $\alpha = .05$, the adjusted alpha level to reject H_0 would be .01 (.05/5 = .01). This is referred to as the per comparison alpha level.

2. The t test does not make use of all available information about the population from which the samples were drawn. The t test is built on the assumption that only two groups have been randomly selected from the population. In the t test, the estimate of the standard error of the difference between means is based on data from two samples only. When three or more samples have been selected, information about the population from three or more samples is available and should be used in the analysis, yet t considers only two samples at a time. Analysis of variance uses all of the available information from all samples simultaneously.

3. Multiple t tests require more time and effort than a simple ANOVA. It is easier, especially with a computer, to conduct one F test than to conduct multiple t tests.

Because of these reasons, analysis of variance is used in place of multiple t tests when three or more groups of data are involved. Analysis of variance can determine whether a significant difference exists among any of the groups represented in the experiment, but it does not identify the group or groups that differ. A significant F value indicates only that at least one significant difference exists somewhere among the many possible combinations of groups. When a significant F is found, additional **post hoc** (after the fact) tests must be performed to identify the group or groups that differ. If F is not significant, no further analysis is needed because we know that no significant differences exist among any of the groups.

Assumptions in ANOVA

The F test is based on the following assumptions:

- The population(s) from which the samples are drawn is normally distributed. Violation of this assumption has little effect on the F value among the samples (Keppel, 1991, p. 97). The F test produces valid results even when the population is not normally distributed. For this reason it is considered to be robust.

- The variability of the samples in the experiment is equal or nearly so (homogeneity of variance). As with the assumption of normality, violation of this assumption does not radically change the F value. However, as a general rule, the largest group variance should not be more than two times the smallest group variance.

- The scores in all the groups are independent; that is, the scores in each group are not dependent on, not correlated with, or not taken from the same subjects as the scores in any other group. The samples have been randomly selected from the population and randomly assigned to various conditions for each group. If a known relationship exists among the scores of subjects in the several groups, use repeated measures ANOVA (see chapter 12).

- The data are based on a parametric scale, either interval or ratio. (For nonparametric data analysis, see chapter 16.)

The F test, like the t test, is considered to be robust. It provides dependable answers even when violations of the assumptions exist. Violations are more critical when sample sizes are small or Ns are not equal. If violations are committed that cannot be controlled and that the researcher thinks may increase the possibility of a type I error, a more conservative alpha value should be used to compensate for the violations (i.e., use $\alpha = .01$ rather than $\alpha = .05$).

Sources of Variance

When several groups of data are compared, each group has a mean (the group mean) and the entire data set, all groups combined, has a mean (the grand mean). The grand mean is computed by summing the scores from all groups and dividing by the total number of subjects in all groups combined (N). Each of the group means may differ from the grand mean. The variance, or deviation, of the group means from the grand mean is called the **between-group variance.** This information determines the numerator of the F ratio.

In addition to variance between the means, each individual score deviates from the mean of its group by a certain amount. This source of variance is called **within-group variance.** This information determines the denominator of the F ratio. These two sources of variance, between-group and within-group, are the basic components used to compute the ANOVA.

A third source of variance, the **total variance,** may be computed by determining the deviation of each score in each group from the grand mean. Total variance is equal to the sum of between-group variance and within-group variance:

$$\text{variance}_{\text{total}} = \text{variance}_{\text{between}} + \text{variance}_{\text{within}}. \tag{11.02}$$

Sum of Squares and Mean Square

The **sum of squares** (*SS*) is the sum of the squares of the deviations of each score from a given mean:

$$SS = \Sigma(X - \overline{X})^2. \tag{11.03}$$

It is computed by subtracting the mean from each score, squaring the deviation scores, and adding them together. Note the analogy with the variance calculation from chapter 5; this is the numerator of the calculation.

The sum of squares within any group (*SS$_W$*) can be computed from the individual scores and the group mean:

$$SS_W = \Sigma\left(X - \overline{X}_{\text{group}}\right)^2. \tag{11.04}$$

The sum of squares between groups (*SS$_B$*) can be computed by subtracting the grand mean from each group mean:

$$SS_B = \Sigma\left(\overline{X}_{\text{group}} - \overline{X}_{\text{grand}}\right)^2. \tag{11.05}$$

The sum of squares for the total (*SS$_T$*) can be computed by subtracting the grand mean from each individual score:

$$SS_T = \Sigma\left(X - \overline{X}_{\text{grand}}\right)^2. \tag{11.06}$$

The total sum of squares is always equal to the between-group sum of squares plus the within-group sum of squares:

$$SS_T = SS_B + SS_W. \tag{11.07}$$

In ANOVA, the size of the sum of squares between groups is compared with the size of the sum of squares within groups. The size of the sum of squares is dependent on the number of scores summed and the size of the squared deviations from the mean, or the variance. To account for the differences in the number of scores that make up the between-group (number of groups) and within-group (number of individual scores) sums of squares, the **mean square** (*MS*) is computed by dividing each sum of squares by the appropriate degrees of freedom (*MS* = *SS/df*). This process makes the mean, or average, variabilities for sums of squares between and within groups comparable.

Degrees of freedom within are determined by subtracting the number of groups (k) from the total number of subjects in all groups (N):

$$df_W = N - k. \tag{11.08}$$

Degrees of freedom between are determined by the number of groups (k) minus 1:

$$df_B = k - 1. \tag{11.09}$$

To determine mean square within, divide the sum of squares within by the degrees of freedom within:

$$MS_W = SS_W/df_W. \tag{11.10}$$

To determine mean square between, divide the sum of squares between by the degrees of freedom between:

$$MS_B = SS_B/df_B. \tag{11.11}$$

These mean square values are now directly comparable and can be used to calculate F:

$$F = MS_B/MS_W. \tag{11.12}$$

The mean square within is the denominator of the F test. Like standard error of the difference in the t test, it represents the amount of variance that can be expected due to chance occurrences alone (the noise). The F ratio compares mean square within with the differences between the means represented by mean square between (the signal). That is, just like the t test, the F ratio is a signal to noise ratio. If the signal is distinguishable from the noise, the F test is statistically significant.

In ANOVA, mean squares within is sometimes referred to as mean square error (MS_E). The term *error* does not mean a mistake; it means the variance that can occur by chance alone, or the variance that is unaccounted for by the effects of treatment. Mean square error is synonymous with mean squares within.

When mean square error is large, it tends to mask small treatment effects. When mean square error is small, it is easier to identify treatment effects. The intent of all research designs is to keep mean square error as small as possible. The value of mean square error depends on the variability in the data and on the number of scores (N) that make up the data. Of these two factors, only N can be directly controlled by the researcher. Therefore, studies with large N are more powerful in detecting mean differences.

Calculating F: The Definition Method

We can calculate F using the formulas presented thus far. The concepts for this calculation of F are easy to understand because they are based on the definition of terms we have already introduced. But, like the definition formulas for standard deviation and for Pearson's correlation, the definition method for ANOVA is tedious to use, especially when raw scores contain decimal values and N is large. For real lab or field research applications, we use a computer. But calculations here are useful to demonstrate the process of ANOVA.

Table 11.1 presents hypothetical strength measurements for five groups (X_1 to X_5) of seven subjects each. The numerical values do not represent any particular unit of measure. They are purposely small and discrete so that the calculations will be easy. Assume that the subjects were randomly selected from a population and randomly assigned to groups. Each subject completed 6 weeks of strength training. In this table, n is the number of subjects in each group and N is the total number of subjects or scores in all groups combined. We test the null hypothesis (H_0) that none of the treatments had an effect. H_0 predicts no significant differences among any of the group means.

Groups 1, 2, 3, and 4 participated in different strength training programs; group 5 was the control group that received no strength training. We want to know whether any differences exist between the mean scores of the groups after the differential training. In this case, using ANOVA to solve the problem takes the place of 10 t tests.

The steps for calculating F by the definition method are as follows:

1. Calculate the sum of each group (ΣX_1, ΣX_2, ΣX_3, . . .), the mean of each group ($\bar{X}_1$, $\bar{X}_2$, $\bar{X}_3$, . . .), the grand sum (ΣX_T), and the grand mean (M_G). These values can be found at the bottom of table 11.1.

2. Calculate the within-group sum of squares for each group.

TABLE 11.1

Data for Simple ANOVA

	X_1	X_2	X_3	X_4	X_5 (control)
	4	5	5	8	5
	5	7	4	4	4
	6	9	6	6	3
	7	8	5	8	4
	4	9	5	5	6
	6	7	6	6	4
	5	10	4	7	5
ΣX	37	55	35	44	31
n	7	7	7	7	7
$\bar{X}$	5.29	7.86	5.00	6.29	4.43

$\Sigma X_T = 37 + 55 + 35 + 44 + 31 = 202$.

$N = 7 + 7 + 7 + 7 + 7 = 35$.

$M_G = 202/35 = 5.77$.

- Determine the deviation scores of each raw score from its group mean $(d = X - \bar{X}_{group})$, as shown in table 11.2.
- Square each deviation and find the sum of squared deviations for each group, as shown in table 11.3.
- Sum the squared deviations from each group. This value is the sum of squares within (also known as sum of squares error):

$$SS_E = \Sigma (X - \bar{X})^2 = 7.40 + 16.86 + 4.00 + 13.40 + 5.71 = 47.37.$$

3. Calculate the between-group sum of squares.

- Find the deviation of each group mean from the grand mean, square each deviation, and sum these deviations (Σd^2_B) (table 11.4).

TABLE 11.2

Calculation of Within-Group Deviations

d_1	d_2	d_3	d_4	d_5
−1.29	−2.86	0.00	1.71	0.57
−0.29	−0.86	−1.00	−2.29	−0.43
0.71	1.14	1.00	−0.29	−1.43
1.71	0.14	0.00	1.71	−0.43
−1.29	1.14	0.00	−1.29	1.57
0.71	−0.86	1.00	−0.29	−0.43
−0.29	2.14	−1.00	0.71	0.57

Note: These values are derived by subtracting the group mean from table 11.1 from the individual score [for the first score (X) in group X_1: 4.00 − 5.29 = −1.29].

TABLE 11.3

Squaring and Summing Within-Group Deviation

$(d_1)^2$	$(d_2)^2$	$(d_3)^2$	$(d_4)^2$	$(d_5)^2$
1.66	8.18	0.00	2.92	0.33
0.08	0.74	1.00	5.24	0.18
0.50	1.30	1.00	0.08	2.05
2.92	0.02	0.00	2.92	0.18
1.66	1.30	0.00	1.66	2.46
0.50	0.74	1.00	0.08	0.18
0.08	4.58	1.00	0.50	0.33
$\Sigma(d_1)^2 = 7.40$	$\Sigma(d_2)^2 = 16.86$	$\Sigma(d_3)^2 = 4.00$	$\Sigma(d_4)^2 = 13.40$	$\Sigma(d_5)^2 = 5.71$

Note: These values are derived by squaring and then summing the deviation scores from table 11.2. This is a sum of squares calculation like we have seen previously (e.g., in calculating the variance and standard deviation).

TABLE 11.4

Calculation of Between-Group Deviations

	$\bar{X}$	M_G	$d_B = (\bar{X} - M_G)$	$(d_B)^2$
Group 1	5.29	5.77	−0.48	0.23
Group 2	7.86	5.77	2.09	4.37
Group 3	5.00	5.77	−0.77	0.59
Group 4	6.29	5.77	0.52	0.27
Group 5	4.43	5.77	−1.34	1.80
				$\Sigma(d_B)^2 = 7.26$

Note: This is another sum of squares calculation. Here, we calculate the difference between each group mean and the grand mean, square the difference, and then sum the squared differences.

- Because n times more values are contributing to the sum of squares within than to the sum of squares between, multiply $\Sigma d^2{}_B$ by n (7) to make the sum of squares between directly comparable with the sum of squares within:

$$SS_B = 7.26 \times 7 = 50.82.$$

4. Determine the degrees of freedom between (df_B) and error (df_E):

$$df_B = 5 - 1 = 4; \, df_E = 35 - 5 = 30.$$

5. Determine the mean square between and the mean square error by dividing the sum of squares values by the appropriate degrees of freedom:

$$MS_B = 50.82/4 = 12.71; \, MS_E = 47.37/30 = 1.58.$$

6. Determine the ratio (F) between the mean square between and the mean square error:

$$F = MS_B/MS_E = 12.71/1.58 = 8.04.$$

Determining the Significance of *F*

With hand calculations, the significance of F is determined by referring to tables A.4, A.5, and A.6 in appendix A, which show the values of F for appropriate degrees of freedom between and within groups. In these tables, degrees of freedom between are read across the top of the table and degrees of freedom error are read down the left side.

Table A.4 (p. 323) shows the values of F for $\alpha = .10$. When $df_B = 4$ and $df_E = 30$, the critical F value is 2.142. Because the obtained F (8.04) exceeds this value, we look in table A.5, which lists the F values for $\alpha = .05$. For $\alpha = .05$, critical $F =$

TABLE 11.5

Tabular Report of ANOVA

Source of variance	SS	df	MS	F	p
Between groups	50.82	4	12.71	8.04*	<.01**
Within groups (error)	47.37	30	1.58		
Totals	98.19	34			

* Slight differences between values in this table and those produced by a computer are due to rounding errors when manually implementing the deviation method.

** This value is only approximate from table A.6. Computer programs can calculate the exact p value for a given F, which in this case is $p = .0002$.

2.690. The obtained F is still larger than critical F, so we look in table A.6, which has F values for $\alpha = .01$. For $\alpha = .01$, critical $F = 4.018$.

The obtained F (8.04) also exceeds this critical F value, so the obtained F is declared significant at $p < .01$. This means that the probability is less than 1 in 100 that an F larger than 4.018 would be obtained by chance alone. Therefore, the level of confidence reached by the obtained F value (8.04) is greater than 99%. We conclude that the treatment had an effect on at least one of the groups and reject H_0. (More precisely, from a computer software analysis, we can determine that with 4 and 30 df and with $F = 8.04$, $p = .0002$).

The results of ANOVA are usually reported in table form. Table 11.5 is an example of a typical ANOVA table.

Post Hoc Tests

As we discussed earlier, a significant F alone does not specify which groups differ from one another. It indicates only that differences exist somewhere among the groups. To identify the groups that differ significantly from one another, a post hoc test must be performed.

A post hoc test is similar to a t test, except post hoc tests have a correction for familywise alpha errors built into them. Some post hoc tests are more conservative than others. Conservative means that the tests are less powerful because they require larger mean differences before significance can be declared. Conservative tests offer greater protection against type I errors, but they are more susceptible to type II errors.

Several post hoc tests may be applied to determine the location of group differences after a significant F has been found. Two of the most commonly used tests, **Scheffé's confidence interval** (I) and **Tukey's honestly significant difference** (HSD), are described here. Scheffé permits all possible comparisons, whereas Tukey permits only pairwise comparisons. Tukey's is easier to apply and is appropriate in most research designs. For information on other post hoc tests, see Keppel (1991, p. 170–177) or other advanced statistical texts.

Scheffé's Confidence Interval

The Scheffé test (Henry Scheffé, 1907–1977) is the most conservative post hoc test. It should be used if all possible comparisons—that is, more than just pairwise comparisons—are to be made. Pairwise comparisons contrast one group mean against another. All possible comparisons include pairwise comparisons plus comparisons of combinations of groups with single groups or other combinations. For example, the average of two treatment effects may be compared with the average of two other treatment effects, or the average of several treatments may be compared with the control group. Scheffé places no restrictions on the number of comparisons that can be made.

Scheffé's confidence interval (I) permits us to compare the raw score means for any two groups or combinations of groups. The formula is

$$I = \sqrt{(k-1)(F_\alpha)\left(\frac{2MS_E}{n}\right)},\qquad(11.13)$$

where k is the number of groups, F_α is the value of F from tables A.4, A.5, or A.6 for a given p value and the df_B and df_E values used in ANOVA. MS_E is mean square error from ANOVA, and n is the size of the groups when all ns are equal.

If groups are not equal in size, equation 11.13 may be modified as follows to accommodate any two groups with unequal values of n:

$$I = \sqrt{(k-1)(F_\alpha)(MS_E)\left(\frac{1}{n_1}+\frac{1}{n_2}\right)}.\qquad(11.14)$$

Because the n values in our example are equal, we apply equation 11.13 to the data from table 11.5. We compute I at $\alpha = .01$ and $\alpha = .05$ as follows:

$$I = \sqrt{(5-1)(4.02)\left(\frac{2(1.58)}{7}\right)} = 2.69, \alpha = .01.$$

$$I = \sqrt{(5-1)(2.69)(\frac{2(1.58)}{7})} = 2.20, \alpha = .05.$$

This interval size is the raw score value by which any two means must differ to be considered significant. Constructing a mean difference table makes it easy to identify the groups that differ from one another.

Table 11.6 shows the differences between all pairwise combinations of means. Scheffé's requires a mean difference of 2.20 for $\alpha = .05$ and 2.69 for $\alpha = .01$. The table identifies groups 2 and 1 as significantly different at $p < .05$ and groups 2 and 3, and 2 and 5, as significantly different at $p < .01$.

TABLE 11.6

Mean Difference Analysis

	Group 1	Group 2	Group 3	Group 4	Group 5
Group 1	0.00	2.57*	0.29	1.00	0.86
Group 2		0.00	2.86**	1.57	3.43**
Group 3			0.00	1.29	0.57
Group 4				0.00	1.86
Group 5					0.00

Note: Values are calculated by taking the absolute value of the difference between two means in table 11.1 (i.e., $\overline{X}_1 - \overline{X}_2 = 5.29 - 7.86 = 2.57$).

$* p < .05; ** p < .01.$

The ordered values of the means provide additional insight.

Group	**Mean**
2	7.86
4	6.29
1	5.29
3	5.00
5	4.43

It is clear that the treatment given to group 2 had a significant effect. Group 2 differs significantly from all other groups except group 4. Groups 1, 3, and 4 do not differ significantly from group 5, the control group.

An alternative method of identifying specific group mean differences using Scheffé's method may be developed by solving equation 11.14 for F. This results in the following:

$$F_{Scheff\acute{e}} = \frac{\left(\overline{X}_1 - \overline{X}_2\right)^2}{(k-1)(MS_E)\left(\dfrac{1}{n_1} + \dfrac{1}{n_2}\right)}. \qquad (11.15)$$

$F_{Scheff\acute{e}}$ is an F value calculated for any two specified groups (in the equation, $\overline{X}_1$ and $\overline{X}_2$ are specified, but any two means could be used). It is interpreted differently than I, which represents the raw score mean difference between any two groups that must be attained for declaration of significance. $F_{Scheff\acute{e}}$ is an actual F value that must be compared with tables A.4, A.5, or A.6 for df_B and df_E in the ANOVA analysis to determine whether the two compared means differ significantly.

Equation 11.15 is sometimes used by computer programs to calculate the F values for all possible pairwise comparisons. If the computer does not also produce the p values, the $F_{Scheff\acute{e}}$ values produced by the computer must be compared with critical values in tables A.4, A.5, or A.6 to determine the significance of the difference between any two groups.

Tukey's HSD

The Tukey's *HSD* test is named after John Tukey (1915–2000), who first developed the procedure. Tukey's test, like Scheffé's, calculates the minimum raw score mean difference that must be attained to declare significance between any two groups. However, Tukey's test does not permit all possible comparisons; it permits only pairwise comparisons. Any single group mean may be compared with any other group mean. The formula for *HSD* is

$$HSD = q_{(k,df_E)}\sqrt{\frac{MS_E}{n}}, \tag{11.16}$$

where q is a value from the Studentized range distribution (see tables A.7, A.8, and A.9 in appendix A) for k and df_E at a given level of confidence (note that k is used, not degrees of freedom between), MS_E is the mean square error value from the ANOVA analysis, and n is the size of the groups.

Equation 11.16 assumes that the n values in each group are equal. It may be modified to compare any two groups with unequal values of n as follows:

$$HSD = q_{(k,df_E)}\sqrt{\frac{MS_E}{2}\left(\frac{1}{n_1}+\frac{1}{n_2}\right)}. \tag{11.17}$$

Because the n values are equal in our example, equation 11.16 is applied for both $\alpha = .01$ ($q = 5.048$) and $\alpha = .05$ ($q = 4.102$) as follows:

$$HSD = 5.05\sqrt{\frac{1.58}{7}} = 2.40, \alpha = .01.$$

$$HSD = 4.10\sqrt{\frac{1.58}{7}} = 1.95, \alpha = .05.$$

These values (2.40 and 1.95, rounded to the nearest 100th) represent the minimum raw score differences between any two means that may be declared significant.

Tukey, a more liberal test, confirms Scheffé but also finds that groups 2 and 1 differ at $p < .01$ (see table 11.6). The values at $p = .01$ and $p = .05$ are both lower in Tukey's *HSD* test than in Scheffé's *I*. This makes *HSD* more powerful (i.e., more likely to reject H_0) than Scheffé's *I*.

Because we started with the null hypothesis and are not making any comparisons other than pairwise (i.e., we are not interested in the combined mean of two or more groups compared with other combined means), Tukey's test is appropriate. Scheffé may be too conservative for this research design. Based on the analysis by Tukey, group 2 is significantly different from groups 1, 3, and 5 at $p < .01$. Figure 11.1 presents the results in bar graph form. The T symbol above each bar represents standard deviation.

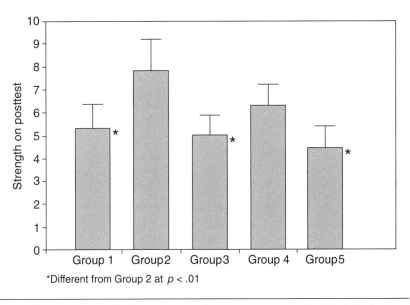

Figure 11.1 Analysis of variance on strength scores (mean ± standard deviation).

Concluding Statement Regarding Post Hoc Tests

When testing the null hypothesis with simple ANOVA, a researcher should first conduct an F test to determine the value of mean square error and whether any differences exist among any of the groups. If F is not significant, no further calculations are needed. If F is significant and comparisons other than pairwise comparisons are needed, Scheffé should be used to investigate differences among various combinations of groups. If F is significant, and only pairwise comparisons are to be made, Tukey should be used to contrast mean differences.

Magnitude of the Treatment (Size of Effect)

A significant F is not the only factor considered when evaluating the importance of a finding. The researcher should also consider the size of the effect of the treatment. A significant F indicates the probability that the differences occurred by chance. Analysis of variance is actually a test of the reliability of the differences. When F is significant, the odds are good that similar mean differences would be found if the experiment were repeated. However, when N is large and mean square error is small, ANOVA may produce a significant F when the mean differences are so small that they have little or no practical value. Determining the size of the effect is a method of comparing treatment effects, independent of sample size.

For example, two groups of 200 subjects each participate in aerobic exercise programs for 4 days a week and 5 days a week. We may find that $\dot{V}O_2max$ values in milliliters per kilogram per minute are 47.6 and 49.1, respectively, and that the difference is significant at $p < .05$. Although the significance test indicates this difference is more than we would expect just due to chance, it is probably not large enough (1.5 milliliters per kilogram per minute) to be worth the effort of an additional workout per week.

Eta Squared

The simplest measure of the effects of treatment in ANOVA is eta squared, symbolized by R^2:

$$R^2 = \frac{SS_B}{SS_T}. \tag{11.18}$$

R squared (R^2) is the ratio of the variance due to treatment and the total variance. Note that this is the same R^2 (and r^2) that we address in chapters 8 and 9. It produces a rough estimate of the size of the effect from a study comparing groups means as it does when assessing the magnitude of effect (variance accounted for) from regression. In the data from the strength training experiment (table 11.5), $R^2 = 50.82/98.17 = .52$. This means that 52% of the total variance can be explained by the treatment effects. The remaining 48% is unexplained.

The value of R^2 varies between 0.00 and 1.00 depending on the relative size of the treatment effects. It is a measure of the proportion of the total variance that is explained by the treatment effects. In this case, a little more than half of the total variance is explained. This is a fairly large proportion and confirms the conclusion that at least one treatment was effective in improving strength. In some statistics books (see Tabachnick and Fidell, 1996, p. 53), R^2 is called **eta squared** (η^2).

A variant of eta squared is called **partial eta squared** (**partial η^2**). Some popular statistical software routinely reports partial η^2. As the name suggests, partial η^2 is similar to η^2. In fact, in simple ANOVA partial η^2 and η^2 are exactly the same. However, in more complicated ANOVA designs (i.e., factorial ANOVA in chapter 14), the values differ. These values are ways of assessing how much variance in the dependent variable is accounted for, or explained by, the independent variable.

Omega Squared

Another common method of determining the size of the effect is **omega squared** (ω^2):

$$\omega^2 = \frac{SS_B - (k-1)(MS_E)}{SS_T + MS_E} \tag{11.19}$$

For the data in table 11.5, omega squared is calculated as follows:

$$\omega^2 = \frac{50.82 - (5-1)(1.58)}{98.19 + 1.58} = .46.$$

Omega squared is a more accurate measure of effect size because it attempts to account for the unexplained variance (mean square error) and usually produces a smaller value than R^2. The value of $\omega^2 = .46$ is smaller than $R^2 = .52$, but it still indicates that a large proportion of the total variance may be attributed to treatment effects. This would still be considered a large effect size. Keppel (1991, p. 66) suggests that for ω^2 a value of .15 is large, .06 is medium, and .01 is small in the behavioral sciences. These guidelines are only advisory; the importance of an effect size depends on the context of the study.

The size of the effect is relatively underreported in kinesiological experimental research because many authors choose to simply report the results of the null hypothesis significance test. However, it may be very important to calculate and report effect sizes after a significant F is found. The null hypothesis significance test provides information regarding our confidence that an effect is different from zero. We are usually more interested in how big an effect the independent variable has on the dependent variable. Further, because of the nature of the relationship between sample size and statistical power, with a large enough N, small and perhaps even trivial effects may be statistically significant.

An Example From Leisure Studies and Recreation

Researchers in the effective use of leisure time have postulated that play activity may be used to reduce stress. It is proposed that play activity is most effective when it is perceived by the subject to be free play (i.e., not directed by others). To test this hypothesis, Finney (1985) randomly divided male and female college-age subjects into four groups: high perceived self-control of the play experience (i.e., low structure), moderate perceived control, low perceived control, and a control group, who performed what they considered to be work, not play. The groups had 19, 20, 20, and 20 subjects, respectively.

All subjects performed a 30-minute stress-producing task—they worked 12 pages of math problems while listening to periodic bursts of 95 decibels of noise delivered through headphones. Upon completing this 30-minute stress period, the subjects engaged for 10 minutes in one of four play activities that varied in the amount of perceived control that the subjects had over their own play behavior. Following this play period, the subjects attempted to solve four geometric puzzles. The subjects were unaware that two of the puzzles were not solvable. Persistence on the two unsolvable puzzles (measured by time in total seconds spent on the two puzzles before giving up) was the dependent variable used to assess the effectiveness of the play period in reducing the stress created by the work task. Whereas H_1 was used to justify the study, the null hypothesis H_0 was tested statistically.

A simple ANOVA applied to the data yielded the information in tables 11.7 and 11.8. To identify specific mean differences, Finney applied $F_{Scheffé}$ (equation 11.15) to the comparisons between each group. This produced a matrix of F values contrasting each group with every other group. Table 11.9 demonstrates this analysis.

The findings of $F_{Scheffé}$ declare that only the group with perceived high self-control of the play period differed from the control group at $p < .05$. (Table A.6 indicates

TABLE 11.7

Effects of Play on Reduction of Stress

Source of variance	SS	df	MS	F	p
Between groups	1,193,210.000	3	397,736.667	4.189	<.01
Within groups (error)	7,121,448.000	75	94,952.640		
Totals	8,314,658.000	78			

TABLE 11.8

Group Means (Time in Seconds) and Standard Deviation by Level of Structured Play

	High	Medium	Low	Control
Mean	705.90	466.30	427.90	385.05
SD	507.64	251.93	235.31	108.80

Note: Data are in total seconds.

TABLE 11.9

$F_{\text{Scheffé}}$ Values for Pairwise Comparisons

	High	Medium	Low	Control
High	0.000	1.964	2.643	3.521*
Medium		0.000	0.052	0.232
Low			0.000	0.064
Control				0.000

* p < .05.

that critical F for $df_B = 3$ and $df_E = 75$ is 4.054 at $\alpha = .01$, and table A.5 indicates critical F is 2.727 at $\alpha = .05$.)

Thinking that perhaps $F_{\text{Scheffé}}$ may be too conservative for this type of data, Finney then applied Tukey's post hoc test to analyze specific raw score mean differences (see table 11.10). Using equation 11.17, Finney found Tukey's HSD was 316.95 at $\alpha = .01$ and 257.79 at $\alpha = .05$. Tukey's confirms Scheffé by indicating that the high group differs from the control group, but using Tukey, the difference is found to be significant at $p < .01$. The Tukey test also found that the high group differed from the low group at $p < .05$. But the low and medium groups

TABLE 11.10

Mean Difference Values for Tukey's Pairwise Comparisons

	High	Medium	Low	Control
High	0.00	239.60	276.00*	320.85**
Medium		0.00	38.40	81.25
Low			0.00	42.85
Control				0.00

$** p < .01; * p < .05.$

still did not show significant differences between each other or from the control group.

The value of ω^2 was .1080, revealing a moderate effect size. About 11% of the differences between the groups can be attributed to the play treatment. Finney concluded that high self-control reduces stress more effectively than low self-control in a play environment.

Summary

Analysis of variance compares the means of three or more groups. When F is significant, it indicates that a significant difference exists somewhere among the several group means. However, the specific mean differences are not identified. A post hoc test must be used to identify which group means differ. After a significant F is found, some measure of the size of the effect of the treatment should be computed to ascertain the relative proportion of the group mean differences that can be attributed to the treatment effects.

Remember that when two means differ significantly at some p value (for example, $p = .05$), the p value indicates the probability (5 in 100) that we would have found mean differences this large (or larger), if the null hypothesis is true. We typically start with the assumption that no difference exists (H_0), and then we test that assumption. If we find significant F values, the null is rejected at the given level of confidence. We never really know whether we are correct in accepting or rejecting the null hypothesis; we know only the odds, or probability, that we are correct in our conclusion.

Problems to Solve

Solve problem 1 by hand with a calculator, then check your hand-calculated answers against computer-generated results. Use a computer to solve problems 2 and 3.

1. A biomechanics researcher wanted to test whether good, average, and poor sprinters differed in horizontal foot speed. She classified the sprinters into three groups based on their sprint times. The horizontal foot speed at touchdown in feet per second was then analyzed with the following results (data is fabricated):

   Poor   4, 5, 8, 6, 7, 6
   Average 7, 8, 9, 6, 7, 10
   Good   10, 13, 12, 8, 11, 12

   a. What are the mean values for each group?
   b. What are the sum of squares values?
   c. What are the degrees of freedom and mean square values?
   d. What is the F value? Is it significant? If so, what is the p value?
   e. Set up a mean difference table.
   f. Apply Scheffé's I to the data. Do any of the means differ?
   g. Apply Tukey's HSD test. Do any of the means differ?
   h. Does Tukey differ from Scheffé in the interpretation of the mean difference table?
   i. What is the size of the effect? Is it small, moderate, or large?

   Hint: If you use a computer to solve this problem, create two columns in the software database. Column 1, the group column, should contain the numbers 1 (rows 1–6, good), 2 (rows 7–12, average), or 3 (rows 13–18, poor). Column 2 (the score column) should contain the foot speed scores.

2. A volleyball coach noted that players who practiced jumping during practice time seemed to be able to jump higher in the games. He wondered whether jumping practice increased vertical jump height more effectively than lower extremity weight training. To test this phenomenon, he proposed the null hypothesis, that no significant differences exist between players who practiced jumping, players who participate in lower extremity weight training, and players who do neither exercise.

For 6 weeks prior to the start of the season, he randomly divided his team into three groups of 10 players each. Group one (control) practiced regularly without any special jumping or weight training. Group 2 (jumping) spent the last 20 minutes of each practice session in jumping exercises, and group 3 (weight training) spent the last 20 minutes of each practice session doing high-resistance leg presses. Following are the results (in inches) of a vertical jump test taken 2 days before the first game.

Control	Weight training	Jumping
23	26	33
21	25	32
25	28	36
26	31	29
31	34	27

Control	Weight training	Jumping
27	35	34
32	29	35
34	27	37
25	28	35
21	25	28

a. What are the means and standard deviations for the three groups?
b. What is the value of F? Is it significant? Is so, what is the p value?
c. Did either of the experimental group means significantly exceed the control group? (Use Tukey's *HSD*.) Which one(s)?
d. Do you accept or reject the null hypothesis?

Review the hint given in problem 1 above for help with setting up the database. The database in problem 2 will be similar to that in problem 1: three groups in column 1 designated by numbers 1, 2, or 3, and vertical jump scores in column 2.

3. A physical education teacher wanted to know the effects of various activity levels on body composition. She surveyed her physical education classes and categorized 30 students into five activity levels based on the amount of daily exercise in which the students participated: inactive, semiactive, normal, active, and very active. She measured percent body fat values using skinfold calipers on all subjects with the following results.

Inactive	Semiactive	Normal	Active	Very active
30.2	29.4	22.9	17.6	10.9
29.6	17.6	25.4	13.4	13.7
35.2	26.4	19.6	20.3	12.8
19.1	25.3	18.7	19.6	14.7
26.3	22.5	21.8	15.1	9.3
22.4	28.6	24.9	10.7	12.7

Use a computer to solve this problem.
a. What is the independent variable in this study? What is the dependent variable?
b. What are the means and standard deviations for each group?
c. Does a significant difference exist among any of the groups? What is the probability value for the test of the null hypothesis?
d. Create a mean difference table and indicate the significance of the differences among the groups using Tukey's *HSD*.

Review the hint given in problem 1 for help with setting up the database. The database in problem 3 will be similar to that in problem 1, but column 1 will have five groups (designated by numbers 1, 2, 3, 4, or 5) and percent body fat values will be in column 2.

See appendix C for answers to problems.

Key Words

analysis of variance
between-group variance
Bonferroni adjustment
eta squared (η^2, also R^2)
F ratio
familywise error rate
mean square
omega squared (ω^2)
partial eta squared

post hoc
R^2 (eta squared)
Scheffé's confidence interval (I)
sum of squares
total variance
Tukey's honestly significant
    difference *(HSD)*
within-group variance

$$H = \left[\frac{12}{18(18+1)} \right]\left[\frac{82.5^2}{6} + \frac{54.5^2}{6} + \frac{34.0^2}{6} \right] - 3(18+1)$$

$$H = 1.0351[1822.09] - 57$$

$$H = 6.77$$

X_4

X_1

X_3

X_2

Analysis of Variance With Repeated Measures

graduate student measured the heart rates (beats per minute) of a set of subjects riding at 5, 7, and 9 miles per hour on a stationary bicycle. Fifteen subjects rode at increasing speeds for 2 minutes at each of the three rates. The mean values for the 15 subjects at the end of each 2-minute exercise bout were determined to be 120 beats per minute at 5 miles per hour, 130 beats per minute at 7 miles per hour, and 150 beats per minute at 9 miles per hour. Does a significant heart rate difference exist among the speeds? Simple analysis of variance assumes that the three groups are independent (that is, they are separate subjects). However, the data in this study are not independent because each mean is based on the same 15 subjects. Using the same subjects in repeated measure research designs produces a relationship (correlation) among the three scores. To analyze this type of data, we must account for this relationship. In this chapter we learn to apply analysis of variance in a repeated-measure design.

One of the most common research designs in kinesiology involves measuring subjects before treatment (pretest) and then after treatment (posttest). A dependent *t* test with matched or correlated samples is used to analyze such data because the same subjects are measured twice **(repeated measures).** In ANOVA, this type of design is referred to as a **within-subjects design** (in contrast to the simple ANOVA presented in chapter 11, which is sometimes referred to as a between-subjects design or between-subjects ANOVA). When three or more tests are given—for example, if a pre-, mid-, and posttest are given with treatment before and after the midtest—ANOVA with repeated measures is needed to properly analyze the differences among the three tests.

The simple ANOVA described in chapter 11 assumes that the mean values are taken from independent groups that have no relationship to each other. In this independent group design, total variability is the sum of

- variability between people in the different groups **(interindividual variability),**
- variability within a person's scores **(intraindividual variability),**
- variability between groups due to treatment effects, and
- variability due to error (variability that is unexplained).

However, in that design we are unable to tease apart how much variability is due to interindividual variability, intraindividual variability, and unexplained variability, so all these sources are lumped together as error. In contrast, when only one group of subjects is measured more than once, the data sets are dependent. The total variability for a single group of subjects measured more than once is expected to be less than if the scores came from different groups of people (that is, if the scores were independent) because interindividual variability has been eliminated by using a single group. This tends to reduce the mean square error term in the denominator of F in a manner similar to the correction made to the standard error of the difference in the dependent t test (equation 10.13, p. 161).

In the repeated measures design, the subjects serve as their own control. If all other relevant factors have been controlled, any differences observed between the means must be due to (a) the treatment, (b) variations within the subjects (intra-individual variability), or (c) error (unexplained variability). Variability between subjects (interindividual variability) is no longer a factor.

Assumptions in Repeated Measures ANOVA

Except for independence of samples, the assumptions for simple ANOVA (between-subjects designs) discussed in chapter 11 also hold true for repeated measures ANOVA (within-subjects designs). But with repeated measures designs, we must also consider the relationships among the repeated measures. Specifically, repeated measures ANOVA must meet the additional assumption of **sphericity.** A detailed explanation of sphericity, and a related assumption referred to as compound symmetry, is beyond the scope of this book. However, a simple way to conceptualize sphericity is to consider a study in which subjects are measured at three time periods: time 1, time 2, and time 3. We can also calculate the difference scores between each time period: time 1 − time 2, time 2 − time 3, and time 3 − time 1. Sphericity requires that the variances of the difference scores are all equal. Violations of sphericity will inflate the type I error rate so that if α is set at, for example, .05, the true risk of committing a type I error will be higher than .05. When only two repeated measures are used (such as pre–post measures for a dependent t test), this assumption is not applicable because only one set of differences can be calculated. Methods of dealing with violations of sphericity are presented later in this chapter.

Calculating Repeated Measures ANOVA

To demonstrate how to calculate ANOVA with repeated measures, we analyze a hypothetical study. A graduate student studying motor behavior was interested in the decrease in balance ability that bicycle racers experience as their fatigue increases. To measure this, the researcher placed a racing bicycle on a roller ergometer. A 4-inch-wide stripe was painted in the middle of the front roller, and the rider was required to keep the front wheel on the stripe. Balance was indicated by wobble in the front wheel and was measured by counting the number of times per minute that the front wheel of the bike strayed off the 4-inch stripe over a 15-minute test period.

As the test progressed, physiological fatigue increased and it became more and more difficult to maintain the front wheel on the stripe. The 15-minute test period was divided into five 3-minute periods for the purpose of collecting data. Data were collected on the number of balance errors during the last minute of each

3-minute period. In this design, the dependent variable is balance errors and the independent variable is time period (we call this variable "time"), which reflects the increase in fatigue.

Table 12.1 presents the raw data in columns and rows. The data (in errors per minute) for the subjects ($N = 10$) are in the rows, and the data for time ($k = 5$ repeated measures) are in the columns. The sum of rows (ΣR) is the total score for each subject over all five time periods, and $\bar{X}_{subjects}$ at the right denotes the mean across time for each subject. The sum of each column (ΣC) is the total for all 10 subjects on a given trial; ΣX_T is presented at the bottom of the table.

Remember that ANOVA stands for analysis of variance. We analyze the variance by breaking the total variance in the data set into the relevant pieces. For the repeated measures ANOVA, we partition the total variance into pieces attributed to (1) differences between measurement periods [for the example in table 12.1, these are the differences between time periods (columns) at minute 3 and minute 6 and so on], which is represented by how the means across time differ; (2) differences between subjects; and (3) unexplained variance (the error or residual). Notice that in contrast to the between-subjects ANOVA presented in chapter 11, where we could partition the total variance into only two pieces [between groups and within groups (error)], in repeated measures ANOVA we have added a third piece. The third piece is the component attributable to differences between subjects. Because each subject provides a score for each time period, we can estimate how much of the total variance is due simply to different abilities of different subjects. As noted previously, in between-subjects ANOVA, interindividual differences are lumped into the error term.

TABLE 12.1

Raw Data: Balance Errors per Minute

Subject	Minute 3	Minute 6	Minute 9	Minute 12	Minute 15	ΣR	$\bar{X}_{subject}$
1	7	7	23	36	70	143	28.6
2	12	22	26	26	20	106	21.2
3	11	6	9	31	30	87	17.4
4	10	18	16	40	25	109	21.8
5	6	12	9	28	37	92	18.4
6	13	21	30	55	65	184	36.8
7	5	0	2	10	11	28	5.6
8	15	18	22	37	42	134	26.8
9	0	2	0	16	11	29	5.8
10	6	8	27	32	54	127	25.4
ΣC	85	114	164	311	365	1,039	
Mean	8.5	11.4	16.4	31.1	36.5		
SD	4.5	8.0	10.8	12.6	21.2		

$\Sigma X_T = 85 + 114 + 164 + 311 + 365 = 1,039$; grand mean ($M_G$) $= 1,039/50 = 20.78$

In repeated measures ANOVA, we partition variance by calculating the sums of squares (SS) for each piece so that

$$SS_{total} = SS_{time} + SS_{subjects} + SS_{error}. \qquad (12.01)$$

We then divide each sum of squares by the appropriate degrees of freedom, which results in mean square values for each piece.

First, the total sums of squares (SS_T) is calculated by subtracting each score from the grand mean (M_G), squaring the differences, and then adding the squared differences:

$$SS_T = \Sigma\,[(X_i - M_G)^2]. \qquad (12.02)$$

From table 12.1, we can see that the sum of all 50 scores (10 subjects times five time periods) is 1,039, resulting in a grand mean of 20.78. Table 12.2 shows the squared differences between each score and the grand mean.

For example, the score for subject one at minute 3 is 7 errors, and therefore the squared difference is $(7 - 20.78)^2 = 189.88$. By adding all the squared differences together, the resulting total sum of squares is 13,356.58.

To assess the variance due to differences between time periods, we must calculate the sum of the squared differences between each time period mean and the grand mean (with an adjustment for sample size):

$$SS_{time} = n\,\Sigma\,[(\bar{X}_{time} - M_G)^2], \qquad (12.03)$$

where n is number of subjects. For the data in table 12.1, $SS_{time} = 10\,[(8.5 - 20.78)^2 + \ldots + (36.5 - 20.78)^2] = 6,115.88$.

TABLE 12.2

Squared Differences Between Each Score
and the Grand Mean

Subject	Minute 3	Minute 6	Minute 9	Minute 12	Minute 15	ΣR
1	189.89	189.89	4.93	231.65	2,422.61	3,038.96
2	77.09	1.49	27.25	27.25	0.61	133.68
3	95.65	218.45	138.77	104.45	85.01	642.32
4	116.21	7.73	22.85	369.41	17.81	534.00
5	218.45	77.01	138.77	52.13	263.09	749.52
6	60.53	0.05	85.01	1171.01	1,955.41	3,272.00
7	249.01	431.81	352.69	116.21	95.65	1,245.36
8	33.41	7.73	1.49	263.09	450.29	756.00
9	431.81	352.69	431.81	22.85	95.65	1,334.80
10	218.45	163.33	38.69	125.89	1,103.57	1,649.92
						$SS_T = 13{,}356.58$

To assess the variance due to differences between subjects, we subtract the mean for each subject across time from the grand mean, square these differences, add them up, and then multiply this sum by the number of trials:

$$SS_{subjects} = T \Sigma (\bar{X}_{subject} - M_G)^2, \qquad (12.04)$$

where T is number of time periods. For the data in table 12.1, $SS_{subjects} = 5 [(28.6 - 20.78)^2 + \ldots + (25.4 - 20.78)^2] = 4{,}242.58$.

To calculate the unexplained variance (SS_{error}), we can rearrange equation 12.01 so that

$$SS_{error} = SS_{total} - SS_{time} - SS_{subjects}. \qquad (12.05)$$

For our example, $SS_{error} = 13{,}356.58 - 6{,}115.88 - 4{,}242.58 = 2{,}998.12$.

We now have the necessary sums of squares values for the repeated measures ANOVA, but as in the between-subjects ANOVA in chapter 11, we need to divide the sums of squares values by their appropriate degrees of freedom. The degrees of freedom for time is

$$df_{time} = T - 1. \qquad (12.06)$$

For the data in table 12.1, $df_{time} = T - 1 = 5 - 1 = 4$.

For the subjects effect, the degrees of freedom are calculated as

$$df_{subjects} = N - 1. \qquad (12.07)$$

In our example, $df_{subjects} = 10 - 1 = 9$.

For the error term, the degrees of freedom are calculated as

$$df_{error} = (T - 1)(N - 1). \qquad (12.08)$$

In our example, $df_{error} = (5 - 1)(10 - 1) = 36$.

Our next step is to calculate the mean square (MS) terms for each component. Recall from chapter 11 that a mean square value is the ratio of the sums of squares term divided by the appropriate df value. The mean square for time is calculated as

$$MS_{time} = SS_{time}/df_{time}. \qquad (12.09)$$

For the data in table 12.2, $MS_{time} = 6{,}115.88/4 = 1{,}528.97$.

The mean square for subjects is calculated as

$$MS_{subjects} = SS_{subjects}/df_{subjects}. \qquad (12.10)$$

In our example, $MS_{subjects} = 4{,}242.58/9 = 471.40$.

Finally, the mean square error term is calculated as

$$MS_{error} = SS_{error}/df_{error}. \qquad (12.11)$$

In our example, $MS_{error} = 2{,}998.12/36 = 83.28$. We now have all the necessary pieces needed for the repeated measures ANOVA.

Recall from chapter 11 that an F ratio is a ratio of mean squares. The F ratio of interest is the F ratio for the time effect, which assesses whether the trial means differ. The F for time is calculated as

$$F_{time} = MS_{time}/MS_{error}. \tag{12.12}$$

For our example, $F_{time} = 1{,}528.97/83.28 = 18.36$ (see table 12.3).

It is also possible to calculate an F value for subjects ($F_{subjects} = MS_{subjects}/MS_{error}$). But this value is not of interest at this time because it simply represents a test of the variability among the subjects. This $F_{subjects}$ value is not important to the research question being considered (Does an increase in fatigue result in an increase in mean balance errors across time?). We use $F_{subjects}$ in chapter 13 when intraclass correlation for reliability is discussed.

To determine the significance of F for time, we look in tables A.4, A.5, and A.6 of appendix A. The df_{time} is the same as df_B from chapter 11. The df_{error} is used in the same way as df_E in chapter 11—to measure within-group variability. Table A.6 shows that for $df\,(4, 36)$ an F of 4.02 is needed to reach significance at $\alpha = .01$. Because our obtained F (18.36) easily exceeds 4.02, we conclude that differences do exist somewhere among the mean values for the five time periods at $p < .01$.

Correcting for Violations of the Assumption of Sphericity

Remember that one of the assumptions of repeated measures ANOVA is that the variances of the differences between the repeated measures, or time periods, is equal. This is called the assumption of sphericity. If the assumption of sphericity is not met, the probability of making a type I error (rejecting H_0 when it is true) is higher than the nominal α. Methods of correcting for a violation of the assumption include the Greenhouse-Geisser adjustment and Huynh-Feldt adjustment. Both correction procedures modify the degrees of freedom values for time and error. Most statistical software automatically checks for sphericity and provides the Greenhouse-Geisser and Huynh-Feldt corrections.

Greenhouse-Geisser Adjustment

The **Greenhouse-Geisser (GG) adjustment** consists of dividing the df_{time} and df_{error} values by $T - 1$ (the number of repeated measures minus 1). Note that $T - 1$ is the same as the value df_{time} (equation 12.06). Hence, in the GG adjustment, the value df_{time} becomes $df_{time}/(T - 1) = 1$. Degrees of freedom for error are adjusted by the same method so that df_{error} becomes $df_{error}/(T - 1)$. In the bicycle research example, the adjusted $df_{error} = 36/(5 − 1) = 9$. We then reevaluate F using tables A.4, A.5, and A.6 and the adjusted df $(1, 9)$. For df $(1, 9)$, table A.6 indicates that F must equal 10.56 to be significant at $\alpha = .01$. Because our obtained value (18.36) exceeds 10.56, we may still conclude that the means of two or more trials are significantly different at $p < .01$.

This application of the GG adjustment assumes maximum violation of the assumption of sphericity. Consequently, when the violation is minimal, this adjustment of degrees of freedom may be too severe, possibly resulting in a type II error (accepting H_0 when it is false).

The Huynh-Feldt Adjustment

An alternate method, the **Huynh-Feldt (HF) adjustment,** attempts to correct for only the amount of violation that has occurred. In the HF adjustment, degrees of freedom for trials and degrees of freedom for error are multiplied by a value (epsilon, ε) that ranges from 0.00 for maximum violation to 1.00 for no violation. As a general rule, if $\varepsilon \geq .75$, the violation is considered to be insignificant. For example, if $\varepsilon = .43$, $df_{time} = 4 \times .43 = 1.72$ and $df_{error} = 36 \times .43 = 15.48$. For df $(1.72, 15.48)$, table A.6 indicates an F value of 6.36 is needed to reach significance at $\alpha = .01$. Because the obtained F (18.36) exceeds 6.36, we are more than 99% confident in declaring F to be significant at $p < .01$ even if sphericity is violated where $\varepsilon = .43$.

Although F is still significant, these adjustments reduce the confidence we can place in our conclusion that the differences among the trial means are significant. If the obtained p value is close to the rejection level of $\alpha = .05$ (suppose we obtain $p = .04$), and the adjustment increases it to .06, H_0 must be accepted.

Calculation of epsilon is a complicated procedure and is not discussed here. Winer et al. (1991, p. 251) present procedures for calculating epsilon (see also Girden, 1992). Most statistical software programs test for violations of the assumption of sphericity and provide both the GG and the HF adjustments, including the epsilon values. (Epsilon values for GG are more conservative than those for HF, thus providing greater protection against making type I errors but increasing the risk of type II errors.)

A strategy for determining the significance of F (when sphericity must be assumed) has been suggested by Keppel (1991, p. 353). A modification of Keppel's strategy follows.

Evaluate F with the GG adjustment (the most conservative condition):

- If F with GG adjustment is significant, reject H_0.
- If F with the GG adjustment is not significant, evaluate F with no adjustment.
- If F with no adjustment is not significant (the most liberal condition), accept H_0.
- If F with GG adjustment is not significant, but F with no adjustment is significant, use HF adjustment (a moderate condition) to make the final determination.

TABLE 12.3

ANOVA: Bicycle Errors

Source	SS	df	MS	F	p
Trials	6,115.88	4	1,528.97	18.36	<.0001
Subjects	4,242.58	9	471.40	5.66	
Error	2,998.12	36	83.28		

Note. Greenhouse-Geisser $\varepsilon = .37$; $p < .0003$. Huynh-Feldt $\varepsilon = .43$; $p < .0001$

An alternate solution if violations are severe is to use multiple analysis of variance (MANOVA) with the repeated measures designated as multiple dependent variables. Under these conditions, the assumption of sphericity is not required. This approach is less powerful and thus provides better protection against type I errors but less protection against type II errors. Further discussion of sphericity may be found in Dixon (1990, p. 504), Keppel (1991, p. 351), Schutz and Gessaroli (1987), Tabachnick and Fidell (1996, pp. 80 and 352), and Winer, Brown, and Michels (1991, p. 251). The ANOVA results for the bicycle test are summarized in table 12.3, as calculated by computer.

Post Hoc Tests

In this study, the researcher was interested in how much fatigue could be tolerated before a significant decrease in balance ability occurred. As the racer pumps up a hill, fatigue increases and balance decreases. It may be wise for the racer to rest at the top of a long hill, or at least proceed down the other side carefully, because balance may be impaired and a serious fall could occur. But at what point does balance ability decrease significantly? One approach to answering that question is to perform post hoc pairwise comparisons between different time levels.

Recall the discussion of familywise α from chapter 11. Briefly, when we perform multiple statistical tests at some α level (e.g., .05), the cumulative risk of committing at least one type I error across the family of tests is greater than the nominal α of .05. The relationship between the single comparison rate and the familywise α rate was given by equation 11.01 on page 180.

In chapter 11, we introduced a variety of post hoc procedures such as Tukey's honestly significant difference test. However, these post hoc tests are also affected by the sphericity assumption of repeated measures ANOVA. Maxwell (1980), using computer simulations, compared different post hoc procedures for conducting pairwise comparisons in repeated measures analyses. The techniques were assessed for their ability to protect against both type I and type II error under different combinations of sample size, number of means, and violations of sphericity. When sphericity is not violated, Tukey's test keeps the type I error rate at the nominal value and results in the most statistical power (Maxwell, 1980). Therefore, when sphericity is not violated, Tukey's honestly significant difference is the post hoc technique of choice. However, when sphericity is violated, the best compromise under these conditions was determined to be the use of dependent t tests with a Bonferroni correction. Recall from chapter 11 that with the Bonferroni correction, the per comparison alpha is calculated by dividing the familywise alpha (FW_α) level by the number of comparisons performed. For the data in table 12.1, five mean values results in a total of 10 possible pairwise comparisons. If we were to perform all possible pairwise comparisons, and assuming a familywise alpha level of .05, the per comparison alpha (PC_α) would be calculated as

$$PC_\alpha = \frac{.05}{10} = .005.$$

A graph of the column means (± standard deviation) indicating the position of the significant differences is presented in figure 12.1. Dependent t tests performed at an α level of .005 (two-tailed) were performed for all 10 possible pairwise comparisons. The results showed that the balance errors at minute 12 were significantly higher than at minutes 3, 6, and 9 ($p = .0001, .0001$, and $.0003$, respectively). Similarly, the balance errors at minute 15 were significantly higher than at minutes 3, 6, and 9 ($p = .0017, .0037$, and $.0023$, respectively). Minutes 3, 6, and 9 did not significantly differ from one another (all $p \geq .0174$), and minute 12 did not differ from minute 15 ($p = .26$). Therefore, between minutes 9 and 12, a significant increase ($p < .005$) is observed in errors committed.

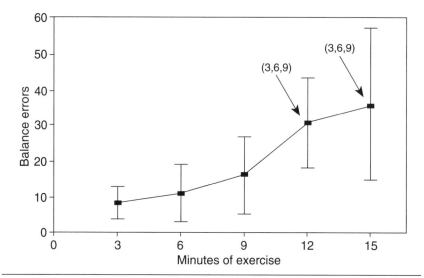

Figure 12.1 Mean (± standard deviation) balance errors over 15 minutes of exercise. Values in parentheses indicate times periods that had significantly ($p < .005$) fewer balance errors than the indicated time period.

The researcher concluded that the increase in work load between minutes 9 and 12 is sufficient to significantly increase balance errors. If this work load can be related to heart rate, racers may be counseled to measure their heart rate after pumping up a long hill. If heart rate at the top of the hill is equal to or higher than a level equivalent to the work load at minute 9, the rider would be wise to proceed down the other side carefully.

Note that just because 10 pairwise comparisons were possible, depending on the research question not all of them need to be performed. Indeed, *a priori* a per comparison alpha of .005 might suggest an overly conservative approach. It may have been more prudent to compare only sequential time periods (i.e., minute 3 vs. minute 6, minute 6 vs. minute 9, minute 9 vs. minute 12, and minute 12 vs. minute 15). Such an approach would have resulted in a per comparison alpha of .05/4 = .0125 instead of .005. For this particular example, the interpretation of the data would not have changed (minute 3 ~ minute 6, minute 6 ~ minute 9, minute 9 < minute 12, minute 12 ~ minute 15). Of course, such an approach would need to be determined before, not after, the fact.

Interpreting the Results

The F for treatment (in the bicycle example) indicates that significant differences exist somewhere among the mean values of the time periods at $p < .01$. The GG and HF adjustments do not alter the p value enough to change the statistical decision. Therefore, we conclude that F is significant at better than $p < .01$ even after adjustment of degrees of freedom for possible violations of the assumption of sphericity. This finding allows us to reject the null hypothesis and supports the research hypothesis that balance errors increase as fatigue increases. Dependent t tests with the Bonferroni correction reveals that a significant difference in number of errors occurred between minutes 9 and 12. These hypothetical results suggest that bicycle riders are less able to maintain balance when they are fatigued.

An Example
From Leisure Studies and Recreation

A graduate student in leisure studies postulated that students' conservation awareness would increase if they participated in a semester-long backpacking class at the university. A questionnaire to measure conservation awareness was developed and found to be valid and reliable. It was administered to 12 students at the beginning (pre), at midterm (mid), and after completion of the class (post). Means and standard deviations for the three measures are reported in table 12.4.

TABLE 12.4

Repeated Measures of Conservation Awareness

	Pre	Mid	Post
Mean	58.6	58.8	62.6
SD	9.3	8.0	6.4

Repeated measures ANOVA produced an F value of 1.94 among the three means. For df (2, 22), table A.4 in appendix A requires a value of 2.56 to reach significance at $\alpha = .10$. Because the obtained F value was less than the lowest critical value from table A.4, the graduate student accepted the null hypothesis and concluded that no difference existed among the three means. Conservation awareness, as measured by the survey, was not affected by participation in the backpacking class. Because F was not significant, sphericity was not an issue, and no further analysis with post hoc tests was needed.

Summary

Repeated measures ANOVA, one of the most common designs used in kinesiological research, permits the researcher to determine the significance of mean differences measured on the same subjects over repeated trials. It is often used in testing designs in which subjects are measured before, during, and after treatment or in which subjects are measured repeatedly over time, as when data is collected minute by minute on a treadmill or bicycle ergometer. It produces an F value that can be evaluated to determine whether any significant differences exist among the mean values of the various trials. When F is significant, a post hoc test must be applied to determine the specific trial means that differ from one another.

Repeated measures ANOVA must meet the additional assumption of sphericity; that is, the variances of the differences among the trials must be approximately equal. When this assumption is not met, adjustments to the degrees of freedom and the p values obtained must be made with either the Greenhouse-Geisser or the Huynh-Feldt correction techniques.

Problems to Solve

Check your hand-calculated answers against a computer-generated result.

1. An exercise physiologist measured $\dot{V}O_2$ (milliliters per kilogram per minute) for five subjects on the treadmill in a graded exercise test (GXT). After 3

minutes of level walking as a warm-up, the work load was increased every 2 minutes for 10 minutes. Five repeated measures of $\dot{V}O_2$ were taken, one at the end of each 2-minute interval, and the following data were obtained:

Experimental Data

Subject	Minute 2	Minute 4	Minute 6	Minute 8	Minute 10
1	20	24	27	31	35
2	17	18	21	28	31
3	19	25	29	36	41
4	12	17	21	27	32
5	15	20	27	31	38

Hint: Set up the computer data file like the data file above: minute 2 in column 1, minute 4 in column 2, and so on. You do not need a column for subject numbers.

    a. What are the mean values for each measured minute?
    b. Calculate the sums of squares for columns (minutes), rows (subjects), and error.
    c. Calculate degrees of freedom and mean square for columns (minutes), rows (subjects), and error.
    d. Calculate F and determine the significance of F.
    e. Set up an ANOVA table to present the results of this problem.
    f. Does a significant difference exist between any of the minutes?
    g. Use post hoc tests to determine which of the minutes differ.

2. A biomechanist compared four commercial abdominal exercise machines with abdominal crunches done without equipment (control). To make the comparison, he attached electromyography electrodes to the upper rectus abdominus muscle in 10 male subjects and recorded the electrical activity (in millivolts) during exercise. Following are the results. Use a computer to solve this problem.

Subject	Machine 1	Machine 2	Machine 3	Machine 4	Control
1	920	913	1,005	811	905
2	566	580	767	833	833
3	328	293	300	290	367
4	568	555	637	484	557
5	248	267	167	312	331
6	815	606	607	462	459
7	313	264	252	295	314
8	538	502	399	354	506
9	1,475	1,563	1,174	1,722	1,581
10	465	434	375	300	352

Hint: Set up this data like the data in problem 1.

Data courtesy of Dr. William Whiting, Director, Biomechanics Laboratory, Department of Kinesiology, California State University Northridge.

    a. What are the mean and standard deviation for each machine and control?
    b. Did any of the machines elicit more electromyographic activity in the upper rectus abdominus muscle than control conditions?

    c. What is the probability that the null hypothesis is true in this experiment?

    d. Is it necessary to perform a post hoc test in this experiment?

See appendix C for answers to problems.

Key Words

Greenhouse-Geisser adjustment
Huynh-Feldt adjustment
interindividual variability
intraindividual variability

repeated measures
sphericity
within-subjects design

$$H = \left[\frac{12}{18(18+1)} \right] \left[\frac{82.5^2}{6} + \frac{54.5^2}{6} + \frac{34.0^2}{6} \right] - 3(18+1)$$

$$H = [.035][1822.09] - 57$$

$$H = 6.77$$

Quantifying Reliability

An exercise physiologist developed a new test to quantify anaerobic capacity in cyclists. Before applying the test to athletes, she wanted to assess the reliability, or reproducibility, of the test. She recruited a sample of high-level cyclists to participate in the reliability study. Each cyclist tested at the laboratory on three separate days and performed the anaerobic capacity test on each test day. How can the reliability data be analyzed and how should she interpret the results?

Chapter 1 introduced the concept of reliability. Reliability is typically thought of in terms of reproducibility and consistency. When an investigator determines the **test–retest reliability** of a test or instrument, the test is administered to a sample of subjects and then repeated at least once at some other time. For example, an athletic trainer may wish to quantify the test–retest reliability of a test of isokinetic strength. Subjects are tested three times within a period of 2 weeks. The data are then analyzed to determine whether the scores are similar to each over the three test periods. Similarly, the **inter-rater reliability** of a test may be examined. Here an investigator assesses whether different raters give similar scores to the same subjects for a given test. For example, a researcher may need to assess ability of different gymnastics judges to give similar scores for gymnastics performances or the ability of different physical therapists to get similar scores when performing manual muscle tests on the same patients. Similarly, **intrarater reliability** assesses the ability of a given rater to give similar scores. Suppose that an investigator is going to use skinfold caliper testing in a study. Before the study, the investigator would test a sample of subjects on multiple occasions and quantify how similar the skinfold values are within the subjects across the test times. If acceptable intrarater reliability is demonstrated, the investigator would have confidence in the quality of the skinfold data.

More generally, the concept of reliability concerns the quantification of **measurement error.** Reliability theory holds that all measurements are made with error. Any time a score is recorded from a subject, that observed score has two components: a true score component and an error component:

$$\text{observed score} = \text{true score} + \text{error.} \qquad (13.01)$$

The true score component is theoretically the average score of a subject recorded from an infinite number of trials. The error component is the difference between the observed score and the true score. The higher the reliability, the smaller the difference between the observed score and the true score. That is, high reliability means that the measurement errors are small.

Measurement error comes in more than one type. At the simplest level, we can break measurement error down into two broad categories: **random error** and **systematic error.** Random error can be conceptualized as noise. As the name suggests, random error is unpredictable. Scores vary from measurement to measurement but over time the error should average zero (positive and negative errors more or less cancel each other). Consider body weight. On a day-to-day basis your weight fluctuates up and down a little bit even if you are not gaining or losing weight over time. In contrast, systematic error means that the observed scores are trending up or down over multiple measures. For example, suppose that you measure the one-repetition-maximum bench press in subjects who have never before lifted weights. If you test those subjects again a week or so later, odds are that most subjects will score a bit higher on the second test because their technique improved.

In chapter 12 we developed the basic tools we need to quantify reliability. We test subjects on at least two occasions (or over at least two different raters if appropriate) and perform a repeated measures ANOVA on the data. From that ANOVA, we derive the needed variance terms to calculate a reliability coefficient. In addition, the ANOVA allows us to test whether significant mean differences exist across the repeated trials. Here we typically root against statistical significance because a significant effect for trials suggests that systematic error is present in the data. If we have systematic error, we need to re-evaluate our measurement schedule, perhaps by including a practice session before the testing in order to remove learning effects or by giving more rest between trials if fatigue or muscle soreness is affecting our scores.

Intraclass Correlation Coefficient

To calculate a reliability coefficient, we can expand the idea conceptualized in equation 13.01 by analyzing variances. Here, the total variance (σ_T) in a set of scores is considered to comprise the variance due to the true scores (σ_t) and the variance due to the error (σ_e):

$$\text{total variance} = \text{true score variance} + \text{error variance.} \qquad (13.02)$$

To calculate the reliability coefficient we make a ratio of the true score variance over the total variance (Weir, 2005):

$$R = \frac{\sigma_t}{\sigma_T} = \frac{\sigma_t}{\sigma_t + \sigma_e}, \qquad (13.03)$$

where R is the reliability coefficient. Here the reliability coefficient quantifies how much variance in the data is due to true score variance. The smaller the error variance, the larger the reliability coefficient. The reliability coefficient can vary from 0.0 (all variance is due to error) to 1.0 (all variance is true variance).

In practice, we estimate reliability using between-subjects variability and the error term from the repeated measures ANOVA as an index of the true score variance as follows:

$$\text{reliability} = \frac{\text{between-subjects variability}}{(\text{between-subjects variability} + \text{error}).} \qquad (13.04)$$

The reliability coefficient is formally quantified using what is called an **intraclass correlation coefficient** (ICC). The Pearson r from chapter 8 is sometimes called an interclass correlation coefficient because two variables are correlated (e.g., height and weight). Here, the intraclass correlation coefficient is "intraclass" because we are calculating an index that comprises the same variable measured on multiple occasions. The intraclass correlation coefficient should look somewhat familiar because we discussed variance ratios in earlier chapters with r squared (chapter 8), R^2 (chapters 9 and 11), and ω^2 (chapter 11).

Table 13.1 contains example data for a small reliability trial of the Wingate anaerobic power test. Ten subjects performed three Wingate tests from which peak anaerobic power (watts) was determined. Each test was performed 1 week apart. Our first step is to perform a repeated measures ANOVA on the data in table 13.1. We examine the effects for trials to determine whether systematic differences appear across the trials. That is, systematic error may be of concern if the means of the trials are significantly different from each other. The ANOVA summary table for the analysis is presented in table 13.2.

From table 13.2, we see that the effect for trials is not significant [F with 2 df in the numerator and 18 df in the denominator: $F(2,18) = 0.26$, $p = .77$]. Further, examination of the means across trials at the bottom of table 13.1 shows relatively small mean differences. Therefore, no systematic error is apparent across the three trial periods.

We now have the necessary variance components from table 13.2 to calculate intraclass correlation coefficients. We use plural here because intraclass correlation coefficients can be calculated in several ways, depending on the situation. However, we must first introduce some terminology to be consistent with what is reported in

TABLE 13.1

Example Data (Peak Power in Watts) for the Wingate Anaerobic Power Test

Trial 1	Trial 2	Trial 3
749.6	702.7	734.0
702.4	732.5	748.2
719.2	714.8	793.8
852	901.7	866.1
772	788.4	713.1
761.6	751.8	790.4
744	761.4	714
837.6	821.9	780.6
871.2	817.5	876.6
726.4	696.7	778.0
$\bar{X} = 773.6$	$\bar{X} = 768.9$	$\bar{X} = 779.5$
$SD = 59.3$	$SD = 64.5$	$SD = 56.8$

the literature. For some intraclass correlation coefficients, it has become common to refer to the information in the "subjects" source as the "between" subjects variability because it reflects how subjects differ from each other. Therefore, for some intraclass correlation coefficients, the mean square for subjects is referred to as mean square between (MS_B). We also need to create a new term called mean square within (MS_W) that is the composite of the mean square for trials and the mean square for error. The sum of squares for trials is 557.9 and the sum of squares for error is 19,265.1, the sum of which is 19,823 (sum of squares within). The degrees of freedom for trials is 2 and the degrees of freedom for error is 18, the sum of which is 20 (degrees of freedom within). Therefore, $MS_W = 19,823/20 = 991.15$. With these terms in mind, we have modified the ANOVA summary table as shown in table 13.3.

Shrout and Fleiss (1979) described six versions of the intraclass correlation coefficient, depending on the situation. The computational formula for each of these models is shown in table 13.4.

The relationship between the computational formulas presented in table 13.4 and the conceptual equations of 13.03 and 13.04 is not intuitively obvious and is beyond the scope of this chapter. The interested reader is referred to in-depth reviews by Shrout and Fleiss (1979), McGraw and Wong (1996), Looney (2000), and Weir (2005).

TABLE 13.2

ANOVA Summary Table for the Data in Table 13.1

Source	SS	df	MS	F	p
Trials	557.9	2	278.97	0.26	.77
Subjects	78,745.2	9	8,749.47	8.17	
Error	19,265.1	18	1,070.28		
Total	98,568.2				

TABLE 13.3

Modified ANOVA Summary Table for the Data in Table 13.1

Source	SS	df	MS	F	p
Between (or subjects)	78,745.2	9	8,749.47 (MS_B: 1-way) (MS_S: 2-way)	8.17	
Within	19,823	20	991.15 (MS_W)		
Trials	557.9	2	278.97 (MS_T)	0.26	.77
Error	19,265.1	18	1,070.28 (MS_E)		
Total	98,568.2				

TABLE 13.4

Formulas for Different Intraclass Correlation Coefficient Models, With Example Calculations

Model	Formula	Computation
1,1	$\dfrac{MS_B - MS_W}{MS_B + (k-1)MS_W}$	$\dfrac{8749.47 - 991.15}{8749.47 + (3-1)991.15} = .72$
1,k	$\dfrac{MS_B - MS_W}{MS_B}$	$\dfrac{8749.47 - 991.15}{8749.47} = .89$
2,1	$\dfrac{MS_S - MS_E}{MS_S + (k-1)MS_E + \dfrac{k(MS_T - MS_E)}{n}}$	$\dfrac{8749.47 - 1070.28}{8749.47 + (3-1)1070.28 + \dfrac{3(278.97 - 1070.28)}{10}} = .72$
2,k	$\dfrac{MS_S - MS_E}{MS_S + \dfrac{k(MS_T - MS_E)}{n}}$	$\dfrac{8749.47 - 1070.28}{8749.47 + \dfrac{3(278.97 - 1070.28)}{10}} = .89$
3,1	$\dfrac{MS_S - MS_E}{MS_S + (k-1)MS_E}$	$\dfrac{8749.47 - 1070.28}{8749.47 + (3-1)1070.28} = .71$
3,k	$\dfrac{MS_S - MS_E}{MS_S}$	$\dfrac{8749.47 - 1070.28}{8749.47} = .88$

However, it is useful to know what the different formulas convey and how to interpret the values. First, note that each model is identified using two terms separated by a comma. The number before the comma is used to identify the general model (model 1, 2, or 3). The model 1 equations are sometimes referred to as one-way models because they lump the trials and error components together into one component called within, as described previously. Models 2 and 3 are sometimes referred to as two-way models because they separate the trials and error terms.

The term after the comma denotes whether the intraclass correlation coefficient will be based on single measurements or mean values. In our reliability study example, each subject has three Wingate scores. If the intent is to assess the reliability of Wingate testing when in practice only one Wingate test will be administered in the future, then use the model of choice denoted with a 1 after the comma. If in practice the Wingate test will be administered three times and the average of the three will be used as the criterion score in future studies, then use the model of choice denoted with k, where k is the number of trials. For example, if using model 3 with the intent of assessing the reliability of the average of the three Wingate tests, then one could describe the intraclass correlation coefficient as 3,3.

The choice of which model to use is not always straightforward. With respect to model 1 (the one-way model), as noted previously this model does not allow for the partitioning of error into separate random and systematic components. However, if one is conducting a rater reliability study, in the one way model the raters are not crossed with subjects, meaning that it can accommodate data where not all subjects are assessed by all raters (Weir, 2005). It is our experience that this situation is rare in kinesiological studies and that all subjects are typically tested by all raters (or tested at each time period in a test–retest study). With respect to model 2 or model 3, the primary distinction centers on whether the intraclass correlation coefficient will include both systematic and random error or will only include random error. Model 2 includes systematic error, whereas model 3 includes only random error.

In the example data of table 13.1, the mean differences are small; therefore, little systematic error is present. Consequently, the differences between intraclass correlation coefficient values are small. Assume that the intent of the analysis is to assess the reliability of a single Wingate test so we can consider the following intraclass correlation coefficients: ICC $(1,1) = .77$, ICC $(2,1) = .72$, and ICC $(3,1)$ $= .71$. Notice that these values are similar to each other, which reflects the fact that because systematic error is small, the mean square within from the one-way model and the mean square error from the two-way models (reflecting random error) are similar (991.15 and 1,070.28, respectively).

Table 13.5 includes another example data set in which we have modified the data in table 13.1 to contain systematic error by adding 50 to 100 watts to each score on trials 2 and 3. Table 13.6 contains the resulting ANOVA summary table from the modified data set. Notice that the mean values from trial 2 and trial 3

TABLE 13.5

Example Data from Table 13.1 Modified by Adding Systematic Error of 50 to 100 Watts

	Trial 1	Trial 2	Trial 3
	749.6	802.7	834.0
	702.4	832.5	848.2
	719.2	814.8	893.8
	852	951.7	966.1
	772	888.4	813.1
	761.6	851.8	890.4
	744	861.4	814
	837.6	871.9	880.6
	871.2	967.5	976.6
	726.4	846.7	878.0
	$\bar{X} = 773.6$	$\bar{X} = 868.9$	$\bar{X} = 879.5$
	$SD = 59.3$	$SD = 54.2$	$SD = 56.8$

TABLE 13.6

ANOVA Summary Table for Data from Table 13.5

Source	SS	df	MS	F	p
Between (or subjects)	72,295.2	9	8,032.80 (MS_B: 1-way) (MS_S: 2-way)	9.79	
Within	82,813.0	20	4,140.65 (MS_W)		
Trials	68,037.9	2	34,018.97 (MS_T)	41.44	<.0001
Error	14,775.1	18	820.84 (MS_E)		
Total	155,108.2				

are both about 100 watts higher than the mean values from trial 1, suggesting the presence of systematic error. Further, the F for trials is now significant [$F(2,18)$ = 41.44; $p < .0001$].

With this added systematic error, the intraclass correlation coefficient values are now ICC (1,1) = .24, ICC (2,1) = .37, and ICC (3,1) = .75 (calculations not shown). Notice that models 1,1 and 2,1 are both markedly depressed with the addition of systematic error, whereas model 3,1 improved slightly. These results reflect the influence of systematic error on models 1 and 2, whereas model 3 is reflective of just random error.

Given the data in table 13.5, the researcher should consider the source of the systematic error. In this example, it appears that subjects markedly improved from trial 1 to trial 2 but likely reached a plateau in performance from trial 2 to trial 3. It might make sense to include a practice session to wash out practice effects. Indeed, performing intraclass correlation coefficient calculations on just trials 2 and 3 (the plateau region) results in the following values: ICC (1,1) = .78, ICC (2,1) = .70, and ICC (3,1) = .69.

Another factor that markedly affects intraclass correlation coefficient values is the amount of between-subjects variability. In general, the more heterogenous the subjects, the higher the intraclass correlation coefficients, and the more homogenous the subjects, the lower the intraclass correlation coefficients. To illustrate, the data in table 13.7 have been modified from the data in table 13.1. Specifically, we added between-subjects variability to the data by adding or subtracting constants (100–200 watts) to all the scores from selected subjects. The ANOVA summary table for the modified data is presented in table 13.8.

First, notice that the mean values do not differ for the data in tables 13.7 and 13.1. However, the standard deviations are larger, reflecting the added between-subjects variability. Second, the effect for trials is the same in tables 13.8 and 13.2. This should be expected because, as addressed in chapter 12, in a repeated measures ANOVA the between-subjects variance does not influence the error term.

The intraclass correlation coefficients for the data in table 13.7 are ICC (1,1) = .95, ICC (2,1) = .95, and ICC (3,1) = .95 (calculations not shown). By adding between-subjects variability but keeping the mean values the same, the intraclass correlation coefficient values have markedly increased relative to the results from the analysis of the data in table 13.1. The sensitivity of the intraclass correlation coefficient calculations to the between-subjects variability illustrates that intraclass correlation coefficient calculations are useful for quantifying the **relative** amount

TABLE 13.7

Example Data from Table 13.1 Modified by Adding Between-Subjects Variability

Trial 1	Trial 2	Trial 3
749.6	702.7	734.0
602.4	632.5	648.2
719.2	714.8	793.8
952	1,001.7	966.1
772	788.4	713.1
761.6	751.8	790.4
744	761.4	714
937.6	921.9	880.6
971.2	917.5	976.6
526.4	496.7	578.0
$\bar{X}$ = 773.6	$\bar{X}$ = 768.9	$\bar{X}$ = 779.5
SD = 146.3	SD = 149.2	SD = 130.3

TABLE 13.8

ANOVA Summary Table for the Data in Table 13.7

Source	SS	df	MS	F	p
Between (or subjects)	526,725.2	9	58,525.02 (MS_B: 1-way) (MS_S: 2-way)	54.68	
Within	19,823.0	20	991.15 (MS_W)		
Trials	557.9	2	278.97 (MS_T)	0.26	.77
Error	19,265.1	18	1,070.28 (MS_E)		
Total	546,548.2				

of measurement error. That is, they quantify the amount of measurement error relative to the amount of total variance (see equation 13.02).

The degree of relative measurement error, as quantified by intraclass correlation coefficients, is useful in estimating statistical power and sample size (in planning a study). Specifically, all else being equal, the greater the intraclass correlation coefficient value, the greater the statistical power when using that instrument (in this case, the Wingate test). Conversely, the lower the intraclass correlation coefficient, the larger the number of subjects needed to attain a desired level of statistical power.

In addition, measurement error will attenuate correlations. In chapter 8 we calculate the Pearson r to estimate the degree of relationship between two variables. The extent of attenuation can be estimated using equation 13.05:

$$\hat{r}_{xy} = \frac{r_{xy}}{\sqrt{ICC_x ICC_y}}, \qquad (13.05)$$

where $\hat{r}_{xy}$ is the correlation between x and y that would be obtained if both x and y were measured without measurement error, r_{xy} is the observed correlation between x and y, ICC_x is the reliability coefficient for variable x, and ICC_y is the reliability coefficient for variable y. In general, the extent of correlation attenuation is minimal if both intraclass correlation coefficient values are greater than .80 (Nunnally and Bernstein, 1994).

Standard Error of Measurement

The intraclass correlation coefficient provides an estimate of the relative error of the measurement; that is, it is unitless and is sensitive to the between-subjects variability. Because the general form of the intraclass correlation coefficient is a ratio of variabilities (see equation 13.04), it is reflective of the ability of a test to differentiate between subjects. It is useful for assessing sample size and statistical power and for estimating the degree of correlation attenuation. As such, the intraclass correlation coefficient is helpful to researchers when assessing the utility of a test for use in a study involving multiple subjects. However, it is not particularly informative for practitioners such as clinicians, coaches, and educators who wish to make inferences about individuals from a test result.

For practitioners, a more useful tool is the **standard error of measurement** (*SEM*; not to be confused with the standard error of the mean). The standard error of measurement is an absolute estimate of the reliability of a test, meaning that it has the units of the test being evaluated, and is not sensitive to the between-subjects variability of the data. Further, the standard error of measurement is an index of the precision of the test, or the trial-to-trial noise of the test. Standard error of measurement can be estimated with two common formulas. The first formula is the most common and estimates the standard error of measurement as

$$SEM = SD\sqrt{1 - ICC}, \qquad (13.06)$$

where ICC is the intraclass correlation coefficient as described previously and *SD* is the standard deviation of all the scores about the grand mean. The standard deviation can be calculated quickly from the repeated measures ANOVA as

$$SD = \sqrt{\frac{SS_{total}}{N - 1}}, \qquad (13.07)$$

where *N* is the total number of scores.

Because the intraclass correlation coefficient can be calculated in multiple ways and is sensitive to between-subjects variability, the standard error of measurement calculated using equation 13.06 will vary with these factors. To illustrate, we use the example data presented in table 13.5 and ANOVA summary from table 13.6. First, the standard deviation is calculated from equation 13.07 as

$$SD = \sqrt{\frac{155,108.2}{29}} = \sqrt{5,348.56} = 73.13.$$

Recall from above that we calculated ICC $(1,1) = .24$, ICC $(2,1) = .37$, and ICC $(3,1) = .75$. The respective standard error of measurement values calculated using equation 13.07 are

$$SEM = 73.13\sqrt{1 - .24} = 73.13\ \sqrt{.76} = 63.75 \text{ watts for ICC } (1,1),$$

$$SEM = 73.13\sqrt{1 - .37} = 73.13\ \sqrt{.63} = 58.05 \text{ watts for ICC } (2,1), \text{ and}$$

$$SEM = 73.13\sqrt{1 - .75} = 73.13\ \sqrt{.25} = 36.57 \text{ watts for ICC } (3,1).$$

Notice that standard error of measurement value can vary markedly depending on the magnitude of the intraclass correlation coefficient used. Also, note that the higher the intraclass correlation coefficient, the smaller the standard error of measurement. This should be expected because a reliable test should have a high reliability coefficient, and we would further expect that a reliable test would have little trial-to-trial noise and therefore the standard error should be small. However, the large differences between standard error of measurement estimates depending on which intraclass correlation coefficient value is used are a bit unsatisfactory.

Instead, we recommend using an alternative approach to estimating the standard error of measurement:

$$SEM = \sqrt{MS_E},$$ (13.08)

where MS_E is the mean square error term from the repeated measures ANOVA. From table 13.6, $MS_E = 820.84$. The resulting standard error of measurement is calculated as

$$SEM = \sqrt{820.84} = 28.65 \text{ watts.}$$

This standard error of measurement value does not vary depending on the intraclass correlation coefficient model used because the mean square error is constant for a given set of data. Further, the standard error of measurement from equation 13.08 is not sensitive to the between-subjects variability. To illustrate, recall that the data in table 13.7 were created by modifying the data in table 13.1 such that the between-subjects variability (larger standard deviations) was increased but the means were unchanged. The mean square error term for the data in table 13.1 (see table 13.2, $MS_E = 1,070.28$) was unchanged with the addition of between-subjects variability (see table 13.8). Therefore, the standard error of measurement values for both data sets are identical when using equation 13.08:

$$SEM = \sqrt{1,070.28} = 32.72 \text{ watts.}$$

Interpreting the Standard Error of Measurement

As noted previously, the standard error of measurement differs from the intraclass correlation coefficient in that the standard error of measurement is an absolute index of reliability and indicates the precision of a test. The standard error of measurement reflects the consistency of scores within individual subjects. Further, unlike the intraclass correlation coefficient, it is largely independent of the population from which the results are calculated. That is, it is argued to reflect an inherent characteristic of the test, irrespective of the subjects from which the data were derived.

The standard error of measurement also has some uses that are especially helpful to practitioners such as clinicians and coaches. First, it can be used to construct a confidence interval about the test score of an individual. This confidence interval allows the practitioner to estimate the boundaries of an individual's true score. The general form of this confidence interval calculation is

$$T = SD \pm Z_{crit} (SEM),$$ (13.09)

where T is the subject's true score, SD is the subject's score on the test, and Z_{crit} is the critical Z score for a desired level of confidence (e.g., $Z = 1.96$ for a 95% CI). Suppose that a subject's observed score on the Wingate test is 850 watts. Because all observed scores include some error, we know that 850 watts is not likely the subject's true score. Assume that the data in table 13.7 and the associated ANOVA summary in table 13.8 are applicable, so that the standard error of measurement for the Wingate test is 32.72 watts as shown previously. Using equation 13.09 and desiring a 95% CI, the resulting confidence interval is

$$T = 850 \text{ watts} \pm 1.96 \ (32.72 \text{ watts}) =$$
$$850 \pm 64.13 \text{ watts} = 785.87 \text{ to } 914.13 \text{ watts.}$$

Therefore, we would infer that the subject's true score is somewhere between approximately 785.9 and 914.1 watts (with a 95% LOC). This process can be repeated for any subsequent individual who performs the test.

It should be noted that the process described using equation 13.09 is not strictly correct, and a more complicated procedure can give a more accurate confidence interval. For more information, see Weir (2005). However, for most applications the improved accuracy is not worth the added computational complexity.

A second use of the standard error of measurement that is particularly helpful to practitioners who need to make inferences about individual athletes or patients is the ability to estimate the change in performance or **minimal difference** needed to be considered real (sometimes called the minimal detectable change). This is typical in situations in which the practitioner measures the performance of an individual and then performs some intervention (e.g., exercise program or therapeutic treatment). The test is then given after the intervention, and the practitioner wishes to know whether the person really improved. Suppose that an athlete improved performance on the Wingate test by 100 watts after an 8-week training program. The savvy coach should ask whether an improvement of 100 watts is a real increase in anaerobic fitness or whether a change of 100 watts is within what one might expect simply due to the measurement error of the Wingate test. The minimal difference can be is estimated as

$$MD = SEM \times Z_{crit} \times \sqrt{2}. \qquad (13.10)$$

Again, using the previous value of $SEM = 32.72$ watts and a 95% CI, the minimal difference value is estimated to be

$$MD = 32.72 \text{ watts} \times 1.96 \times \sqrt{2} = 90.7 \text{ watts.}$$

We would then infer that a change in individual performance would need to be at least 90.7 watts for the practitioner to be confident, at the 95% LOC, that the change in individual performance was a real improvement. In our example, we would be 95% confident that a 100-watt improvement is real because it is more than we would expect just due to the measurement error of the Wingate test. Hopkins (2000) has argued that the 95% LOC is too strict for these types of situations and a less severe level of confidence should be used. This is easily done by choosing a critical Z score appropriate for the desired level of confidence.

It is not intuitively obvious why the use of the $\sqrt{2}$ term in equation 13.10 is necessary. That is, one might think that simply using equation 13.09 to construct the true score confidence interval bound around the preintervention score and then seeing whether the postintervention score is outside that bound would provide the answer we seek. However, this argument ignores the fact that both the preintervention score and the postintervention score are measured with error, and this approach considers only the measurement error in the preintervention score. Because both observed scores were measured with error, simply observing whether the second score falls outside the confidence interval of the first score does not account for both sources of measurement error.

We use the $\sqrt{2}$ term because we want an index of the variability of the difference scores when we calculate the minimal difference. The standard deviation of the difference scores (SD_d) provides such an index, and when there are only two measurements like we have here, $SEM = SD_d/\sqrt{2}$. We can then solve for the standard deviation of the difference scores by multiplying the standard error of measurement by $\sqrt{2}$. Equation 13.10 can be re-conceptualized as

$$MD = SD_d \times Z_{crit}.$$

As with equation 13.09, the approach outlined in equation 13.10 is not strictly correct, and a modestly more complicated procedure can give a slightly more accurate confidence interval. However, for most applications the procedures described are sufficient.

Putting It All Together

We recommend a three-layered approach to quantifying measurement error. First, a repeated measures ANOVA should be performed to assess the presence of systematic error. If sufficient systematic error is present, the measurement schedule should be re-evaluated and modified by perhaps including practice sessions or modifying rest intervals. If systematic error is present when assessing inter-rater reliability, perhaps one or more of the raters is scoring performance incorrectly. Second, the repeated measures ANOVA provides the necessary mean square terms to calculate the intraclass correlation coefficient of choice. Third, the square root of the mean square error term can be used to quantify the standard error of measurement.

Summary

Reliability refers the measurement error of a test or instrument. The relative degree of reliability can be quantified using different intraclass correlation coefficients. The primary differences between coefficients center on whether the investigator wishes the intraclass correlation coefficients to reflect just random error or to include the influence of systematic error. The intraclass correlation coefficient is sensitive to the degree of between-subjects variability in the data such that, all else being equal, the larger the between-subjects variability, the larger the magnitude of the intraclass correlation coefficient.

The standard error of measurement is an index that reflects the degree of absolute measurement error. That is, the standard error of measurement is in units of the dependent variable and is not sensitive to the characteristics of the population from which the sample data were drawn. For example, increases in between-subjects variability (with all else being equal) will not directly affect the standard error of measurement but will amplify the intraclass correlation coefficients.

Problems to Solve

A physical therapist tests the reliability of the active knee extension test for hamstring flexibility (degrees). She administers the active knee extension test to 20 subjects on three separate days. The following table contains the resulting data.

Day 1	Day 2	Day 3
50	47	55
24	22	27
37	40	40
24	25	22
45	43	41
25	24	21
37	33	39
30	29	27
39	37	38
30	26	25
20	24	20
32	28	33
33	29	30
44	41	41
34	33	33
35	32	36
33	36	30
38	42	40
40	45	36
47	52	46

1. Perform a repeated measures ANOVA on the data in the table and construct an ANOVA summary table. Does the ANOVA suggest that systematic differences exist across trials?

2. Calculate the intraclass correlation coefficient for models 2,1 and 3,1.

3. Calculate the standard error of measurement.

4. Assume a person has an observed score of 35°. Construct a 95% CI about this subject's observed score to estimate the interval in which we think that person's true score lies.

5. Calculate the minimal difference using a 95% CI.

See appendix C for answers to problems.

Key Words

inter-rater reliability
intraclass correlation coefficient
intrarater reliability
measurement error
minimal difference

random error
relative
standard error of measurement
systematic error
test–retest reliability

$$H = \left[\frac{12}{18(18+1)} \right] \left[\frac{82.5^2}{6} + \frac{54.5^2}{6} + \frac{34.0^2}{6} \right] - 3(18+1)$$

$$H = (.035)(1822.09) - 57$$

$$H = 6.77$$

Factorial Analysis of Variance

A researcher in motor learning wants to study the effect of practice time on learning a novel task of throwing accuracy. The researcher designs an experiment in which college students practice for 1, 3, or 5 days per week, 20 minutes per day, for 6 weeks. The researcher also wants to know whether subjects with athletic experience (varsity athletes) benefit differently from the practice conditions than do students with no experience in athletic competition. Nine skilled athletes and nine students with no competitive experience are selected and assigned to the three practice conditions. How can these data be analyzed?

In chapters 11 and 12, ANOVA was conducted on one factor. In chapter 11, the effects of differing exercise regimens (one factor with five groups of subjects) on strength were considered. In chapter 12, the effect of physiological fatigue (one factor with five repeated measures on the same subjects) on bicycle balance skill was considered.

However, researchers often want to know the simultaneous effects of several **factors** on the dependent variable. When we use ANOVA to simultaneously analyze two or more factors, we call this a **factorial ANOVA.** For example, one could study the effects of five exercise regimens (five levels of factor A) and sex (two levels of factor B) on strength. This design would answer the question, Do different exercise regimens have the same effect on the development of strength in men as they do on strength in women? This type of factorial ANOVA is often referred to as a between–between design because it analyzes differences on the dependent variable *between* groups of subjects assigned to each of the five levels of factor A and between males and females (two levels of factor B) simultaneously.

Studies may also involve a *within* factor, where subjects are measured repeatedly. Later in this chapter we consider a design in which three groups of subjects learning under different conditions (factor A—between) are measured three times (factor B—within) on a novel motor skill. This design would be classified as between–within, between groups and within trials. Sometimes designs that include at least one between factor and one within factor are referred to as **mixed designs.**

It is also possible to have a within–within design. For example, the same subjects are tested on different modes of exercise (e.g., treadmill vs. cycle ergometer—factor A—within) and measured on $\dot{V}O_2$ in milliliters per kilogram of body weight per minute (dependent variable) repeatedly minute by minute (factor B—within) until exhaustion. In this design, the same subjects are measured both within modes and within minutes.

In kinesiology, we are frequently interested in the effects of various treatments, training, teaching methods, or other factors as they affect different subgroups of the population such as males versus females, athletes versus nonathletes, young versus old, fit versus unfit, and so forth. In these cases, a factorial ANOVA may be used to determine the simultaneous effects of two or more factors on a dependent variable.

The designs of these studies are sometimes identified by the number of factors and the levels of each factor being studied. If two factors, such as levels of intensity of training regimens (low, medium, high) and levels of sex (male, female), were applied to strength development (the dependent variable), the design would be 3×2 (three levels of factor A combined with two levels of factor B). If levels of the factor of age (15–30 years, 31–50 years, and 51–70 years) were added to this study, the design would become $3 \times 2 \times 3$ (training $\times$ sex $\times$ age). Although it is theoretically possible to have any number of factors, usually no more than three are used.

Factorial ANOVA produces an *F* value for each factor and an *F* value for the interaction of the factors. **Interaction** is the combined effect of the factors on the dependent variable. Two factors interact when the differences between the mean values on one factor depend upon the level of the other factor. It may be thought of as a measure of whether the lines describing the effects of the two factors are parallel (see figures 14.1 to 14.6). If the slopes of the lines are not significantly different (i.e., the lines are parallel), then interaction is not significant. Interaction would not be significant if it were found that training regimens affect men and women in the same way, but it would be significant if it were found that the training regimens affect men differently than they do women. Further discussion of interaction follows later in this chapter.

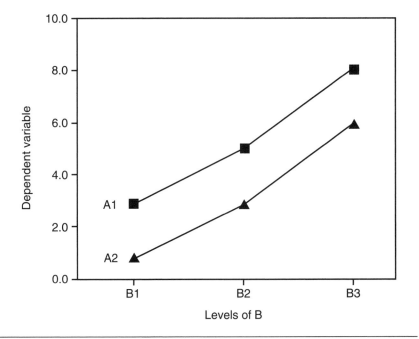

Figure 14.1 No significant interaction—lines parallel.

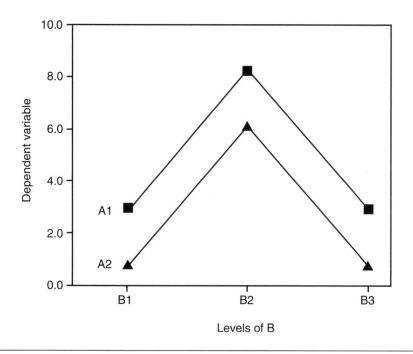

Figure 14.2 No significant interaction—lines parallel.

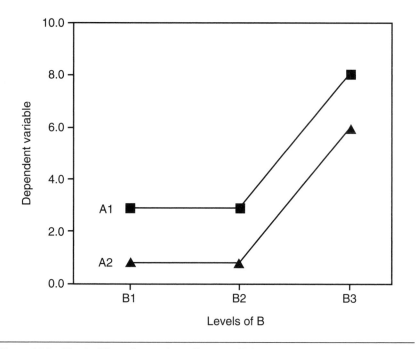

Figure 14.3 No significant interaction—lines parallel.

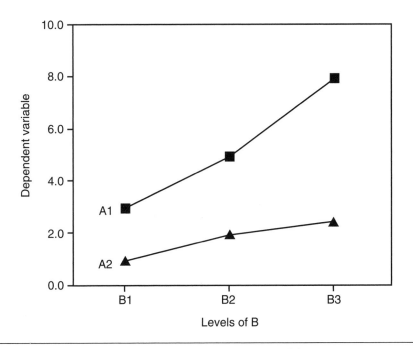

Figure 14.4 Significant interaction (ordinal)—lines not parallel.

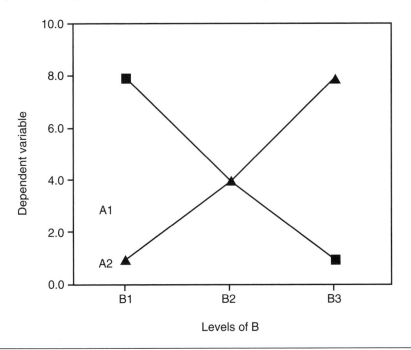

Figure 14.5 Significant interaction (not ordinal)—lines not parallel.

A Between–Between Example

To demonstrate **between–between** factorial ANOVA, a hypothetical study of learning motor skills is described. The data are fabricated and are intentionally simple so that the concepts may be easily understood and followed throughout the analysis without the confounding effect of complicated data. The assumptions for between–between designs are the same as for a simple ANOVA as described in chapter 9.

In this design, the researcher wanted to study the effect of 1 day, 3 days, or 5 days of practice per week (20 minutes per day) on the learning of a novel task of nondominant arm throwing accuracy (the dependent variable). In addition, the researcher wanted to know whether subjects with athletic experience (college-level varsity athletes) benefit differently from the various practice conditions than do college students with no experience in athletic competition.

This between–between design (3 practice conditions × 2 groups of athletic experience) with equal numbers of subjects in each group is presented for demonstration purposes. However, more complicated designs with additional factors and unequal N in each cell could be analyzed. Factorial ANOVA is typically performed by a computer that can deal with any number of factors, any number of levels, and with equal or unequal N. When the concepts are understood for a simple design, more complicated designs will be easy to comprehend and the computer can do

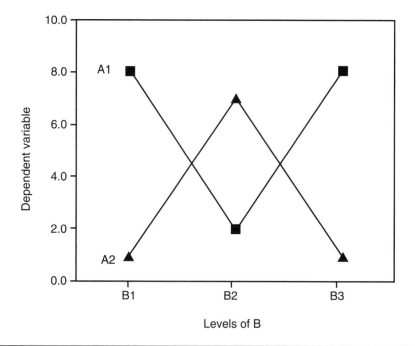

Figure 14.6 Significant interaction (not ordinal)—lines not parallel.

the work. The actual formulas for calculating the results are not presented here. It is assumed that the student has access to a computer that can perform this analysis. The emphasis in this chapter is on understanding the concepts of the analysis and how to interpret the results.

Nine highly skilled varsity athletes from a basketball team were selected as the experienced group, and nine college students with no experience in high school or college competitive sport were randomly selected from the student body and identified as the no experience group. Athletic experience versus no experience was identified as factor A.

Three subjects in each group were then randomly assigned to one of the three practice conditions (1, 3, or 5 days of practice per week for 20 minutes per day). Number of days of practice was identified as factor B. Subjects' scores on the throwing task at the end of 6 weeks were recorded as the dependent variable. The scores ranged from one to 10 (10 = high). Table 14.1a presents the raw data from the end of the 6-week period.

Steps in the Analysis

1. Arrange the data into cells by factors (see table 14.1b). In this example, factor A is athletic experience (A_1 = experience, A_2 = no experience). Factor B is days of practice (B_1 = 1 day per week, B_2 = 3 days per week, and B_3 = 5 days per week).

2. Determine the cell means and the marginal mean values (averages across factors) for A_1, A_2, B_1, B_2, and B_3 (see table 14.2). For example, the cell mean for athletes with 1 day of practice per week is the mean of scores 1, 2, and 1 because that cell contains the scores in the intersection of A_1 (athletes) and

TABLE 14.1a

Raw Data and Factor Sums

| | | Factor A (experience) | | |
		A_1 (athletes)	A_2 (nonathletes)	Sums of B
Factor B (practice)	B_1 (1 time/wk)	1	2	9
		2	2	
		1	1	
	B_2 (3 times/wk)	4	3	27
		6	2	
		8	4	
	B_3 (5 times/wk)	9	5	41
		8	6	
		9	4	
Sums of A		48	29	Grand sum 77

B_1 (1 day per week). Similarly, the marginal mean for all individuals with 1 day of practice per week (marginal mean for B_1) is the mean of scores 1, 2, 1, 2, 2, and 1.

3. Notice that the marginal values represent the mean values down each column of factor A or across each row of factor B. In other words, the average for all experienced athletes (A_1) regardless of the practice schedule in which they were placed is 5.22. Likewise, the average for nonathletes (A_2) is 3.22.

TABLE 14.1*b*

Between-Between Data Setup

Factor A (experience)	Factor B (practice)	Throwing score
1	1	1
1	1	2
1	1	1
1	2	4
1	2	6
1	2	8
1	3	9
1	3	8
1	3	9
2	1	2
2	1	2
2	1	1
2	2	3
2	2	2
2	2	4
2	3	5
2	3	6
2	3	4

TABLE 14.2

Cell Means and Marginal Means

| | | Factor A (experience) | | |
		A_1	A_2	Marginals for B
Factor B (practice)	B_1	1.33	1.67	1.50
	B_2	6.00	3.00	4.50
	B_3	8.33	5.00	6.67
	Marginals for A	5.22	3.22	4.28 (grand mean)

To determine whether athletes benefited from practice more than nonathletes, we can use a computer to calculate an F value (simple ANOVA) between these two **marginal mean** values. This is called the **main effect** of factor A. When we analyze a main effect, we are examining the effect of a factor while ignoring the effect of the other factor. Main effects are assessed by analyzing the differences between the marginal means.

4. Note that the average for all subjects (athletes and nonathletes combined) in the 1-day group (B_1) is 1.5. Likewise the mean for the 3-day group (B_2) is 4.50 and for the 5-day group (B_3) is 6.67.

To determine whether days of practice had an effect on learning, we can compute an F value (simple ANOVA) on these three marginal mean values. This is called the main effect of factor B.

5. Using a computer, compute the F value for interaction. Remember that interaction is the combined effect of the factors A and B on the dependent variable. If athletes learn at the same rate as nonathletes, then the lines in figure 14.7 will be parallel; that is, they will have the same or nearly the same slope. If athletes benefit more from practice than nonathletes, then the slope of the line for athletes will be greater (steeper) than the line for nonathletes. The interaction analysis computes an ANOVA between the slopes of the lines.

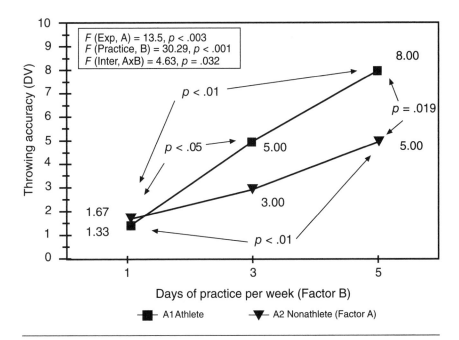

Figure 14.7 Factorial between–between.

Interaction may be the most important finding in the study. Interaction can take several forms. Figures 14.1 to 14.6 represent a design like the one we have been studying—the effects of two levels of A and three levels of B on a dependent variable.

Note that we will use computer software to calculate the three F ratios (one for each main effect and the interaction) because calculating the ratios by hand is very cumbersome and time consuming. Statistical software that can perform factorial ANOVA will provide a printout that includes all the of the needed F ratios.

When interaction is not significant (figures 14.1, 14.2, and 14.3), it may be appropriate to conduct further analyses on the main effects by using simple ANOVA at each level of A and B followed by post hoc tests. If interaction is significant and main effects are significant but ordinal (the mean values for A_1 and A_2 are in the same order at all levels of B), as is exemplified by figure 14.4, it may be appropriate to perform further analysis on the main effects by using simple ANOVA and post hoc tests.

When interaction is significant and disordinal, as in figures 14.5 and 14.6, main effects may be clouded because the average of A_1 and A_2 will be similar across all levels of B. The average of B_1, B_2, and B_3 may also be similar across all levels of A. This can result in a nonsignificant main effect even though the mean values for the cells are considerably different. Under these conditions, a careful review of the data and the potential causes of disordinal interaction effects as they relate to the purpose of the study is needed before a decision to further analyze main effects is made.

In general, when interaction effects are significant (whether ordinal or disordinal), we must be cautious in interpreting the main effects. That is, a significant interaction tells us that the effects of factor A differ across levels of factor B (and vice versa). Consider the pattern of results in figure 14.4. We might surmise that both main effects are significant because, on average, $B_3 > B_2 > B_1$ and, on average, $A_1 > A_2$. However, the interaction (ordinal) is also significant. We could interpret the pattern in table 14.4 as indicating that the effect of going from B_1 to B_2 to B_3 is more pronounced in A_1 than it is in A_2.

For the practice versus experience example (figure 14.7), this would suggest that practice has a pronounced effect on athletes but has a lesser effect on nonathletes. If we were to interpret the main effect of factor B, we would conclude that more practice improves throwing accuracy, but this is incomplete because the effect is large for the athletes but smaller for the nonathletes. At a minimum, when the interaction is significant, the data are more interesting than is reflected in the main effects.

Interpreting the Results

The interpretation of a factorial ANOVA is a **step-down process**. First, the main effects and the interaction are evaluated. If these F values are not significant, the analysis stops here. If any of them are significant, further analysis is warranted. The purpose of the step-down procedure is to protect against type I errors. The familywise error rate (see chapter 11) would be increased considerably if all possible cell mean differences were analyzed without first checking for main effects on each factor.

When none of the main effects or interaction is significant, to proceed further (someone called it data snooping) is to risk making a type I error by encountering the difference that occurs 1 out of 20 times at $\alpha = .05$ or 1 out of 100 times at $\alpha = .01$ by chance alone. When main effects or interaction are significant, however, further analysis is appropriate because we are confident that additional differences exist. It is now just a matter of finding them.

When these three F values (F for factor A, F for factor B, and F for interaction) have been calculated, a factorial ANOVA table (see table 14.3) is prepared to summarize the results.

The F values for factors A and B each represent an ANOVA on the marginal means (the main effects). The significant F ($F = 13.50$, $p = .003$) for factor A (athletic experience) indicates that for all three levels of B combined, athletes (A_1 marginal mean = 5.22) scored significantly better than nonathletes (A_2 marginal mean = 3.22; see table 14.2 and figure 14.7). Because there are only two levels of factor A, post hoc tests are not required because there is only one mean difference.

The significant F ($F = 30.29$, $p < .001$) for factor B (practice) indicates that for both levels of A combined, a significant difference exists somewhere among the three levels of B (B_1 marginal mean = 1.50, B_2 marginal mean = 4.50, and B_3 marginal mean = 6.67). The F for practice does not indicate which of the three levels of B differ from each other. A post hoc test is needed to identify the location of significant cell mean differences. Tukey's HSD post hoc comparisons indicate that B_2 and B_3 were significantly higher than B_1, but the difference between B_2 and B_3 was not significant.

The significant F ($F = 4.63$, $p = .03$) for the interaction tells us that the effects of practice on throwing accuracy differ in athletes and nonathletes. Figure 14.7 suggests that the effect of more days of practice on throwing accuracy is larger in the athletes than in the nonathletes. The significant F for the interaction tells us that the lines in figure 14.7 are not statistically parallel. We could stop the interaction interpretation right here and simply note that the significant interaction indicates that athletes improve more with more practice than do nonathletes. However, we are often interested is delineating the specific mean differences in the cell means.

TABLE 14.3

Factorial ANOVA

Source	SS	df	MS	F	p
Factor A (experience)	18.00	1	18.00	13.50	.003
Factor B (practice)	80.78	2	40.39	30.29	<.001
A × B (interaction)	12.33	2	6.17	4.63	.032
Error	16.00	12	1.33		
Total	127.11	17			

To find the differences in the cell means we can decompose the data into smaller pieces by conducting a simple ANOVA across A at each of the levels of B, and across B at each of the levels of A. These tests are referred to as tests of **simple effects**. The results of the decomposition of the interaction effect are presented in table 14.4.

Simple ANOVAs across A at B_1, B_2, and B_3 reveal that the only significant difference between athletes (A_1) and nonathletes (A_2) occurs with 5 days of practice (B_3) ($F = 14.29$, $p = .019$). Because there are only two groups in this ANOVA (A_1 and A_2), no further analysis with a post hoc test is necessary. Notice that a t test would be sufficient here because only two mean values are being compared. Given that $t^2 = F$ when $k = 2$, both t and F will produce the same determination of significance in this case.

The simple effects across A indicate that the consequence of classification as athlete or nonathlete is not remarkable until 5 days of practice are conducted. Classification by athletic experience does not produce significant differences at 1 or 3 days of practice. The simple effects for A also reveal that the interaction effect (while beginning to take effect at 3 days of practice) becomes significant only at 5 days of practice.

A second set of simple effects tests are performed across B at A_1 and A_2. For B_1 versus B_2 versus B_3 at A_1 (athletes), $F = 20.18$, $p = .002$, and Tukey's $HSD = 3.44$

TABLE 14.4

Factorial ANOVA

Main effects
$F_A = 13.50$, $\quad p = .003$
$F_B = 30,29$, $\quad p < .001$
$F_{AB} = 4.63$, $\quad p = .032$

Simple effects (A)
F across A at $B_1 = \quad .50$, $\quad p = .52$.
F across A at $B_2 = \quad 5.40$, $\quad p < .081$
F across A at $B_3 = \quad 14.29$, $\quad p = .019$

Simple effects (B)
F across B at $A_1 = 20.18$, $p = .002$, $HSD* = 3.44$ at $\alpha = .05$; $HSD = 5.03$ at $\alpha = .01$
F across B at $A_2 = 10.86$, $p = .010$, $HSD = 2.21$ at $\alpha = .05$; $HSD = 3.22$ at $\alpha = .01$

* HSD = Tukey's honestly significant difference.

Note: p values determined by computer are considered to be exact (designated by =, unless p is less than the level of precision of the computer calculation, in which case we denote $p < .001$); p values determined by tables are designated as less than a given value ($<$).

at $\alpha = .05$ and 5.03 at $\alpha = .01$. Applying this information to figure 14.7, it can be observed that the athletes who practiced 3 days are significantly better ($p < .05$) than are the athletes who only practiced 1 day ($6.00 - 1.33 = 4.67 > 3.34, p < .05$). Because the interaction is ordinal, it follows that the athletes who practiced 1 day per week also differ significantly from the athletes who practiced 5 days per week ($8.33 - 1.33 = 7.00, p < .01$), but 3 days per week does not sufficiently differ from 5 days per week to exceed HSD ($8.33 - 6.00 = 2.33$ does not meet the minimal requirement for HSD at $\alpha = .05$ of 3.44).

The F value for simple ANOVA across B_1, B_2, and B_3 at A_2 (nonathletes) is 10.86, $p = .010$, and $HSD = 2.21$ at $\alpha = .05$ and 3.22 at $\alpha = .01$. Figure 14.7 indicates that it takes 5 days of practice to produce a significant effect on throwing accuracy in nonathletes ($5.00 - 1.67 = 3.33, p < .01$); 3 days is not enough ($3.00 - 1.67 = 1.33$). In addition, 3 days of practice does not differ from 5 days.

Magnitude of the Treatment (Size of Effect)

The proportion of the total variance that may be attributed to each factor and interaction can be computed with a redefinition of the formula for omega squared (see equation 11.19 on page 192). The values for sums of squares, mean square, and degrees of freedom may be obtained from the factorial ANOVA in table 14.3.

For each factor, omega squared is

$$\omega_{\text{fac}}^2 = \frac{SS_{\text{fac}} - df_{\text{fac}}(MS_E)}{SS_T + MS_E} \tag{14.01}$$

and for interaction, omega squared is

$$\omega_{AB}^2 = \frac{SS_{AB} - (k_A - 1)(k_B - 1)(MS_E)}{SS_T + MS_E}. \tag{14.02}$$

Calculations:

For factor A (athletic experience),

$$\omega_A^2 = \frac{18.00 - 1(1.33)}{127.11 + 1.33} = .13.$$

For factor B (practice frequency),

$$\omega_B^2 = \frac{80.78 - 2(1.33)}{127.11 + 1.33} = .61.$$

For interaction,

$$\omega_{AB}^2 = \frac{12.33 - (2-1)(3-1)(1.33)}{127.11 + 1.33} = .07.$$

These calculations indicate that practice is by far the most important factor contributing to the differences in the scores. Practice contributes 61% of the total variance, athletic experience contributes only 13%, and the combined effect of experience and practice adds another 7%. The remainder of the variance is unexplained (error). Other indices such as eta squared (η^2) and partial eta squared may also be reported by various software programs. These are conceptually similar to omega squared in that they estimate variance in the dependent variable accounted for by a factor in the analysis. See Keppel (1991, pp. 221–224) for further discussion of the process for evaluating the magnitude of the treatments.

Conclusions

Factorial ANOVA is a method of simultaneously determining the effects of two or more factors on a dependent variable. In our hypothetical study, the effects of prior athletic experience (factor A; athletes vs. nonathletes) were compared with the effects of 1, 3, or 5 days of practice per week (factor B) on accuracy in a novel throwing task (the dependent variable).

It was determined that athletic experience did have an effect on performance on the throwing task; athletes scored significantly higher than nonathletes (5.22 vs. 3.22, $F = 13.50$, $p = .003$). This interpretation is derived from the significant main effect for factor A (table 14.3).

The amount of practice (factor B) was also significant ($F = 30.29$, $p < .001$). Specifically, 3 and 5 days of practice were significantly better than 1 day of practice. This interpretation is derived from the significant main effect for factor B and the subsequent post hoc comparisons.

Finally, the interaction was significant ($F = 4.63$, $p = .032$), which tells us that the effects of practice on throwing performance differ between athletes and nonathletes. The lines depicting the ability of athletes and nonathletes to benefit from practice are not parallel. The line for athletes has a greater slope than does the line for nonathletes. Athletes appear to make more effective use of practice than do nonathletes. For the athletes, 3 days ($p < .05$) or 5 days ($p < .01$) are better than 1 day. For the nonathletes, it takes 5 days of practice per week to observe a significant difference ($p < .01$) in throwing accuracy. Three days per week was no different from 1 day per week.

Equivalently, the interaction tells us that the effects of athletic experience on throwing performance differ at different levels of practice. With 1 and 3 days of practice per week, differences between athletes and nonathletes are not significant. However, with 5 days of practice per week, athletes perform significantly better than nonathletes.

The relative importance (effect size) of each main effect and interaction was determined with omega squared. Practice was found to be the major factor affecting the dependent variable, contributing 61% of the total variance in the throwing scores. Athletic experience and interaction contributed 13% and 7%, respectively.

You may wonder why no difference exists between athletes and nonathletes at 3 days of practice per week (figure 14.7). The lines diverge at 3 days, but the difference between these two points is not significant ($5.00 - 3.00 = 2.00$). This is the result of the artificially small sample size used in the example, resulting in insufficient statistical power. A power analysis to determine the minimum N needed per cell to have an 80% chance of detecting real differences as small as 2.0 could have been performed before the study began. With a larger N, the difference between A_1 and A_2 at B_2 likely would be significant. See chapter 7 for a discussion of power.

A Between–Within Example

Factorial ANOVA with repeated measures on one factor is often referred to as a **between–within,** or mixed model design. "Between" signifies that the analysis on factor A is an ANOVA between independent groups. Each of the groups in factor A consists of different subjects randomly assigned to the groups. "Within" signifies that the analysis on factor B is a repeated measures analysis. The subjects in each of the groups are measured two or more times.

The mixed model is a very common design in kinesiology and many other physical and behavioral sciences. For example, exercise physiologists often want to know the effects of various training schedules on physiological variables during the training program. Subjects may be placed into treatment groups (e.g., high, medium, and low intensity) and then trained for a period of weeks. Measurements of $\dot{V}O_2$, heart rate max, respiratory exchange ratio, and other physiological variables are measured weekly throughout the training period to determine the effects of treatment over time.

Motor behavior researchers also use this design to evaluate the effects of various learning schedules on speed and amount of learning over multiple trials or over an extended period of time. This design calls for two or more groups of subjects to be tested two or more times. Hence, the design is a combination of a simple ANOVA (between) and a repeated measures ANOVA (within). Both ANOVA are conducted simultaneously.

As in the between–between example presented earlier, this analysis also produces three F values. One for the between analysis (groups), one for the within (repeated measures) analysis (trials), and one for interaction (groups × trials). Interaction is interpreted in this design in the same manner that it was interpreted in the factorial ANOVA (between–between) design. Often, interaction is the most interesting and important finding in a between–within design.

You will recall that interaction identifies the combined effects of group assignment and repeated measurement. Frequently, groups will respond to treatment in a different manner over time. High-intensity training may produce a more rapid improvement in physiological variables than will moderate- or low-intensity training. Certain teaching methods may produce more rapid learning of a specific skill than do other methods. Because of its utility, this design is commonly reported in the professional literature.

The *assumptions* for a between–within design are the same as for the between analysis in a simple ANOVA (chapter 11) and the within analysis in ANOVA repeated measures (chapter 12). However, the sphericity assumption (discussed in chapter 12) for the repeated measures factor must now be applied to the pooled data (across all of the groups) as well as to each individual group. This pooled condition is referred to as multisample sphericity, or **circularity** (Schutz and Gessaroli, 1987). Advanced computer programs test for this assumption and provide the epsilon values for the Greenhouse-Geisser (GG) and Huynh-Feldt (HF) corrections.

To demonstrate this design, another study with simple, hypothetical data is described. When the concepts from this simple study are understood, more complicated studies with unequal N and a different number of levels of each factor may be considered. Analysis of this design is almost never computed by hand. The conceptual analysis is presented here (without detailed equations) to guide you through the process. An analysis of this nature should always be done by computer to reduce the possibility of calculation error and to save time and effort.

A researcher wanted to investigate the relative effects of distributed and massed practice (factor A—between groups) on the learning over time (factor B—within trials) on a novel motor skill as measured by time on target on the pursuit rotor apparatus (dependent variable). Fifteen college-age students were randomly placed into three practice groups with five subjects in each group. The groups were then randomly assigned to one of the following three practice conditions for the 10-day study.

1. Distributed practice: the subjects practiced in 30-second trials with a 30-second break between trials for a total of 10 trials (5 minutes total practice time each day for 10 days).

2. Massed practice: the subjects practiced for 5 minutes per day without a break for 10 days.

3. Control: the subjects took the pre-, mid-, and posttests, but practiced on an unrelated gross motor task for 5 minutes each day for 10 days.

The dependent variable for the pre-, mid-, and posttests was time on target during the pursuit rotor task for a 15-second period. The dependent variable was measured prior to the first practice day (pre), after the fifth day (mid), and after the tenth day (post). This design with three groups measured over three trials would be categorized as a 3×3 between–within factorial ANOVA.

The factorial ANOVA will produce an F value for the differences *between* groups over the pre-, mid-, and posttests and an F value for the differences *within* the three tests (trials) averaged over group. An F value for interaction, which will determine the combined effects of the practice conditions and the three trials, will also be calculated. The interaction will compare the relative speed of learning of the three groups.

Steps in the Analysis

1. Arrange the raw data (X) into a matrix with factor A (groups) on the vertical axis and factor B (trials) on the horizontal axis. Compute the sum of each group and for each trial and the grand sums (see table 14.5a).

2. Set up a matrix with subjects in factor A arranged by group (1, 2, 3) and factor B in columns 2, 3, and 4. This is the format in which the data will be entered into the computer (table 14.5b).

3. Set up a matrix for the mean values of factor A crossed with factor B (table 14.6.).

4. Using a computer, compute the between F value across rows on the marginal means for factor A (groups). This is called main effects for factor A.

5. Using a computer, compute the within F value across columns on the marginal means for factor B (trials). This is called main effects for factor B.

6. Using a computer, compute the F value to determine interaction among groups and trials.

7. Create a factorial ANOVA table to summarize the results (table 14.7).

8. Conduct a step-down analysis from main effects to simple effects to cell differences and present in table form (table 14.8).

When data from table 14.5b is entered into the computer, the information from tables 14.6 and 14.7 will be provided in the computer output.

TABLE 14.5a

Raw Data and Factor Sums

	B_1 (pre)	B_2 (mid)	B_3 (post)	A sum
A₁ (distributed)	2	5	10	17
	3	7	11	21
	5	9	13	27
	4	8	9	21
	3	6	12	21
B sum	17	35	55	107
A₂ (massed)	2	7	9	18
	4	5	8	17
	3	3	7	13
	3	5	8	16
	2	8	7	17
B sum	14	28	39	81
A₃ (control)	2	4	6	12
	3	3	5	11
	2	5	6	13
	3	2	8	13
	3	3	4	10
B sum	13	17	29	59
Grand sum	44	80	123	247

Factor A = groups (between); factor B = trials (within).

TABLE 14.5*b*

Between–Within Data Setup

Group	Pretest	Midtest	Posttest
1	2	5	10
1	3	7	11
1	5	9	13
1	4	8	9
1	3	6	12
2	2	7	9
2	4	5	8
2	3	3	7
2	3	5	8
2	2	8	7
3	2	4	6
3	3	3	5
3	2	5	6
3	3	2	8
3	3	3	4

TABLE 14.6

Cell Means

	B_1 (pre)	B_2 (mid)	B_3 (post)	Marginals for A
A_1 (distributed)	3.4	7.0	11.0	7.13
A_2 (massed)	2.8	5.6	7.8	5.4
A_3 (control)	2.6	3.4	5.8	3.93
Marginals for B	2.93	5.33	8.2	5.49 (grand mean)

Factor A = groups (between); factor B = trials (within).

The F value for main effects on factor A ($F = 19.03, p < .001$) is a simple ANOVA on the marginal means (7.13, 5.40, and 3.93) for factor A (groups) (see table 14.6). This F value indicates that, ignoring any differences between the three time periods, a difference exists somewhere among the marginal means of the three groups. If we wish to further explore these differences, we would need to perform a post hoc analysis. If the F value for main effects for A is not significant, the analysis for factor A stops here. Because this is not a repeated measures analysis, we do not have to worry about the assumption of circularity.

The F value for main effects on factor B ($F = 67.77, p < .001$) is a repeated measures ANOVA on the marginal means (2.93, 5.33, and 8.20) for factor B (see table 14.6). This is called the main effects for B. This significant F indicates that,

TABLE 14.7

Factorial ANOVA

Source	SS	df	MS	F	p
Factor A (Groups)	76.98	2	38.49	19.03	<.001
Error A	24.27	12	2.02		
Factor B (Trials)	208.58	2	104.29	67.77	<.001
A × B (Interaction)	26.49	4	6.62	4.30	.009
Error B	36.93	24	1.54		
Totals	373.25				

ε: GG = .909, HF = 1.00.

Adjustment for trials

| GG adjusted values | $F = 67.78$, $df = 1.82$, 21.81, $p < .001$ |
| HF adjusted values | $F = 67.78$, $df = 2.00$, 24.00, $p < .001$ |

Adjustment for interaction

| GG adjusted values | $F = 4.30$, $df = 3.64$, 21.81, $p = .012$ |
| HF adjusted values | $F = 4.30$, $df = 4.00$, 24.00, $p = .009$ |

GG = Greenhouse-Geisser; HF = Huynh-Feldt.

TABLE 14.8

Step-Down Analysis of Factorial ANOVA

Main effects

$$F_A = 19.03, p < .001$$
$$F_B = 67.77, p < .001$$
$$F_{AB} = 4.30, p = .009$$

Simple effects (A)

F across A at B_1 = 1.13, Not significant
F across A at B_2 = 6.50, $p = 0.12$; HSD^* = 3.59, $p < .01$.
F across A at B_3 = 19.11, $p < .001$; HSD = 3.02, $p < .01$.

Simple effects (B)

F across B at A_1 = 46.96, $p < .001$; HSD = 2.52, $p < .01$.
F across B at A_2 = 20.40, $p < .001$; HSD = 2.52, $p < .01$.
F across B at A_3 = 9.01, $p < .001$; HSD = 2.52, $P < .01$.

* HSD = Tukey's honestly significant difference.

ignoring differences between the three groups, a difference exists somewhere among the marginal means of the three time periods. If we wish to further explore these differences, we would need to perform a post hoc analysis. If the F value for main effects for B is not significant, the analysis for factor B stops here. Because this is a repeated measures analysis, we must check for a possible violation of the assumption of circularity (sphericity) by determining the size of the epsilon value.

Epsilon values (GG = .909, conservative; HF = 1.00, liberal) confirm that the violation was minimal. Remember that an epsilon value of 1.00 indicates no violation, and a value > .75 indicates minimal violation. The HF adjustment indicates no violation, and the GG adjustment indicates very little violation. The GG and HF adjustments do not change the p values for the trials analysis. This indicates that the small violation that occurred is not remarkable.

The significant F value for interaction ($F = 4.30$, $p = .009$) reveals that the three groups responded differently over the three trials. Figure 14.8 suggests that the distributed practice group improved their scores over the three trials faster than either the massed or control group. In other words, the lines on the graph of groups by trials are not parallel. Simple ANOVA across groups at the pre-, mid-, and posttests reveals that the differences are significant at the mid- and posttests but not at the pretests (table 14.8). After GG or HF adjustments for violation of circularity (table 14.7), p values are still highly significant (HF indicates no change

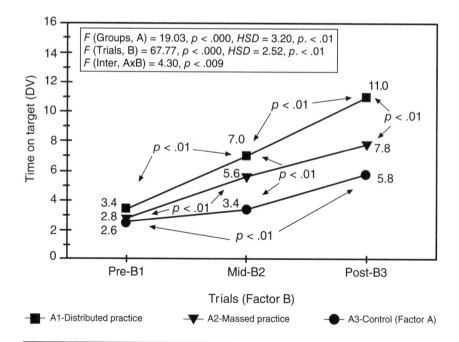

Figure 14.8 Factorial between–within.

because HF $\varepsilon = 1.00$) corroborating that minimal, if any, violation of the assumption occurred.

Note that if the GG or HF adjustments reduce the p values below the preset alpha level for rejection of the null, the data may be reanalyzed using multiple ANOVA (MANOVA; chapter 18). In this case, the repeated measures are treated as multiple dependent variables and the assumptions of sphericity and circularity are not required. If the F value for trials and interaction is still significant under MANOVA analysis (which is less powerful than the repeated measures design), the null hypothesis may be rejected with the confidence indicated by the MANOVA analysis. See Schutz and Gessaroli (1987) for further discussion of this issue.

The ANOVA for simple effects and post hoc tests are conducted in the manner described for a single factor ANOVA in chapter 11 (independent groups) and chapter 12 (repeated measures). You will notice that the values for Tukey's *HSD* test for determination of cell differences are the same for all simple effects on factor B (repeated measures). This is because the pooled mean square error term for main effects was used to calculate *HSD* in each case, as is done in most computer software programs. Some difference of opinion exists among statisticians as to whether the main effects error term should be used or whether a separate error term for each simple effect should be calculated. Arguments exist on both sides of this issue. The interested reader is referred to Tabachnick and Fidell (1996, p. 45) for further discussion of this matter. Note that the *HSD* values could be calculated for $p < .05$ if this level had been set as an acceptable rejection level before data collection. In addition, note that because the sphericity violation was minimal, we used Tukey's *HSD* post hoc procedure. As described in chapter 12, if sphericity was substantively violated, a better approach would be to employ paired t tests with the Bonferroni correction.

Magnitude of the Treatment (Size of Effect)

Applying equations 14.01 and 14.02 for omega squared to the data in table 14.7 results in a determination of the relative importance of the factors of groups and trials as they affect the dependent variable (time on target).

For groups,

$$\omega_G^2 = \frac{76.98 - 2(2.02)}{373.25 + 2.02} = .19.$$

For trials,

$$\omega_T^2 = \frac{208.58 - 2(1.54)}{373.25 + 1.54} = .55.$$

For interaction,

$$\omega_{GT}^2 = \frac{26.48 - 2(3-1)(3-1)(1.54)}{373.25 + 1.54} = .05.$$

Conclusions

When table 14.8, figure 14.8, and ω^2 are interpreted based on the main effects, simple effects, and post hoc tests, the following conclusions seem justified.

1. No difference exists among the groups on the pretest. This confirms that the distribution of subjects into groups by random assignment was effective.

2. A significant difference exists in the amount of learning among the three groups. On the midtest, the distributed group is significantly better than the control group ($7.0 - 3.4 = 3.6$). On the posttest, the distributed group is significantly better than the massed or control groups ($11.0 - 7.8 = 3.2$) but the massed group is not significantly better than the control group ($7.8 - 5.8 = 2.0$). Group assignment accounts for 19% of the variance in time on target.

3. The scores of the distributed group improved significantly between the pre- and the midtest ($7.0 - 3.4 = 3.6$) and between the mid- and the posttest ($11.0 - 7.0 = 4.0$). The scores of the massed group improved between the pre- and the midtest ($5.6 - 2.8 = 2.8$) but did not continue to improve between the mid- and the posttest ($7.8 - 5.6 = 2.2$). The scores of the control group improved between the pre- and the posttest ($5.8 - 2.6 = 3.2$) but not between the pre- and the midtest ($3.4 - 2.6 = 0.8$) or between the mid- and the posttest ($5.8 - 3.4 = 2.4$). Because the control group did not practice the skill, this improvement probably represents learning that took place during the testing sessions. Factor B, trials, is the most important factor to affect the dependent variable. It accounts for more than one-half (55%) of the variance in time on target scores.

4. The groups did not learn at the same rate. Significant interaction indicates that the lines are not parallel. The distributed group learned at a rate faster than the control or massed groups because it differs ($p < .01$) from the control at the midtest, and it differs from both other groups at the posttest ($p < .01$). Given that the massed group is not significantly different from the control group at either the mid- or the posttest, we cannot conclude that they were learning faster than the control group. Interaction took place only between the distributed and the other two groups and it accounts only for 5% of the total variance.

A Within–Within Example

A researcher wanted to know whether physiological differences existed in response to exercise on a traditional treadmill compared with exercise on a stair-step treadmill. Thirteen healthy male college students were randomly selected and asked to report to the laboratory on two different days with at least one rest day in between. On the first day, subjects were asked to perform a graded submaximal exercise test on a treadmill. The test consisted of four increasing stages of work. Heart rate responses were recorded at each stage as dependent variables. The amount of physical work performed on the treadmill at each stage was calculated.

On the second day, the same subjects performed an identical amount of physical work over four increasing stages on a stair-step treadmill device. Hence, the same subjects performed on two different modes of exercise (factor A—within) and over four equivalent stages of work (factor B—within). This design may be categorized as a 2 × 4 **within–within** factorial ANOVA. The data layout for the example is shown in table 14.9. (*Note:* Set up the computer database for this problem as follows in table 14.9.)

The analysis of this design is similar to the analysis of between–within, but because both factors are repeated measures on the same subjects, the assumptions of sphericity must be met for both factors. The analysis produces three F values: main effect of mode of exercise (factor A), main effect for stages of exercise (factor B), and interaction. If interaction is not significant, then subjects respond in a similar manner on both modes of exercise to the identical graded exercise stimulus. If the interaction is significant, then physiological responses are different on the two modes of exercise. The mean values for heart rate are presented in table 14.10 and the factorial ANOVA summary is presented in table 14.11.

TABLE 14.9

Within–Within Data Setup

Treadmill stage 1	Treadmill stage 2	Treadmill stage 3	Treadmill stage 4	Stair stage 1	Stair stage 2	Stair stage 3	Stair stage 4
98	117	133	143	99	113	131	142
102	114	135	139	102	108	132	137
95	121	130	148	95	117	126	145
97	116	134	142	100	114	135	144

This is an incomplete data set intended as an example only.

TABLE 14.10

Cell Means

	1 (B_1)	2 (B_1)	3 (B_3)	4 (B_4)	Marginals for A
A_1 (treadmill)	98	117	133	143	122.25
A_2 (stairs)	99	113	131	142	121.25
Marginals for B	98.5	115.0	132.0	142.5	121.75 (grand mean)

TABLE 14.11

Factorial ANOVA

Source	SS	df	MS	F	p
Factor A (Mode)	51.24	1	51.24	.34	.570
Error A	1,800.63	12	150.05		
Factor B (Stages)	29,403.26	3	9,801.09	177.81	<.001
Error B	1,984.37	36	55.12		
A × B (Interaction)	94.72	3	31.57	1.23	.313
Error A × B	923.90	36	25.66		

Check for sphericity:

Mode
Because there are only two values for mode, sphericity cannot be violated. Epsilon values are not produced; therefore, no adjustment is made. Further, because unadjusted F is not significant, no adjustments would be necessary.

Stages
ε: GG = .7804, HF = .9813

Adjustment for stages

GC adjusted values	$F = 177.81$,	$df = 2.34, 28.09$,	$p < .001$
HF adjusted values	$F = 177.81$,	$df = 2.94, 35.33$,	$p = .001$

Interaction
ε: GG = .6775, HF = .8168.

Adjustment for interaction
Because unadjusted F is not significant, no adjustments are needed.

GG = Greenhouse-Geisser; HF = Huynh-Feldt.

Step-Down Analysis

The F for the main effect for mode (factor A) is not significant; therefore, no further analysis is justified. We can conclude that the heart rate response, averaged across the four stages, is similar between treadmill and stair exercise. The F value for the main effect for stages (factor B) is significant ($p < .001$). This indicates that the heart rate values, averaged across the two modes of exercise, vary across the different stages. A post hoc test may be used to determine which stages differ. Tukey's *HSD* at $p < .01$ is ~ 10. Because every stage for both modes is more than 10 beats per minute higher than the previous stage, we can conclude that heart rate increases significantly at every stage. The F value for interaction is not significant, indicating no differences in the slopes of the lines. Figure 14.9 presents the data in graphic format.

Conclusions

1. The F value for mode is not significant; therefore, we concluded that the heart rate did not differ on the two exercise modes.

2. The F value for stages is highly significant. Although a small violation of circularity existed for stages, it was not sufficient to alter the p value. The increase in heart rate was expected because physical work increased stage by stage. Indeed, if we did not see a significant increase in heart rate over stages, we would suspect an error. Tukey's *HSD* indicates that a statistically significant increase in heart rate occurred between each of the four stages.

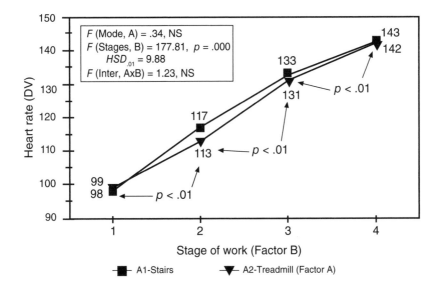

Figure 14.9 Factorial within–within.

3. The *F* value for interaction is not significant. This confirms the insignificant *F* value for mode and provides further evidence that the subjects responded physiologically in the same manner on both modes of exercise over all four stages.

Note: If more than one physiological variable is included (i.e., heart rate, $\dot{V}O_2$, blood pressure, respiratory exchange ratio, and so on), the data should be analyzed using MANOVA (see chapter 18).

Summary

Factorial ANOVA is a valuable tool for evaluating the simultaneous effects of more than one factor on a dependent variable. It is commonly used in kinesiology and related disciplines. Theoretically any number of factors can be considered, but typically no more than three are used.

When different subjects are involved in either factor, the design is referred to as a between design. This indicates that the groups are independent. When the same subjects are measured more than once, the design is referred to as a within design. Under these conditions, the assumptions of sphericity must be met. Epsilon values to determine the level of violation are calculated by a computer. When circularity is not met, the degrees of freedom are multiplied by epsilon to arrive at an adjusted *p* value, which will be higher (less significant) than the original *p* value.

F values for each factor and for the interaction of the factors are calculated. When the *F* value for main effects on either factor is significant, further analysis with post hoc tests across the marginal means may be warranted. However, the interaction effect may be the value of most interest in these studies because it indicates the combined effect of the factors. Further, when the interaction is significant we must be cautious in interpreting main effects because the interaction indicates that the effect of one factor on the dependent variable differs across levels of the other factor.

Key Words

between–between
between–within
circularity
factor
factorial analysis of variance
interaction

main effect
marginal mean
mixed design
simple effect
step-down process
within–within

$$H = \left[\frac{12}{18(18+1)}\right]\left[\frac{82.5^2}{6} + \frac{54.5^2}{6} + \frac{34.0^2}{6}\right] - 3(18+1)$$

$$H = [.035][1822.091] - 57$$

$$H = 6.77$$

Analysis
of Covariance

A sports nutrition researcher conducts a study comparing a protein supplement with placebo on training responses to a 12-week bench press regimen. The subjects are given a bench press pretest to determine one-repetition maximum; the subjects then perform the 12-week training program. One group of subjects is given a protein supplement drink immediately after each training session whereas the placebo group is given an isocaloric carbohydrate drink. After the training program, the subjects are again given a one-repetition-maximum bench press test. A colleague has suggested that the researcher consider using analysis of covariance to analyze the results. Why might this be a good analytical strategy for the study?

In chapter 9, the technique of multiple regression was presented, which allows us to quantify the relationship between a dependent variable and two or more independent variables. Similarly, factorial ANOVA was introduced in chapter 14. Factorial ANOVA allows us to examine mean differences on a dependent variable across levels of two or more independent variables. [Recall that the t test is a special case of ANOVA ($F = t^2$), so our comments regarding ANOVA apply to the t test also.]

Relationship Between ANOVA and Regression

We have not yet explicitly shown that regression and ANOVA are based on the same underlying process. To illustrate, consider the data in table 15.1, which presents simulated data from two groups measured on a reaction time task. One group was given caffeine and the other was given a placebo. Imagine that the subjects were tested for reaction time in milliseconds 1 hour after administration of caffeine or placebo, and that all testing was conducted under double blind conditions.

We could easily perform an independent t test on the data in table 15.1 using the procedures in chapter 10, but to illustrate that ANOVA and regression are the same process, we instead performed an ANOVA. The ANOVA summary table is shown in table 15.1 (calculations not shown).

Now let us analyze the data in table 15.1 using a regression approach. The independent variable is the experimental group (caffeine vs. placebo). The trick is to turn this variable into a number. A simple approach is to use what is called dummy coding. In a simple study like this, we can dummy code the data so that each subject in the placebo group is given a score of 0 and each subject in the caffeine group is given a score of 1. With three or more groups, the technique of dummy coding is more complicated and is beyond the scope of our discussion. For a more thorough discussion, see the classic paper by Cohen (1968).

Here we use bivariate regression because we have one independent variable ($X =$ dummy code for experimental group) and the dependent variable is reaction time ($Y =$ milliseconds). The regression analysis of the data (calculations not shown) using

TABLE 15.1

Example Reaction Time (Milliseconds) Data for Simple Reaction Time Study

Placebo	Caffeine
51	49
53	48
49	51
51	47
54	50
$\bar{X} = 51.6$	$\bar{X} = 49.0$
$SD = 1.95$	$SD = 1.58$

TABLE 15.2

Factorial ANOVA

Source	SS	df	MS	F	p
Between	16.9	1	16.9	5.37	.049
Within	25.2	8	3.15		
Total	42.1	9			

the above dummy code reveals that the y-intercept is 51.6 and the slope coefficient is −2.6. The resulting regression equation is

$$Y = 51.6 - 2.6X.$$

The Pearson $r = -.634$ ($r^2 = .40$) and $p = .049$. Notice the p value is the same for the regression analysis as for the ANOVA. Because the X values are the dummy codes, we would predict that reaction time of subjects in the placebo group (dummy code = 0) would be $51.6 - 2.6(0) = 51.6$ milliseconds. Similarly, for individuals in the caffeine group we would predict that reaction time would be $51.6 - 2.6(1) = 49.0$ milliseconds. Notice that these are equal to the mean values for the respective groups. That is, if all we know about a subject is that subject's group affiliation, our best prediction of that person's reaction time is the mean value for their group. Also, from the ANOVA summary table (table 15.2) we see that the sum of squares between is 16.9 and the total sum of squares is 42.1. The ratio of sum of squares between to sum of squares total is $16.9/42.1 = .40$. This is equal to the coefficient of determination from regression and illustrates that group affiliation (caffeine vs. placebo) accounts for about 40% of the variance in reaction time.

The purpose of the previous discussion is to illustrate that regression and ANOVA are based on the same underlying model. Given that regression and ANOVA are the same underlying process, it is not surprising that we have a statistical tool that allows us to combine regression and ANOVA. This technique is called **analysis of covariance** (ANCOVA). Researchers typically use regression when the independent variable(s) is measured on an interval or ratio scale. Conversely, ANOVA is typically used when the independent variables(s) is nominal data (e.g., experimental group vs. control group). What do we do when we have two or more independent variables, with at least one measured on an interval or ratio scale and at least one measured on a nominal scale? Analysis of covariance allows us to analyze data in such a situation. Analysis of covariance is typically used in situations in which the researcher wishes to conduct a study comparing groups but wants to adjust the scores on the dependent variable for the influence of some other variable. Collectively then, we have two independent variables: (1) the different groups and (2) the other variable that we want to make adjustments for.

As an example, suppose that a researcher in sports nutrition wishes to conduct a study comparing the effect of a dietary supplement and a placebo on resting energy expenditure (REE). Subjects are randomly assigned to receive either the supplement or the placebo. Subsequently, the REE of the subjects is quantified using a metabolic cart for measurement of oxygen consumption and carbon dioxide production (in liters per minute). The researcher also knows that body mass influences REE: on average, heavier people use more energy than lighter people. Because the subjects were assigned to the two groups randomly, it is unlikely that the two groups systematically differ in body weight. However, it is also unlikely that the two groups are exactly equal in body weight. The researcher wishes to adjust the REE so that the groups are statistically similar.

Analysis of covariance allows us to compare the supplement and placebo groups on REE while adjusting for differences in body weight. Here, we have an independent variable that is a categorical variable (and has two levels: supplement and placebo) and an independent variable that is measured on a continuous (ratio) scale (body weight). The dependent variable is REE (in kilocalories; derived from the gas exchange measurements). In the language of ANCOVA, the variable that we adjust for is called the **covariate;** in this example, the covariate is body weight.

Conceptually, ANCOVA is a two-step process. First, we perform a regression analysis between the covariate and the dependent variable. The dependent variable is adjusted based on the regression, and the adjusted scores are then compared between groups using ANOVA. For the REE example, picture a simple linear regression analysis in which we regress body weight (the covariate, X) against REE (the dependent variable, Y). The scores on REE are adjusted for differences in body weight based on the regression analysis. The adjustment can be performed using equation 15.01:

$$Y' = Y - b_{\text{pooled}}\left(X - \bar{X}\right), \tag{15.01}$$

where Y' is the adjusted score, Y is the unadjusted score on the dependent variable, b_{pooled} is the common regression slope for all groups combined, and X is the score on the covariate. Notice that equation 15.01 is just a variant of the bivariate regression equation where Y is the intercept and b_{pooled} is the slope. The scores adjusted for body weight are then compared using ANOVA (or equivalently an independent t test because there are only two groups).

The typical motive for a researcher to use ANCOVA is to adjust groups so that when they are compared, the comparison is fairer. However, researchers must be careful when trying to correct for group differences using ANCOVA. When groups are created by randomization, using ANCOVA to adjust scores is a reasonable approach. However, it is tempting to try to equate groups that were not created by randomization. For example, one may wish to compare males and females on some variable. To make a fair comparison, the researcher may wish to use ANCOVA to adjust the male and female scores based on some covariate (e.g., body weight). The researcher cannot randomize subjects to the male and female groups because these are pre-existing characteristics. Using ANCOVA to statistically control for group differences that pre-exist due to factors beyond chance is controversial, and this discussion is beyond the scope of this chapter. For a more thorough discussion see Christenfeld and colleagues (2004). Here we simply note that the researcher should be wary of the use of ANCOVA to statistically equate groups that were not created by randomization.

ANCOVA and Statistical Power

Beyond the desire to statistically equate groups, a better motive for the use of ANCOVA is that, if we have a good covariate, we will have more statistical power. By adjusting scores on the dependent variable we remove unexplained variance, or noise. This increases our signal-to-noise ratio, which essentially decreases the denominator of our F ratio. Consider the example data presented in table 15.3, which reflects the REE example described earlier. If we ignore the body mass covariate and simply compare the two groups on REE, we find that no significant differences exist between groups (mean $\pm$ SD; supplement = 1,800.9 $\pm$ 367.2 kilocalories, placebo = 1,695.8 $\pm$ 297.3 kilocalories; $F(1,28) = 0.74$; $p = .40$; see table 15.4 for ANOVA summary table).

In contrast, if we include body mass as a covariate, we can reduce some of the noise (unexplained variance) in the data and likely increase our statistical power. First consider the relationship between body mass and REE, which is shown in figure 15.1. Body mass is nicely correlated with REE, so by covarying for body mass, we may be able to remove enough noise to markedly enhance our statistical power. Table 15.5 shows the ANCOVA summary table from the analysis.

Compare the ANOVA summary tables in tables 15.4 and 15.5. The total sums of squares are the same for both analyses, which is to be expected because they reflect the total variance of the same raw data. However, they differ in how that

TABLE 15.3

Example Resting Energy Expenditure Data

Body mass (kg) supplement	Resting energy expenditure (kcal), supplement	Body mass (kg) placebo	Resting energy expenditure (kcal), placebo
71.4	1,636.4	52.9	1,401.4
89.9	2,487	74.3	1,589.4
52.3	1,595.4	75	1,922.2
61.4	1,503.8	93.9	1,716
80.8	1,918	88.6	1,710
46.6	1,556	94.2	2,313.4
80.5	2,176	64.2	1,329.8
63.4	1,481.2	60.2	1,456.6
62.6	1,753.4	87.3	1,926.6
62.6	1,763.6	65	1,561
104.8	2,240.8	81.8	2,095.4
63.9	2,259.6	67.7	1,585.4
48	1,278.2	59.1	1,201.8
98	2,023.2	77.3	1,807.4
62.1	1,339.8	75	1,819.8

TABLE 15.4

ANOVA Summary Table
for Resting Energy Expenditure Example

Source	SS	df	MS	F	p
Between	82,813.6	1	82,813.6	0.74	.396
Within	3,125,803.3	28	111,635,8		
Total	3,208,616.8	29			

variance is partitioned. In the ANCOVA, much of the total variance is credited to the covariate (body mass). In fact, the ratio of sums of squares covariate to total sums of squares is 1,808,487.7/3,208,616.8 = ~.56; about 56% of the variance in REE is accounted for by body mass. We have thus removed a lot of the variability in REE by covarying for body mass. Body mass is a statistically significant covariate as indicated by significant F ratio [$F(1,27) = 37.07$, $p < .001$]. Note that now the difference between groups (supplement vs. placebo) is statistically significant [$F(1,27) = 4.88$, $p = .036$]. This is in contrast to the effect shown in table 15.4. By covarying for body mass, we have removed unexplained variance due to body mass

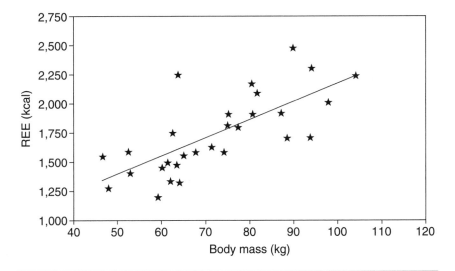

Figure 15.1 Bivariate regression between body mass and REE.

TABLE 15.5

ANOVA Summary Table
for Resting Energy Expenditure Example

Source	SS	df	MS	F	p
Covariate	1,808,487.7	1	1,808,487.7	37.07	<.001
Group	237,891.55	1	237,891.55	4.88	.036
Error	1,317,315.5	27	48,789.47		
Total	3,208,616.8	29			

and thereby increased statistical power, resulting in a much different outcome than with the simple ANOVA.

As noted earlier, ANCOVA increases our statistical power if we have a good covariate. A good covariate is highly correlated with the dependent variable but is not correlated with the independent variable. If our groups were created by randomization, then no systematic basis exists for the groups to differ on the covariate (i.e., no relationship should exist between the covariate and the independent variable). It is therefore purely bad luck if our covariate is correlated with the independent variable. However, if we compare groups not created by randomization, we may have some trouble. Suppose that you are comparing strength between males and females, but you know that males are on average heavier than females, and you know that, on average, heavier people are stronger than lighter people. The impulse is to covary for body weight, but body weight is correlated with the dependent variable (which is a good) and is correlated with the independent variable sex (which is not so good). Astute readers may note that this is the issue of collinearity that we encountered in multiple regression in chapter 9.

Assumptions in ANCOVA

As with all statistical procedures, certain assumptions apply for the calculations to be valid. In addition to the typical assumptions, ANCOVA has an assumption known as **homogeneity of regression.** This is the assumption that the slope of the regression line that describes the relationship between the covariate and the dependent variable is the same among the groups. That is, the slopes must be parallel to each other. Figures 15.2 and 15.3 show two situations. In figure 15.2, the slopes of the regression line between the covariate and the dependent variable are parallel in group A and group B. In this case, the homogeneity of regression assumption is not violated and we are safe in our use of ANCOVA. (Notice that the lines do not have the same y-intercept values; this is okay because it is the slopes that matter.) In figure 15.3, the slopes of the regression lines are not parallel; the slope is markedly steeper for group A than it is for group B. In this case, the use of ANCOVA may give us misleading results. If groups A and B were created by randomization, it is unlikely (although not impossible) that the homogeneity of regression assumption will be violated.

The Pretest–Posttest Control Group Design

One of the most common research designs in kinesiology is the pretest–posttest control group design. In this experimental design, subjects are randomly assigned to one of at least two groups. One group is a control group (or better yet, a placebo group), and one or more other groups receive a treatment. For example, table 15.6 contains simulated data from a flexibility training study. Three stretching programs [proprioceptive neuromuscular facilitation (PNF), static stretching, and control] were compared on their effectiveness in improving hamstring range of motion as assessed by the 90/90 test, which is a common test in physical therapy for hamstring length. In this test, the subject lies supine with the thigh in a vertical position and is asked to fully extend the knee. The 90/90 test score is the difference in degrees between the measured knee angle and full extension; a lower score reflects greater flexibility. Subjects were given the 90/90 test and then randomized to one of the three groups and given the appropriate intervention. The 90/90 test was administered post-treatment. To prevent bias in the flexibility measurements, the therapist performing the 90/90 test was blinded to the subjects' group assignments.

For illustrative purposes, we ignore the pretest scores and perform a simple 1×3 between-subjects ANOVA on the posttest 90/90 scores. The means ± standard

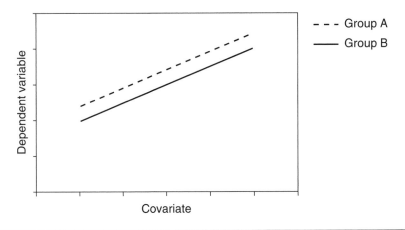

Figure 15.2 Slopes between covariate and dependent variable showing homogeneity of regression.

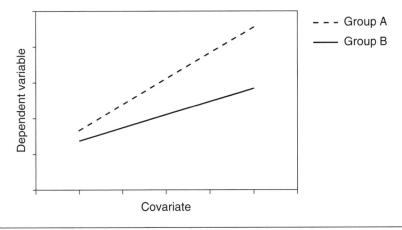

Figure 15.3 Slopes between covariate and dependent variable showing violation of homogeneity of regression.

deviations are shown in figure 15.4, and the ANOVA summary table is shown in table 15.7. Although some differences seem to exist between the group means, the results are far from significant [$F(2,33) = 0.61, p = .55$].

In contrast, we can perform a simple ANCOVA in which we include the pretest 90/90 data as the covariate. The pretest serves as a nice covariate because we would expect it to be highly correlated with the dependent variable (the posttest). That is, subjects who tend to score high on the posttest are likely to tend to score high on the pretest. Similarly, those who tend to score low on the posttest are likely to tend to score low on the pretest. This can be seen in figure 15.5; indeed, the coefficient of determination for this relationship is .90. When we perform the

TABLE 15.6

Example 90/90 Scores for the Stretching Intervention Study

	PNF pre	PNF post	Static pre	Static post	Control pre	Control post
	33	28	42	43	30	27
	35	33	36	32	31	30
	43	38	53	49	30	31
	40	37	31	31	34	38
	45	42	45	43	34	39
	37	35	23	23	49	51
	27	23	45	45	42	44
	28	25	42	41	29	27
	31	29	23	21	40	41
	33	31	38	39	51	53
	48	45	27	29	38	40
	51	47	40	36	36	37
Mean	37.58	34.42	37.08	36.00	37.00	38.17
SD	7.83	7.69	9.40	8.87	7.34	8.53

PNF = proprioceptive neuromuscular facilitation

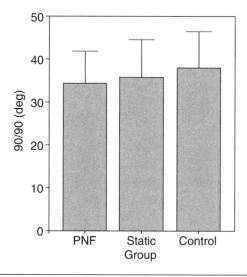

Figure 15.4 Mean ± standard deviation for the posttest scores on the 90/90 data from table 15.6.

ANCOVA on the data in table 15.6, we get a much different outcome than from the simple ANOVA on the posttest scores. The ANCOVA summary is shown in table 15.8. The first row is the effect for the covariate; as expected, the covariate (the pretest) is significant [$F(1,32) = 585.56, p < .001$]. The second row shows the

TABLE 15.7

ANOVA Summary Table for the Posttest 90/90 Scores

Source	SS	df	MS	F	p
Between	85.06	2	42.53	0.61	.55
Within	2,316.58	33	70.20		
Total	2,401.64	35			

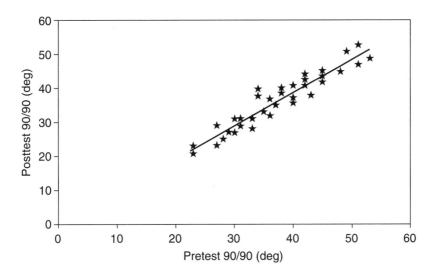

Figure 15.5 Relationship between the pretest and posttest for the 90/90 example data.

TABLE 15.8

ANOVA Summary Table for the 90/90 Scores

Source	SS	df	MS	F	p
Covariate	2,196.55	1	2,196.55	585.56	<.001
Group	112.33	2	56.17	14.97	<.001
Error	120.04	32	3.75		
Total	4,103.25				

effect for the independent variable (group). This effect is the assessment of whether differences exist among the groups on the dependent variable, which is the effect of interest. But with ANCOVA, the dependent variable is now the adjusted post-test 90/90 score. The 90/90 scores have been adjusted based on the pretest 90/90 performance as shown in table 15.6. The effect for group is significant [$F(2,32)$ = 14.97, $p < .001$], in contrast to the analysis of the unadjusted posttest means as shown in table 15.7. By including a covariate in the analysis, we have markedly increased the statistical power as evidenced by the increase in F (0.61 vs. 14.97). Notice that we do pay a penalty, however: we lose 1 df in the denominator (33 vs. 32). This penalty is acceptable as long as the correlation between the covariate and the dependent variable is high. After covarying for the pretest, we can say that a significant effect exists for group, which is equivalent to saying that differences exist among the stretching conditions.

Pairwise Comparisons

Recall from chapter 11 when ANOVA was introduced that when there are three or more groups in the analysis, a significant omnibus F test indicates only whether differences exist between the groups. It does not tell us where the specific differences occur. As before with ANOVA, with ANCOVA we can perform pairwise comparisons to delineate the specific group differences. However, with ANCOVA we must modify our procedures.

First, it is important to note that the comparisons between means need to be performed on the **adjusted means.** The adjusted means are calculated as

$$\overline{Y}'_j = \overline{Y}_j - b_{\text{pooled}} \left(\overline{X}_j - \overline{X} \right), \qquad (15.02)$$

where $\overline{Y}'_j$ is the adjusted group mean for the jth group, $\overline{Y}_j$ is the unadjusted group mean for the jth group, b_{pooled} is the common regression slope for all groups, $\overline{X}_j$ is the mean of the covariate for the jth group, and $\overline{X}$ is the grand mean for the covariate. Notice the similarity between equation 15.02 and equation 15.01. These values should be available from standard ANCOVA statistical software. Note that b_{pooled} is not simply the regression slope for a bivariate regression between the covariate and the dependent variable. Rather, it is the regression coefficient for the covariate from multiple regression including the group code(s) and the covariate as independent variables. Standard ANCOVA software should provide you with the common regression slope.

For our example data in table 15.6, the pooled regression coefficient is .99 and the covariate grand mean is 37.22 (calculations not shown). The adjusted means for each group are calculated as

$$\overline{Y}'_{\text{PNF}} = 34.42 - .99(37.58 - 37.22) = 34.06,$$

$$\overline{Y}'_{\text{static}} = 36.00 - .99(37.08 - 37.22) = 36.14,$$

$$\overline{Y}'_{\text{control}} = 38.17 - .99(37.00 - 37.22) = 38.39.$$

To perform pairwise comparisons on the adjusted group means, we outline the Bryant-Paulson generalization of Tukey's *HSD* procedure (Bryant and Paulson, 1976; Huitema, 1980). The equation for the **Bryant-Paulson procedure** is

$$Q = \frac{\bar{Y}'_i - \bar{Y}'_j}{\sqrt{\dfrac{MS_{E_{ANCOVA}}\left[1 + \left(\dfrac{MS_{B_{cov}}}{SS_{W_{cov}}}\right)\right]}{n}}}, \tag{15.03}$$

where $\bar{Y}'_i$ and $\bar{Y}'_j$ are the adjusted means to be compared, $MS_{E_{ANCOVA}}$ is the mean square error term from the ANCOVA, $MS_{B_{cov}}$ is the between-groups mean square from an ANOVA procedure comparing the groups on the covariate, $SS_{W_{cov}}$ is the within-groups sums of squares term from an ANOVA procedure comparing the groups on the covariate, and n is the sample size per group (if the sample size per group is not equal, then n is calculated using the harmonic mean).

As the previous description suggests, to get the necessary pieces to perform this calculation we must also perform a between-subjects ANOVA with the covariate as the dependent variable. The ANOVA summary table for the ANOVA comparing the three groups on the covariate is shown in table 15.9.

Suppose that we wish to perform a comparison between the adjusted means for the PNF and control groups. Entering the appropriate values into equation 15.03 yields the following:

$$Q = \frac{34.06 - 38.39}{\sqrt{\dfrac{3.75\left[1 + \left(\dfrac{1.19}{2,239.83}\right)\right]}{12}}} = \frac{-4.33}{.559} = -7.75$$

We ignore the sign of Q and compare it with the critical value of Q (Q_{crit}) obtained from table A.10. With three groups and $df = N - J - 1 = 36 - 3 - 1 = 32$ (where N = total number of subjects across all groups and J = number of groups), the critical Q is between 3.501 (for $df = 40$) and 3.555 (for $df = 30$). We will use the more conservative value of 3.555. The critical Q is less than the calculated Q of 7.75, and we conclude that PNF stretching improved hamstring range of motion relative to the control treatment.

TABLE 15.9

ANOVA Summary Table on the Covariate

Source	SS	df	MS	F	p
Between	2.39	2	1.19	0.02	.98
Within	2,239.83	33	67.87		
Total	2,242.22	35			

We can perform similar comparisons for the other two pairwise comparisons. Note that the denominator of Q (.559) is the same for these comparisons, so that for the static versus control comparison,

$$Q = \frac{36.14 - 38.39}{.559} = -4.03.$$

Similarly, for the PNF versus static comparison,

$$Q = \frac{34.06 - 36.14}{.559} = -3.72.$$

When we ignore the sign, both of these values exceed the critical value of 3.555. Collectively, the pattern of comparisons is PNF > static > control, and we can conclude that both stretching techniques improved hamstring range of motion relative to the control condition, but that the PNF technique was more effective than the static technique.

We can use the same Bryant-Paulson model to create pairwise confidence intervals for the ANCOVA-adjusted group means (Huitema, 1980). Equation 15.03 can be modified for this purpose as

$$\left(\overline{Y}'_i - \overline{Y}'_j \right) \pm Q_{\text{crit}} \sqrt{\frac{MS_{E_{\text{ANCOVA}}} \left[1 + \dfrac{MS_{B_{\text{cov}}}}{SS_{W_{\text{cov}}}} \right]}{n}}. \tag{15.04}$$

Notice that these are the same terms that were used in equation 15.03. To apply this to the pairwise comparison between the PNF and control groups, the difference between the adjusted means is $38.39 - 34.06 = 4.33$. Recall that $Q_{\text{crit}} = 3.555$ (for $\alpha = .05$, i.e., the 95% CI) and that the square root term is .559. The 95% CI is then

$$4.33 \pm (3.555)(.559) = 4.33 \pm 1.99 = 2.34 \text{ to } 6.32.$$

We can then conclude that we are 95% confident that the true or population difference in adjusted means is somewhere from 2.35 to 6.31 degrees. The fact that the confidence interval does not contain zero is equivalent to concluding that the difference is statistically significant at $\alpha = .05$. The 95% CI for static versus control is

$$38.39 - 36.14 \pm (3.55)(.559) = 2.25 \pm 1.98 = 0.27 \text{ to } 4.23.$$

Similarly, for static versus PNF,

$$36.14 - 34.06 \pm (3.55)(.559) = 2.08 \pm 1.98 = 0.10 \text{ to } 4.06.$$

Notice again that both of these 95% CIs exclude zero, which is perfectly consistent with the previous tests of statistical significance. This should be expected because the confidence interval and null hypothesis significance tests are based on the exact same underlying statistical model.

Summary

The ANCOVA procedure is the intersection between ANOVA and regression. The essence of ANCOVA is to adjust group means using regression so that the groups are statistically equivalent on the covariate. Analysis of covariance is a useful procedure for increasing statistical power and we suggest that investigators consider incorporating a covariate when possible. It is especially useful in pretest–posttest control group designs when the pretest is used as the covariate. The Bryant-Paulson modification of the Tukey procedure can be used to perform pairwise comparisons of statistical significance and to create simultaneous confidence intervals.

Problems to Solve

An investigator compares the effects of a low-intensity and a high-intensity exercise program on the power output at the ventilatory threshold. For comparison purposes, a control group is also evaluated. A total of 30 subjects are randomized to the three groups. All subjects are given a pretest to determine the power output (watts) at which the ventilatory threshold occurs. The two exercise groups then undergo 8-week training programs, whereas the control group does not exercise. After the 8-week period, the subjects undergo a posttest. The data are as follows in the table. Using statistical software, perform an ANCOVA on the data where group is the independent variable, pretest ventilatory threshold (watts) is the covariate, and posttest ventilatory threshold is the dependent variable.

Control pre	Control post	Low-intensity pre	Low-intensity post	High-intensity pre	High-intensity post
175.00	150.00	125.00	125.00	125.00	225.00
200.00	200.00	150.00	150.00	75.00	175.00
150.00	175.00	100.00	125.00	150.00	175.00
125.00	100.00	100.00	125.00	75.00	125.00
100.00	100.00	175.00	175.00	125.00	175.00
150.00	225.00	150.00	200.00	150.00	175.00
175.00	200.00	175.00	250.00	75.00	100.00
175.00	175.00	125.00	150.00	150.00	200.00
225.00	200.00	200.00	225.00	150.00	175.00
200.00	225.00	175.00	175.00	225.00	275.00

1. Construct an ANCOVA summary table from the analysis. Does the analysis suggest that the training programs have different effects on the ventilatory threshold?

2. Do the data violate the homogeneity of regression assumption?

3. Compare the three groups using the Bryant-Paulson generalization of the Tukey *HSD* procedure.

See appendix C for answers to problems.

Key Words

adjusted means

analysis of covariance

Bryant-Paulson procedure

covariate

homogeneity of regression

$$H = \left[\frac{12}{18(18+1)}\right]\left[\frac{82.5^2}{6} + \frac{54.5^2}{6} + \frac{34.0^2}{6}\right] - 3(18+1)$$

Analysis of Nonparametric Data

Aphysical education teacher believed that the ability to serve well was related to the success or failure of beginning tennis students. To test this belief, she conducted a ladder tournament with beginning players to determine her students' abilities to win games against their peers. When the tournament was over, she ranked the students according to the order they finished in the tournament. Next, she administered a serving test that ranked the students from best to worst on serving skill. The resulting two sets of rank-order scores represent ordinal data that are nonparametric (that is, the assumptions of normality do not apply). How can the teacher analyze these data?

As discussed in chapter 1, data can be classified into four categories: nominal, ordinal, interval, and ratio. When ratio or interval data are collected, analysis by parametric statistical techniques is appropriate. Pearson's correlation coefficient, the t test, and ANOVA in all of its varieties are parametric statistical techniques. But when data are of the nominal or ordinal type, the assumptions of normality are not met and **nonparametric** statistical techniques must be used.

This chapter presents several of the most commonly used nonparametric statistical techniques:

- **Chi-square** (χ^2) compares two or more sets of nominal data that have been arranged into categories by frequency counts.
- Spearman's **rank order correlation** coefficient determines the relationship between two sets of ordinal data.
- The **Mann-Whitney U test** determines the significance of the difference between ordinal rankings of two groups of subjects ranked on the same variable. It is similar to an independent t test.
- **Kruskal-Wallis ANOVA** by ranks compares the ranking of three or more independent groups. It is similar to simple ANOVA.
- **Friedman's two-way ANOVA** by ranks is similar to repeated measures ANOVA. It determines the significance of the difference between ranks on the same subjects.

Chi-Square (Single Classification)

Chi-square is used to compare two or more sets of nominal data that have been arranged into categories by frequency counts. Sex is an example of nominal data. A person is classified as male or female, and we simply count the number in each category. No variability exists within the category; all subjects are of equal value. Classification must be mutually exclusive; a subject may be classified into only one of the categories.

Grades are another example of this type of classification. After considering all of a student's test scores, the teacher assigns the student a grade for the course. The grade is just a name (nominal data) for performance that has a certain meaning. The grades do represent ordered values (A, B, C, D, and F), but we treat them here as nominal by counting the number of students in each category.

A certain number of students (the frequency) receive A, a certain number receive Bs, and so on for Cs, Ds, and Fs. All the As are equally valuable, all the Bs equally valuable, and so on for Cs, Ds, and Fs. The chi-square test should be used to compare two sets of data, classified by category, to determine whether the frequencies of the categories differ by amounts larger than would be expected by chance. It reveals the significance of the differences in the frequency counts. The following example shows how chi-square may be applied.

An Example From Education Administration

A physical education department established a policy that teachers should award certain percentages of the various grades in their classes. After much discussion, the teachers agreed to keep the grade distribution to approximately 10% As, 30% Bs, 35% Cs, 20% Ds, and 5% Fs. The teachers acknowledged that for a small class these percentages might not hold true, but when all classes for a given teacher are combined, the final grade percentages should closely approximate the established policy.

Everyone accepted the policy, and for a year or two the faculty followed it closely. But then one teacher began to consistently give higher grades than would be expected by the policy. In a roster of 141 students, this teacher gave 30 As, 57 Bs, 32 Cs, 15 Ds, and 7 Fs. The chairperson wanted to know whether such a distribution differed significantly from the accepted policy. Based on 141 students, departmental policy would call for 14 As (10%), 42 Bs (30%), 49 Cs (35%), 28 Ds (20%), and 8 Fs (5%).

The null hypothesis (H_0) presumes that the differences occurred by pure chance. If H_0 is demonstrated to be false at an acceptable level of confidence, H_1 is accepted and it must be concluded that an influence other than chance caused the differences in the frequency counts.

To determine the odds that the teacher's distribution could occur by chance alone (and was thus not a violation of policy), the department chair applied the chi-square technique. The formula for chi-square is as follows:

$$\chi^2 = \Sigma \left[\frac{(O-E)^2}{E} \right],$$ (16.01)

where O is the observed frequency and E is the expected frequency.

To apply this formula, we must find the difference between the observed (teacher's grades) and expected (department policy) frequencies for each category and square each of these differences. Each squared difference is then divided by the expected frequency for its category. Finally we obtain the sum of the

TABLE 16.1

Calculation of Chi-Square

Grades	A	B	C	D	F	Total
Observed (teacher's grades)1	30	57	32	15	7	141*
Expected (departmental policy)	14	42	49	28	8	141*
$O - E$	16	15	-17	-13	-1	
$(O–E)^2$	256	225	289	169	1	
$(O–E)^2/E$	18.29	5.36	5.90	6.04	0.13	35.72

* The total of the observed and the expected frequencies must be equal.

squared differences divided by expected frequency. Table 16.1 demonstrates the process.

Table 16.1 is called a one by five (1×5) classification table because the observed frequencies occupy only one row and it has five columns.

In this problem, $\chi^2 = 35.72$. The degrees of freedom in chi-square are the number of categories, k (in this case k = number of grade categories), minus one ($df = k - 1$, or $5 - 1 = 4$). Table A.11 in appendix A indicates that for df = 4, chi-square must be 13.277 to reach the 99% LOC ($\alpha = .01$). Because 35.72 easily exceeds 13.277, the probability that the two distributions differ by chance alone is less than one in 100. The department chair rejected H_0 and concluded that the teacher was awarding grades that were significantly higher than permitted by departmental policy.

Chi-Square (Two or More Classifications)

Chi-square may also be used to solve more complicated problems. Our second example uses two classifications in a contingency table, which is a distribution of frequencies into both rows and columns. One category of frequency occupies the rows and the other category occupies the columns. The following problem involves a 2×3 (two rows by three columns) contingency table.

An Example From Motor Learning

A teacher wanted to know which of two methods of teaching the front crawl stroke was more effective in producing swimmers with good stroke form. The first method was the traditional part method, in which the kick was taught first, then the arm stroke, then the breathing, and finally all the parts were put together as a complete stroke. The second method was the whole method, in which after the prone float and glide had been learned, the students were introduced to all parts of the stroke simultaneously and practiced the entire stroke for the remainder of the semester.

After teaching 55 students the part method and 52 students the whole method for 15 weeks, twice weekly for 1 hour each class, the teacher asked another swimming expert to classify each student's stroke form into one of three categories: good, average, or poor. The results are shown in table 16.2. The values in the cells represent the number of students in each class who were classified as good, average, or poor.

We now ask, because more whole-method students are in the good category and more part-method students in the average and poor categories, Is the whole method better or could these frequencies have occurred by chance? We use the chi-square technique to obtain the answer.

First, we must assume that the total group of 107 students represents the best estimate that can be obtained for the expected frequencies in each category. Because no control group exists with which each method can be compared, we calculate how many students would be expected to fall into each category had they all been taught alike by either method (i.e., without the influence of whole or part instruction methods). The totals for each group and for each category must be computed as shown in table 16.3.

The percentage of the total number of students expected in each category is calculated by dividing the total number in that category by the grand total.

TABLE 16.2

Chi-Square Contingency Table: 2 × 3 Classification

Type of Instruction	Good	Skill categories Average	Poor
Part	15	27	13
Whole	21	19	12

TABLE 16.3

Determination of Percent of Total

Type of Instruction	Good	Skill categories Average	Poor	Total
Part	15	27	13	**55**
Whole	21	29	12	**52**
Total	36	46	25	**107**
% of grand total	**33.64**	**42.99**	**23.36**	

Totals are represented in bold.

Good	36/107 = 33.64%
Average	46/107 = 42.99%
Poor	25/107 = 23.36%

With this information, we can calculate how many students would fall into each category of each group if all students were taught alike. This is the expected frequency for the chi-square calculation. It is determined by multiplying the percentage of students expected in each category by the total number in that group.

For the part group:

Good	.3364 × 55 = 18.50
Average	.4299 × 55 = 23.64
Poor	.2336 × 55 = 12.85

For the whole group:

Good	3364 × 52 = 17.49
Average	.4299 × 52 = 22.35
Poor	.2336 × 52 = 12.15

An alternative, and shorter, method of determining expected frequencies per cell is to apply the following formula:

$$\text{expected frequency} = \frac{(\Sigma_{row})(\Sigma_{column})}{N}. \tag{16.02}$$

For example, the expected frequency for the good students in the part group is

$$(55 \times 36)/107 = 18.50.$$

The expected frequencies are then entered into a table along with the observed frequencies (table 16.4).

The differences between the observed and expected frequencies are now calculated, squared, divided by the expected value, and summed to compute the chi-

TABLE 16.4

Observed and Expected Frequencies

	Good Observed	Good Expected	Average Observed	Average Expected	Poor Observed	Poor Expected
Part	15	18.50	27	23.64	13	12.85
Whole	21	17.49	19	22.35	12	12.15

TABLE 16.5

Calculation of Chi-Square's Contingency Table

Observed	Expected	$O-E$	$(O-E)^2$	$(O-E)^2/E$
15	18.50	−3.50	12.25	0.662
21	17.49	3.51	12.32	0.702
27	23.64	3.36	11.29	0.478
19	22.35	−3.35	11.22	0.502
13	12.85	0.15	0.02	0.002
12	12.15	−0.15	0.02	0.002
				$\chi^2 = 2.35$

square statistic. Table 16.5 demonstrates the process; for this example, chi-square is found to be 2.35.

The degrees of freedom for a chi-square contingency table are number of rows minus one times number of columns minus one (see table 16.2):

$$df = (R-1)(C-1). \tag{16.03}$$

In this example, $df = (2-1)(3-1) = (1)(2) = 2$.

For $df = 2$, table A.11 in appendix A indicates that a chi-square value of 4.605 is needed for the 90% LOC. Because the calculated value for chi-square (2.35) is less than 4.605, we must accept H_0. We conclude that the group frequencies are not significantly different and that students learn equally well with either method. The small differences noted are the result of chance occurrences.

Limitations of Chi-Square

Chi-square does not apply well to small samples, especially those in a 2 × 2 table. The total number of frequencies (N) should be at least 20, and the value of each cell in the expected frequencies row should not be less than 1. Also, in the 2 × 2 table, some statisticians suggest the Yates correction for continuity by subtracting 0.5 from each of the $(O-E)$ values before they are squared to protect against type I errors (Thomas, Nelson, & Silverman, 2010).

The chi-square procedure can be very helpful for analyzing frequency counts by categories and is applicable to many problems in kinesiology.

Rank Order Correlation

Spearman's rank order correlation coefficient is used to determine the relationship between two sets of ordinal data. It is the nonparametric equivalent of Pearson's correlation coefficient. Often data in kinesiology result from experts ranking subjects. For example, a teacher may rank students by skill: 1 (highest skill), 2 (next highest), and so forth on down to the last rank (lowest skill). In recreational sports, ladder and round-robin tournaments may result in a rank order of individuals or teams. Even when data have been collected on a parametric variable, the raw data may be converted to rankings by listing the best score as 1, the next best as 2, and so on.

Ranked data are ordinal and do not meet the criteria for parametric evaluation by Pearson's product moment correlation coefficient. To measure the relationship between rank order scores, we must use Spearman's rank order correlation coefficient (rho, or ρ). The formula for **Spearman's rho** is

$$\rho = 1 - \frac{6 \Sigma d^2}{N(N^2 - 1)}, \tag{16.04}$$

where d is the difference between the two ranks for each subject and N is the total number of subjects (i.e., the number of pairs of ranks). The number 6 will always be in the numerator.

The degrees of freedom for rho are the same as for Pearson's r: $df = N_{pairs} - 2$. The significance of rho is determined by looking up the value for the appropriate degrees of freedom in table A.12 in appendix A.

An Example From Physical Education

We use the tennis example introduced at the beginning of the chapter to demonstrate how to apply Spearman's rho. Recall that the students were ranked from highest (1) to lowest (25) on the serving test. These ranks were then compared with the final placements on the ladder tournament. The results are presented in table 16.6.

When two scores are tied in rank data, each is given the mean of the two ranks. The next rank is eliminated to keep N consistent. For example, if two subjects tie for fourth place, each is given a rank of 4.5, and the next subject is ranked 6. In table 16.6, there are no ties because a ladder tournament in tennis does not permit ties; one person must win. The instructor also ranked the students on serving skills without permitting ties.

The difference between each student's rank in the ladder tournament and rank in serving ability was determined, squared, and summed (see table 16.6). The signs of the difference scores are not critical because all signs become positive when the differences are squared. With these values we can calculate rho by applying equation 16.04:

$$\rho = 1 - \frac{6(362)}{25(25^2 - 1)} = 1 - 0.14 = .86.$$

Table A.12 (p. 339) indicates that for $df = 23$, a value of .510 is needed to reach significance at $\alpha = .01$. The obtained value, .86, is greater than .510, so H_0 is rejected

TABLE 16.6

Spearman's Rank Order Correlation Coefficient

Subject	Place on ladder	Rank on serve	d	d^2
1	8	10	−2	4
2	16	17	−1	1
3	7	4	3	9
4	24	25	−1	1
5	2	5	−3	9
6	15	9	6	36
7	1	3	−2	4
8	23	24	−1	1
9	6	6	0	0
10	22	20	2	4
11	17	13	4	16
12	5	7	−2	4
13	9	18	−9	81
14	14	19	−5	25
15	25	21	4	16
16	3	1	2	4
17	18	15	3	9
18	13	12	1	1
19	10	16	−6	36
20	19	11	8	64
21	4	2	2	4
22	21	23	−2	4
23	12	8	4	16
24	20	22	−2	4
25	11	14	−3	9
				$\Sigma d^2 = 362$

and it is concluded that tennis serving ability and the ability to win tournament games are related at better than the 99% LOC.

Remember, a significant correlation does not prove that being a good server is the cause of winning the game. Many factors are involved in successful tennis ability; serving is just one of them. Other factors related to both serving and winning may cause the relationship.

Mann-Whitney *U* Test

The Mann-Whitney *U* test is used to determine the significance of the difference between rankings of two groups of subjects who have been ranked on the same variable. The *U* value indicates whether one group ranks significantly higher than the other. It is the nonparametric equivalent of an independent, two-group t test. It may be used instead of the t test when the assumptions of the t test cannot be met, such as when the data are ordinal or highly skewed.

When interval or ratio data are highly skewed, we may want to create one rank order list (on the dependent variable) for all subjects from both groups. For example, the highest scoring person from both groups is ranked 1, the second highest scorer is ranked 2, and so on until all subjects in both groups have been ranked. Then we compare the ranks in group 1 to the ranks in group 2 using the Mann-Whitney U test.

All subjects from both groups are ranked on the same variable and placed in order from highest to lowest, and the subjects in group 1 and group 2 are then listed by their ranks. The sums of the ranks for each group are compared to determine whether the median rankings between the groups differ by more than would be expected by chance alone. The formulas in equations 16.05 and 16.06 are modified from Bruning and Kintz (1977, p. 224).

The formula for U_1 is

$$U_1 = n_1 n_1 + \left[\frac{n_1(n_1+1)}{2} \right] - \Sigma R_1 \tag{16.05}$$

and the formula for U_2 is

$$U_2 = n_1 n_2 - U_1. \tag{16.06}$$

In equations 16.05 and 16.06, n_1 and n_2 are the number of subjects in each group and ΣR_1 is the sum of the rankings for group 1. (It does not matter which group is designated as group 1.)

When $n_1 + n_2 \geq 20$, a Z score may be computed to determine the significance of the differences in ranks. When $n_1 + n_2 < 20$, we can judge the significance of the smaller U value using tables A.13, A.14, and A.15 in appendix A.

An Example From Motor Learning

A student in a motor learning class was assigned a term project to ascertain whether gymnasts had better balance skills than the general population. The student measured 10 gymnasts and 15 nongymnasts on the Bass stick test for upright balance. The test results of all 25 subjects were then ranked (1 for best, 25 for worst) in a single list. The results are shown in table 16.7.

For these data, U_1 and U_1 are computed as follows:

$$U_1 = (10)(15) + \frac{10(10+1)}{2} - 103 = 102,$$
$$U_2 = (10)(15) - 102 = 48.$$

Because $n_1 + n_2 \geq 20$, a Z score is used to determine significance. The formula to compute Z for either group is

TABLE 16.7

Balance Rankings

Gymnasts	Nongymnasts
1	3
2	5
4	6
7	10
8	12
9	13
11	14
17	15
21	16
23	18
	19
	20
	22
	24
	25
$\Sigma R_1 = 103$	$\Sigma R_2 = 222$

$$Z = \frac{U - \dfrac{n_1 n_2}{2}}{\sqrt{\dfrac{n_1 n_2 (n_1 + n_2 + 1)}{12}}}. \qquad (16.07)$$

Note: The number 12 is a constant and will always be in the denominator within the square root sign. The number 2 in the numerator represents the number of groups.

It does not matter which U value is used to calculate Z. The Z values for U_1 and U_2 will have the same absolute value: One will be positive and one will be negative. We shall use U_1:

$$Z_1 = \frac{102 - \dfrac{(10)(15)}{2}}{\sqrt{\dfrac{(10)(15)(10 + 15 + 1)}{12}}} = 1.50.$$

Because we are testing H_0, we use a two-tailed test to interpret Z. For a two-tailed test, we need $Z = 1.65$ for $\alpha = .10$, $Z = 1.96$ for $\alpha = .05$, and $Z = 2.58$ for $\alpha = .01$ (see table A.1 in appendix A). In this problem, Z does not reach the limits for $\alpha = .10$, so we accept H_0 and conclude that no significant difference exists between gymnasts and nongymnasts in balance ability as measured by the Bass stick test.

Comparing Groups With Small Values of N

If $n_1 + n_2 < 20$, the Z test may be biased, so a table of U (tables A.13 to A.15 in appendix A) must be used to determine the significance of U. The critical U value is found in the table and compared with the smaller calculated U. If the smaller U value is equal to or less than the table value, the rank difference is significant. In the balance problem, the smaller U (U_2) is 48. Table A.13 shows that for $\alpha =$ .10 (a two-tailed test with $n_1 = 10$ and $n_2 = 15$), the smaller U must be 44 or less. Because $U_1 = 48$, the difference between the ranks is not significant. This agrees with our conclusion based on Z.

Kruskal-Wallis ANOVA for Ranked Data

If data are ranked and there are more than two groups, a nonparametric procedure analogous to simple ANOVA is available called the Kruskal-Wallis ANOVA for ranked data. This procedure produces an H value that, when $N > 5$, approximates the chi-square distribution. Once we have calculated H, we can determine its significance by using the chi-square table A.11 in appendix A for $df = k - 1$, where k is the number of groups to be ranked.

An Example From Athletic Training

Athletic trainers and coaches want to return athletes to competitive condition as soon as possible after a debilitating injury. Anterior cruciate ligament (ACL) tears repaired with surgery require extensive rehabilitation. An athletic trainer wanted to know whether accelerated rehabilitation (closed kinetic chain activities using weight-bearing exercises) was superior to normal rehabilitation activities (knee extension and flexion exercises) as compared with no special rehabilitation exercises (control).

Eighteen subjects, each of whom had undergone recent ACL reconstruction with the patellar tendon replacement technique, were selected and divided into three groups: control, normal, and accelerated. After 6 months, three orthopedic physicians evaluated each subject and jointly ranked all 18 according to their level of rehabilitation. Following are the rankings classified according to the type of rehabilitation technique (see table 16.8). Ties are given the average of the two-tied ranks.

Clearly, differences exist in the sums and the means. The question we must ask is, Are the differences large enough to be attributed to the treatment effects, or are they chance differences that we would expect to occur even if the treatments had no effect?

To solve this problem we apply the following formula for the Kruskal-Wallis H value where N is the total of all subjects in all groups, n is subjects per group, and k is number of groups.

TABLE 16.8

Rankings According to Rehabilitation Technique

	Control (R_1)	Normal (R_2)	Accelerated (R_3)
	15	13	3.5
	11	8	16.5
	18	7	1
	16.5	3.5	5
	12	14	6
	10	9	2
Sums	82.5	54.5	34.0
Means	13.75	9.08	5.67

$$H = \left[\frac{12}{N(N+1)} \right] \left[\frac{\Sigma R_1^2}{n_1} + \frac{\Sigma R_2^2}{n_2} + \ldots \frac{\Sigma R_k^2}{n_k} \right] - 3(N+1). \qquad (16.08)$$

Substituting values from the example in table 16.8, we find

$$H = \left[\frac{12}{18(18+1)} \right] \left[\frac{82.5^2}{6} + \frac{54.5^2}{6} + \frac{34.0^2}{6} \right] - 3(18+1)$$

$$H = [.035][1822.09] - 57 = 6.77.$$

From the chi-square table A.11, the critical value for $\alpha = .05$ for $df = 3 - 1 = 2$ is 5.991. Because our obtained value of 6.77 exceeds 5.99, we reject the null hypothesis and conclude that significant differences exist somewhere among the three groups.

The formula for H assumes that no ties have occurred. If more than a few ties occur, a correction for H has been suggested by Spence and colleagues (1968, p. 217). There is usually little practical value in calculating H_C unless the number of ties is large and the value of H is close to the critical value. The correction formula rarely changes the conclusion.

$$H_C = \frac{H}{1 - \frac{(t_1^3 - t_1 + t_2^3 - t_2 + \ldots t_k^3 - t_k)}{N^3 - N}}, \qquad (16.09)$$

where t is the number of scores tied at a given rank and k is the number of times ties occur. In table 16.8, two scores are tied at 3.5 and two scores are tied at 16.5; therefore, $t_1 = 2$, $t_2 = 2$, and $k = 2$.

In our example,

$$H_C = \frac{6.77}{1 - \frac{\left(2^3 - 2\right) + \left(2^3 - 2\right)}{18^3 - 18}} = \frac{6.77}{.9979} = 6.78.$$

To differentiate among the groups, we can calculate the standard error (SE) of the difference for any two values using a procedure suggested by Thomas and Nelson (2001, p. 205):

$$SE = \sqrt{\frac{N(N+1)}{12} \frac{2}{n}}. \tag{16.10}$$

When n are unequal, use $1/n_1 + 1/n_2$ in place of $2/n$, and calculate a separate standard error for each pair of groups to be compared.

In our example,

$$SE = \sqrt{\left[\frac{18(18+1)}{12}\right]\left[\frac{2}{6}\right]} = 3.08.$$

Because we are making three comparisons, we must use a Bonferroni adjustment (i.e., divide by 3) of our rejection α value of .05 (.05/3 = .017). Now we look for pairwise differences at $\alpha = .017$. This protects against type I errors.

For a two-tailed test at $\alpha = .05$, we expect 2.5% of the area under the normal curve to be in each tail. Using the Bonferroni correction to the p value for three comparisons results in 2.5/3 = .83 at each end of the curve. This leaves 50 − .83, or 49.17% of the curve between the mean and the critical value. Using table A.1 we note that the Z score for 49.17% under the curve is ± 2.39. Multiplying our standard error value, 3.08 × ± 2.39 = ± 7.36 gives us the critical value for pairwise comparisons at $\alpha = .017$. To apply this value, it is helpful to create a mean difference table (see table 16.9).

Therefore, we conclude that the accelerated group is significantly different from the control group at $p < .017$, but the normal group is not different from either the control or the accelerated group.

TABLE 16.9

Rankings According to Rehabilitation Technique

Group	Control	Normal	Accelerated
Control	0.00	4.67	8.08*
Normal		0.00	3.41
Accelerated			0.00

*Any value in the table that exceeds 7.36 is significant at $p < .017$.

Friedman's Two-Way ANOVA by Ranks

When subjects are measured three or more times using ranked data, or if interval or ordinal data are converted to ranks, a nonparametric procedure similar to repeated measures ANOVA is used. Friedman's two-way ANOVA by ranks computes a chi-square value for the differences between the sum of the ranks for each repeated measure. This test was developed by statistician and economist Milton Friedman (Friedman, 1937), who received the Nobel Prize in economics in 1976. This test is analogous to the single factor repeated measures ANOVA but is called two-way because "subjects" is a factor in repeated measures ANOVA.

An Example From Physical Education

A researcher wanted to know whether physical education was judged by students to be more or less popular than selected academic classes. Using middle school students as subjects, the researcher asked 10 students to rank physical education, math, and English according to how well they liked the classes; 1 represented the most-liked class, 2 the class in the middle, and 3 the least-liked class. Table 16.10 presents the fabricated results of the hypothetical survey.

Friedman's formula to compute chi-square among the sums of the ranks is

$$\chi_R^2 = \left[\frac{12}{Nk(k+1)}\right]\left[\Sigma R_1^2 + \Sigma R_2^2 + \ldots \Sigma R_k^2\right] - 3N(k+1) \qquad (16.11)$$

where N is number of subjects and k is number of repeated measures.

Substituting values from table 16.10, we compute

$$\chi_R^2 = \left[\frac{12}{10(3)(3+1)}\right]\left[17^2 + 25^2 + 20^2\right] - 3(10)(3+1)$$

$$\chi_R^2 = [.1][1314] - 120 = 11.4.$$

TABLE 16.10

Results of Hypothetical Study

Student	Physical education	Math	English
A	1	3	2
B	2	3	1
C	3	2	1
D	1	2	3
E	1	3	2
F	2	3	2
G	1	2	3
H	2	3	2
I	1	3	2
J	3	1	2
Totals	17	25	20

Using $df = k - 1 = 3 - 1 = 2$ and $\alpha = .01$, we note from table A.11 that the critical chi-square = 9.210. Because the obtained value (11.4) is greater than the critical value, we reject the null hypothesis and conclude at $p < .01$ that students like physical education best. *Note:* Data are fabricated.

Summary

When data are collected that do not meet the assumptions of parametric statistics, alternative techniques must be used to reach conclusions about the relationships or the differences among the variables.

Chi-square, which can be used in both single and double classification problems, is the appropriate technique to use to analyze differences in frequency counts of nominal data. Spearman's rank order correlation coefficient may be used to determine the relationship between two variables of ranked data. It is interpreted in a manner similar to Pearson's correlation coefficient for parametric data.

The Mann-Whitney U test can be used to determine the significance of the difference between ordinal rankings of two groups of subjects on the same variable. It is the nonparametric equivalent of an independent t test. Kruskal-Wallis ANOVA by ranks is used to compare the ranking of three or more independent groups. It is similar to simple ANOVA. Friedman's two-way ANOVA by ranks is similar to repeated measures ANOVA. It determines the significance of the difference between ranks on the same subjects.

Problems to Solve

If your computer software includes nonparametric statistics, first solve the problem by hand, then compare your results with the computer output.

1. A principal in a high school asked a random sample of students if they thought an equal amount of money should be allocated to the boys and girls athletic programs. The boys responded 27 yes and 42 no; the girls responded 35 yes and 36 no. If the total group is the best estimate of the expected response, does a significant difference exist between the opinions of the boys and the girls?

2. The following listing shows the rank order of finish of 10 gymnasts in the compulsory and optional competitions for the all-around event. What is the correlation between the two rankings?

Gymnast	Compulsory	Optional
1	8	6
2	5	7

Gymnast	Compulsory	Optional
3	4	5
4	6	4
5	9	9
6	2	1
7	7	8
8	10	10
9	3	2
10	1	3

3. In the national collegiate football rankings, the Eastern teams seemed to have an advantage. The following listing shows the rankings for the top 25 teams by area of the country. Does a significant difference exist in favor of the East?

West—5, 6, 8, 12, 15, 19, 20, 24, 25

East—1, 2, 3, 4, 7, 9, 10, 11, 13, 14, 16, 17, 18, 21, 22, 23

4. To determine the effectiveness of learning a cartwheel by watching a video, a gymnastics teacher randomly divided 15 novice gymnasts into three groups of five students each. One group (control) viewed a live demonstration of a cartwheel one time and then was tested. A second group (video) watched a video of a cartwheel 10 times and then was tested. The third group (live) watched 10 live demonstrations of a cartwheel and then was tested. All 15 subjects were ranked on their ability to perform the cartwheel with the following results (1 is highest rank, 15 is lowest rank).

Control	Video	Live
8	2	1
12	5	3
13	7	4
14	9	6
15	11	10

Does a significant difference exist among the groups in their ability to perform the cartwheel?

5. A physical education teacher wanted to know which sport is most popular with high school students. He asked 14 students to rank football, baseball, and basketball; 1 represented the most-liked sport, 2 the sport in the middle, and 3 the least-liked sport. Following are the results.

Student	Football	Baseball	Basketball
A	1	2	3
B	2	1	3
C	2	1	3
D	1	3	2
E	2	1	3
F	2	1	3

(continued)

Student	Football	Baseball	Basketball
G	3	2	1
H	2	1	3
I	1	2	3
J	3	1	2
K	3	2	1
L	1	2	3
M	3	1	2
N	2	1	3

Does a significant difference exist in the rankings of the three sports?

See appendix C for answers to the problems.

Key Words

chi-square

Friedman's two-way analysis of variance

Kruskal-Wallis analysis of variance

Mann-Whitney U test

nonparametric

rank order correlation

Spearman's rho

$$H = \left[\frac{12}{18(18+1)} \right] \left[\frac{82.5^2}{6} + \frac{54.5^2}{6} + \frac{34.0^2}{6} \right] - 3(18+1)$$

$$H = [.035][1622.09] - 57$$

$$H = 6.77$$

Clinical Measures
of Association

A n athletic trainer wishes to know whether a new type of football helmet decreases the number of concussions compared with the old helmet design. A physical therapist develops a new test to diagnose rotator cuff injuries. These types of inquiry involve dependent variables that are measured on a nominal scale (e.g., injured vs. noninjured). A variety of statistical techniques can help analyze these clinically oriented questions.

Many clinical measures are scored as categorical data (e.g., injured vs. noninjured, dead vs. alive). Similarly, we are often interested in risk factors that are categorical (e.g., smokers vs. nonsmokers). In this chapter, we introduce some common clinical measures of association. We refer to these as clinical because they are most typically used in clinical disciplines. In kinesiology, athletic training, physical therapy, and clinical exercise physiology are clinical disciplines in which these techniques are often applicable. We use the term association to denote the idea that the procedures quantify the degree of relationship between the independent and dependent variables. Here, we limit our discussion to situations that can be cast into what is referred to as a 2 × 2 contingency table (see table 17.1). The data in each cell are simply a count of the number of cases that are in each cell. For consistency, we follow the convention where the first vertical column contains cases in which the response is present (e.g., cases who are injured) and the second vertical column contains cases in which the response is absent (cases who are noninjured). Similarly, the top horizontal row contains cases in which the exposure is present (e.g., performs warm-up exercises) and the bottom horizontal row contains cases in which the exposure is absent (performs no warm-up exercises). Cell A (top left) then contains cases in which both the exposure and the response are present (e.g., injured subjects who warmed up), cell B (top right) contains cases in which the exposure is present but the response is absent, cell C (lower left) contains cases in which the exposure is absent but the response is present, and cell D (lower right) contains cases in which the exposure and response are both absent.

Relative Risk

Relative risk (RR), also known as the risk ratio, is a ratio of proportions. Specifically, it is a ratio of the rate of exposure in the individuals who have the condition (or response) divided by the rate of exposure in the individuals who do not have the condition. Consider the example in table 17.2, in which an experimental warm-up program is examined to determine whether it could decrease the risk of knee injuries. Subjects were randomized to either the warm-up group or the control group. Here, the warm-up program is the exposure (the independent variable) and knee injury is the condition (the dependent variable). If the warm-up program is

TABLE 17.1

Layout for a 2 × 2 Contingency Table

	Response present	Response absent
Exposure present	A: exposure present, response present	B: exposure present, response absent
Exposure absent	C: exposure absent, response present	D: exposure absent, response absent

TABLE 17.2

Relative Risk Example

	Knee injured	Knee not injued	Totals
Performed warm-up	A: 14	B: 86	100
Performed no warm-up	C: <u>27</u>	D: <u>73</u>	<u>100</u>
Totals	41	159	200

Example inspired by Olsen and colleagues (2005). Data in table are fabricated.

effective, we would expect the rate of knee injuries to be lower in those who do the warm-up (those exposed to the warm-up) than is those in the control group (those who were not exposed to the warm-up). (It may be that the warm-up is dangerous and therefore makes knee injuries more likely, in which case we would expect the rate of knee injury to be higher in the warm-up subjects than in the controls.) Conversely, if the warm-up is ineffective, the rate of knee injuries should be about the same in the exposed and nonexposed groups.

If the rate of knee injuries is about the same in the two groups, then the relative risk should be around 1.0. This means that the null hypothesis value for relative risk is 1.0, and we need to statistically assess how different from 1.0 our sample data have to be for us to be confident that the exposure influences the rate of the condition. In practice, the formal null hypothesis significance test is not typically performed on relative risk calculations; instead, confidence intervals are constructed about the calculated relative risk. If the 95% CI does not include 1.0, that is equivalent to rejecting H_0 at an α level of .05.

The relative risk value is an effect size value. If the relative risk is 3.0, then we are estimating that the rate of condition is three times higher in the exposed group than in the unexposed group and, therefore, the exposure is associated with increased risk. If the relative risk is 0.33, then we are estimating that the rate of the condition is about one-third of that in the unexposed group and, therefore, the exposure is associated with decreased risk.

In the example in table 17.2, 100 subjects were randomized to a structured program in which they warmed up before practice (the exposure), and 100 did not warm up. Of those in the exposed group, 14 experienced a knee injury (cell A) and 86 did not experience a knee injury (cell B). In the unexposed group, 27 experienced a knee injury (cell C) and 73 did not experience a knee injury (cell D). To calculate the numerator for relative risk, we determine the proportion of knee injuries in the exposed group, which in this example is 14/100 = 0.14 (14% of the subjects in the warm-up group experienced a knee injury). This is simply the ratio of the number of subjects in the exposed group with a knee injury (cell A) divided by the total number of subjects in the exposed group (cell A + cell B), so that numerator for relative risk is A/(A + B). Similarly, to calculate the denominator of relative risk, we calculate the proportion of knee injuries in the unexposed group. Here, this is equal to 27/73 = 0.27 (27% of subjects in the unexposed group experienced a knee injury), which is the ratio of the number of subjects in the unexposed group with a knee injury (cell C) divided by the total number of subjects in the unexposed group (cell C + cell D). Therefore the denominator of relative risk is C/(C + D).

Collectively, the equation for relative risk is

$$RR = [A/(A + B)]/[C/(C + D)], \tag{17.01}$$

which for the data in table 17.2 is $[14/(14 + 86)]/[27/(27 + 73)] = 0.14/0.27 = 0.519$ ~ 0.52. What does a relative risk of 0.52 tell us? One way to think of this value is that the warm-up program was associated with a reduction of risk of approximately 48%. Another is that the warm-up program was associated with about 52% of the risk of the program in which subjects did not warm up.

The calculation of the confidence interval is not shown here because it a bit more involved than confidence intervals based on the normal distribution, but statistical software calculates the confidence interval automatically. From the data in table 17.2, the 95% CI is 0.29 to 0.93. Because the interval excludes 1.0, this is tantamount to rejecting the null hypothesis at $\alpha = .05$. Our best estimate of the relative risk is 0.52, and we are 95% confident that the true relative risk is somewhere between 0.29 and 0.93. Collectively, we would conclude that the warm-up program appears to reduce the risk of injury by somewhere between 7% and 71%. Note that the cause–effect inference stems from the design of the study (an experiment) and not from the statistical outcome per se.

From the data in table 17.2, we can calculate two other indices. First is absolute risk reduction (ARR; also known as attributable risk). With absolute risk reduction, instead of making a ratio we take the difference between the proportion of injured subjects in the exposed group and the proportion of injured subjects in the unexposed group. In equations form,

$$ARR = [A/(A + B)] - [C/(C + D)]. \tag{17.02}$$

We simply take the difference between the numerator and denominator of the relative risk calculation. From the example data, ARR = 0.14 − 0.27 = −0.13. We can interpret this as meaning that about 13 injuries will be prevented for every 100 individuals who perform the warm-up program. The other index we can calculate is the more commonly reported value of the number needed to treat (NNT). The number needed to treat is simply the inverse of the absolute risk reduction. Here we ignore the sign and NNT = 1/0.13 = 7.69. A number needed to treat value of 7.69 indicates that we need to treat about 7.7 individuals with a warm-up program to prevent one injury.

Odds Ratio

The **odds ratio** (OR), also known as the relative odds, is an index that is similar to relative risk. However, instead of calculating a ratio of proportions as in relative risk, with the odds ratio we calculate a ratio of odds. Remember from chapter 6 (equation 6.03, p. 78) that odds are mathematically defined as

$$\text{odds} = p/(1 - p),$$

where p is the probability of the outcome in question. The probability of the flip of a fair coin turning up heads is $1/2 = .5$. The odds then of a coin flip turning up heads is $.5/(1 - .5) = 1.0$, or more commonly we would say that the odds are 1/1 or one to one. The odds should be equal to the total number of possible outcomes.

Why calculate the odds ratio instead of relative risk? The odds ratio is needed in a specific research design called the case control study. A case control study is a common design in epidemiological research and is typically used in the early stages of an area of inquiry. In a case control study, the researcher recruits subjects who have the condition under study (these are the cases) and subjects who do not have the condition (the controls). The researcher then looks retrospectively (back in time) to see which subjects experienced the exposure.

For example, for the data in table 17.3 the researchers enrolled 100 individuals who have Achilles tendon pathology (the cases) and 100 subjects who do not have Achilles tendon pathology (the controls). Note that the ratio of cases to controls (50% of all subjects) does not reflect the incidence of Achilles tendon pathology in

TABLE 17.3

Odds Ratio Example

	Achilles tendon pathology	Controls	Totals
Polymorphism present	A: 20	B: 40	60
Polymorphism absent	C: 80	D: 60	140
Totals	100	100	200

Example inspired by Mokone and colleagues (2006). Data in table are fabricated.

the population at large, so the relative risk would result in an inaccurate assessment of risk. In this example, the exposure is the presence of a specific polymorphism in a gene that encodes for a collagen protein. The hypothesis was that individuals with Achilles tendon pathology would be less likely to express the polymorphism.

We interpret the odds ratio in a similar manner to the relative risk. The null hypothesis value is OR = 1.0 because an odds ratio of 1.0 indicates that the odds of exposure are the same in both groups. An odds ratio less than 1.0 suggests that the exposure is protective because the controls have higher odds of being exposed. An odds ratio greater than 1.0 suggests the exposure is deleterious because the cases have higher odds of experiencing it. The construction of a confidence interval provides information on the precision to the estimate of the odds ratio. If $\alpha = .05$, then construction of the 95% CI will allow the researcher to determine whether H_0 should be rejected based on whether 1.0 is inside (accept H_0) or outside (reject H_0) the confidence interval.

To calculate the odds ratio, we can cast the data as we have done for relative risk. Now, however, we want to calculate the odds that those who have the condition were exposed and the odds that those who do not have the condition were exposed. That is, what are the odds of exposure in the two groups? The odds of exposure for the cases are A/C = 20/80 = 0.25 and the odds of exposure for the controls are B/D = 40/60 = 0.67. The odds ratio is the ratio of these odds so that

$$OR = (A/C)/(B/D). \tag{17.03}$$

For the example data, OR = 0.25/0.67 = 0.375. As with relative risk, we do not go into detail on the calculation of the confidence interval; statistical software will perform the calculation for you. For the data in table 17.3, the 95% CI is 0.20 to 0.71. Because the 95% CI does not include 1.0, this is tantamount to rejecting the null hypothesis at $\alpha = .05$. Therefore, the odds of a case having the polymorphism are about one-half the odds that a control subject will have the polymorphism. This suggests that the polymorphism is protective because the controls have higher odds of expressing it than do the cases.

Diagnostic Testing

Diagnostic tests result in binary outcomes. A variety of diagnostic tests are associated with kinesiology. For example, clinicians may perform a graded exercise test to diagnose coronary artery disease. Athletic trainers may perform the Lachman test to determine whether an athlete has a torn anterior cruciate ligament (ACL). To assess the validity of diagnostic tests, we can cast the data once again into a 2×2 contingency table. The typical layout for a diagnostic test is illustrated in table 17.4. In cell A is the count of individuals who do have a torn ACL and have a positive Lachman test; these are the true positives. In cell B is the count of individuals who do not have a torn ACL but have a positive Lachman test; these are

TABLE 17.4

Layout of a 2 × 2 Contingency Table for Diagnostic Tests

	Condition present	Condition absent
Test positive	A: true positive	B: false positive
Test negative	C: false negative	D: true negative

false positives. In cell C is the count of individuals who do have a torn ACL but have a negative Lachman test; these are the false negatives. In cell D is the count of individuals who do not have a torn ACL and who have a negative Lachman test; these are the true negatives.

Sensitivity and Specificity

Sensitivity and specificity are the two most commonly reported indices of diagnostic accuracy. **Sensitivity** is the true positive rate and is an index that quantifies how well a diagnostic test identifies those with the condition. **Specificity** is the true negative rate and is an index of how well a diagnostic test identifies those without the condition.

To calculate sensitivity, the true positive rate, we create a ratio of the true positives in the numerator (cell A) and the total number of individuals with the condition in the denominator (A + C) so that

$$\text{sensitivity} = A/(A + C). \tag{17.04}$$

For the data in table 17.5 the true positive count is 80 individuals with a torn ACL and who were correctly identified as having a torn ACL by the Lachman test. The total number of individuals with a torn ACL is 100. The sensitivity then is 80/100 = .80. Therefore, of all the individuals with a torn ACL, the Lachman test correctly identified 80% of them as having a torn ACL.

TABLE 17.5

Diagnostic Validity of Lachman Test

	Torn ACL	Intact ACL	Totals
Positive Lachman	A: 80	B: 45	125
Negative Lachman	C: 20	D: 55	75
Totals	100	100	200

Example inspired by Cooperman, Riddle, and Rothstein (1990). Data in table are fabricated.

To calculate specificity, the true negative rate, we create a ratio with the true negatives in the numerator (cell D) and the total number of individuals who do not have the condition in the denominator (B + D) so that

$$\text{specificity} = D/(B + D). \qquad (17.05)$$

For the data in table 17.5, the true negative count is 55 individuals who do not have a torn ACL and who were correctly identified as such by the Lachman test. The total number of individuals who do not have a torn ACL is 100. The specificity then is 55/100 = .55. Therefore, of all the individuals who do not have a torn ACL, the Lachman test correctly identified 55% of them as not having torn the ACL.

Positive and Negative Predictive Values

Unfortunately, sensitivity and specificity have limited usefulness for practitioners in terms of making probabilistic statements about a given individual. Specifically, they are the reverse of the clinical decision making progress (Fritz and Wainner, 2001). Sensitivity and specificity start from knowing the diagnosis and then involve calculating the probability of a positive and negative test. In practice, a clinician will know the outcome of a diagnostic test (e.g., Lachman test) and will then want to know the probability of having the condition (e.g., torn ACL). For example, imagine that an athletic trainer performs the Lachman test on an injured athlete and the test is positive. Because the test is imperfect, the trainer does not know with certainty whether the athlete truly has a torn ACL. The Lachman test may have been correct or it may have given a false positive. Similarly, if the Lachman test is negative, the test may have been correct and the ACL is intact, or the test may have given a false negative. To help practitioners make probabilistic statements about individuals, instead of using sensitivity and specificity, we use what are referred to as positive and negative predictive values.

The **positive predictive value** is the proportion of individuals with a positive test who really do have the condition. Compare this with sensitivity, which is the proportion of individuals with the condition who have a positive test. Mathematically,

$$\text{positive predictive value} = A/(A + B). \qquad (17.06)$$

Notice that the positive predictive value has the same numerator as sensitivity. The difference is the denominator. For positive predictive value, the denominator is the count of all individuals with a positive test (not the count of all individuals with the condition, as in sensitivity). For the data in table 17.5, the total number of individuals correctly identified by the Lachman test as having a torn ACL is 80. The total number of positive Lachman tests is 125. The resulting positive predictive value is then 80/125 ~ .64. Therefore, about 64% of the individuals who have a positive Lachman test really do have a torn ACL. This allows you to make a probabilistic statement about a person being evaluated. If that person has a positive Lachman test, the probability that that person has a torn ACL is about 64%.

TABLE 17.6

Diagnostic Validity of Lachman Test
With Altered Prevalence

	Torn ACL	Intact ACL	Totals
Positive Lachman	A: 240	B: 45	285
Negative Lachman	C: 60	D: 55	115
Totals	300	100	400

The **negative predictive value** is the proportion of individuals with a negative test who really do not have the condition. In contrast, specificity is the proportion of individuals who do not have the condition who have a negative test. Mathematically,

$$\text{negative predictive value} = C/(C + D). \tag{17.07}$$

Negative predictive value has the same numerator as specificity but the denominator is different. For negative predictive value, the denominator is the count of all individuals with a negative test (not the count of all individuals without the condition, as in specificity). For the data in table 17.5, the total number of individuals correctly identified by the Lachman test as not having a torn ACL is 55. The total number of negative Lachman tests is 75. The resulting negative predictive value is 55/75 ~ .73. In other words, about 73% of individuals with a negative Lachman test do not have a torn ACL. For an individual with a suspected ACL tear, if the subsequent Lachman test is negative, then that person has about a 73% probability of not having a torn ACL.

Both positive and negative predictive values are influenced by the prevalence of the condition (sensitivity and specificity are not). To illustrate, the data in table 17.6 have been modified by tripling the values for cells A and C so that the number of torn ACL is about twice as large as the number of ACL that are not torn. If we recalculate sensitivity (240/300 = .80) and specificity (55/100 = .55), we get the same values as before. However, the positive predictive value has increased from 64% to about 84% (240/285 ~ .84). The negative predictive value has decreased from about 73% to about 48% (55/115 ~ .478). In general terms, a higher prevalence increases the positive predictive value and decreases the negative predictive value. In other words, "…the more common a disease is, the more likely it is that a positive test result is right and a negative test is wrong" (Whiting et al., 2008). Decreasing the prevalence decreases the positive predictive value and increases the negative predictive value.

Positive and Negative Likelihood Ratios

Recently, likelihood ratios have become increasingly used in diagnostic assessment. We can calculate both a positive likelihood ratio and a negative likelihood ratio.

Likelihood ratios have come into favor because, like predictive values, the direction of clinical reasoning is consistent with their practical use (i.e., test outcome to statement about individual) but is not sensitive to prevalence as are predictive values.

The **positive likelihood ratio** is a ratio of probabilities in which the numerator is the probability of a positive test in a person with the condition and the denominator is the probability of a positive test in a person without the condition. Mathematically,

$$\text{positive likelihood ratio} = [A/(A + C)]/[B/(B + D)]. \qquad (17.08)$$

The numerator is equal to the sensitivity and the denominator is equal to 1 minus specificity. Large positive likelihood ratios are indicative of a test that effectively rules in the condition. For the data in table 17.5, the likelihood ratio is $[80/(80 + 20)]/[45/(45 + 55)] \sim 1.78$. Therefore, individuals with a torn ACL are 1.78 times more likely to have a positive Lachman test than are individuals who do not have a torn ACL. For comparison purposes, positive likelihood ratios greater than 10 are considered strong evidence that the condition is present.

The **negative likelihood ratio** is a ratio of probabilities in which the numerator is the probability of a negative test in an individual with the condition and the denominator is the probability of a negative test in an individual without the condition. Mathematically,

$$\text{negative likelihood ratio} = [C/(A + C)]/[D/(B + D)]. \qquad (17.09)$$

The numerator is equal to 1 minus sensitivity and the denominator equals specificity. Small negative likelihood ratios are indicative of a test that effectively rules out the condition. For the data in table 17.5, the negative likelihood ratio is $[20/(50 + 20)]/[55/(45 + 55)] \sim 0.36$. Therefore, individuals with a torn ACL are about one-third as likely to have a negative Lachman test as are those who do not have a torn ACL.

The use of likelihood ratios allows clinicians to make probabilistic statements about an individual's test results in manner that is different than statements that can be made with predictive values. Specifically, if the pretest probability of the condition is known, then based on the test result (positive or negative) and the applicable likelihood ratio, a posttest probability of the condition can be estimated. This is a form of Bayesian analysis. The trick is to know the pretest probability, which may be gleaned from the research literature. As an example, we use data for the Lachman test from a recent meta-analysis by Benjaminse and colleagues (2006). They report that the positive likelihood ratio for the Lachman test is 9.4 and the negative likelihood ratio is 0.1. (Note that these are markedly different than the values used in our example.) Further, the pretest probability for a torn ACL is .44. That is, if an individual experiences a knee injury that is visually consistent with a torn ACL but no testing has yet been conducted, then the probability that the person truly does have a torn ACL is .44.

Imagine that after landing from a jump, an athlete falls to the ground and grabs her right knee. We can estimate the pretest probability of having a torn ACL to be .44. The athletic trainer performs a Lachman test and the test is positive. If the positive likelihood ratio is 9.4, what is the posttest probability of a torn ACL?

The basic equation to make this estimate is

$$\text{posttest odds} = \text{pretest odds} \times \text{likelihood ratio.} \qquad (17.10)$$

The first step is to convert the pretest probability to pretest odds. Because odds $= p/(1 - p)$, then the pretest odds for our example are $.44/(1 - .44) = 0.786$. If we assume a positive likelihood ratio of 9.4, then the posttest odds $= 0.786 \times 9.4 = 7.39$. That is, for every one injured athlete with a positive Lachman test that does not have a torn ACL, about 7.4 injured athletes with a positive Lachman test do have a torn ACL. To convert the posttest odds to a posttest probability, we can use the following equation:

$$p = \text{odds}/(1 + \text{odds}). \qquad (17.11)$$

For our example, $p = 7.39/(1 + 7.39) = .88$. The information added by the positive Lachman test has increased the estimated probability from .44 to .88.

Now assume the injured athlete had a negative Lachman test. Once again, the pretest probability of a torn ACL is .44 and therefore the pretest odds are 0.786. The negative likelihood ratio is 0.10, so the posttest odds are $0.786 \times 0.10 = 0.0786$ and the posttest probability is $0.0786/(1 + 0.0786) = .073$. Therefore, the posttest probability of a torn ACL in an injured athlete with a negative Lachman test is 7.3%.

Summary

Common indices of clinical association often have binary outcome measures (e.g., response present or not present) and can often be cast into 2×2 contingency tables. Similarly, diagnostic tests have binary outcomes: condition present or not present. The efficacy of a diagnostic test can be characterized using (a) sensitivity and specificity, (b) positive and negative predictive values, and (c) positive and negative likelihood ratios. For clinicians, predictive values and likelihood ratios allow for probabilistic estimates for individual athletes and patients.

Problems to Solve

1. An athletic trainer wishes to study the effectiveness of knee braces in preventing tears of the ACL in college football linemen. Over several years, linemen are randomized into two groups. One group wears the knee brace and the other group (control) wears only the standard protective gear. The linemen are categorized as to whether they experienced a torn ACL. The results of the study are as follows.

	Injured ACL	No injury to ACL	
Knee brace	A = 5	B = 40	45
No knee brace	C = 20	D = 40	60
	25	80	Total = 105

    a. Calculate and interpret the relative risk.
    b. Calculate and interpret the attributable risk.
    c. Calculate and interpret the number needed to treat.

2. A different investigator also examined the relationship between knee brace use and ACL tears. However, a case control study design was used. This study enrolled 50 linemen who had an ACL tear (the cases) and another 50 linemen who did not have an ACL tear (the controls). The investigator then retrospectively determined which subjects had worn a knee brace and which had not. The results showed that 21 of the cases had worn a knee brace and 37 of the controls had worn a knee brace. Cast the data into a 2 × 2 contingency table and calculate the odds ratio.

3. An investigator studies the diagnostic efficacy of a new field test to diagnose a tear of the labrum. The gold standard for the diagnosis of a labral tear is an arthroscopic examination. The performance of the field test relative to the arthroscopic examination is summarized in the following table.

	Positive labral tear	Negative labral tear	
Positive field test	A = 112	B = 27	139
Negative field test	C = 22	D = 154	176
	134	181	Total = 315

Calculate the following:
    a. Sensitivity
    b. Specificity
    c. Positive predictive value
    d. Negative predictive value
    e. Positive likelihood ratio
    f. Negative likelihood ratio

See appendix C for answers to problems.

Key Words

negative likelihood ratio
negative predictive value
odds ratio
positive likelihood ratio

positive predictive value
sensitivity
specificity
relative risk

$$H = \left[\frac{12}{18(18+1)} \right]\left[\frac{82.5^2}{6} + \frac{54.5^2}{6} + \frac{34.0^2}{6} \right] - 3(18+1)$$

$$H = [.035][1822.09] - 57$$

$$H = 6.77$$

Advanced Statistical Procedures

Discrepancies between studies in the research literature may be due to insufficient statistical power in some research studies. Meta-analysis is a procedure used to systematically review the literature and statistically combine the effect sizes of multiple studies into one average effect size that reflects the influence of all the subjects in the combined studies.

Researchers often want to analyze more than one dependent variable, yet performing multiple ANOVAs on many dependent variables invites a type I error. Multiple analyses of variance (MANOVA) can help solve this problem.

What are the factors that make up a complex motor skill? What are the components of physical fitness? Factor analysis can identify the factors that are inherent in a general concept, such as skill or fitness. How can we classify subjects into groups based on common characteristics? Discriminant analysis allows us to do this. These advanced statistical techniques are discussed in this chapter on an introductory basis.

Meta-Analysis

Meta-analysis is a procedure that allows an investigator to statistically combine the results of multiple studies. Here we present the essential concepts of the statistical approach to meta-analysis, but no specific calculations or examples are shown. For a more complete discussion see, for example, by Lipsey and Wilson (2001).

Meta-analysis is intimately tied to the process of a systematic review. Historically, literature reviews have been narrative reviews. In a narrative review, the author(s) are experts who interpret the research literature concerning a specific topic. However, the process is not systematic because the studies reviewed are chosen by the author(s) and may reflect their bias.

In a systematic review, the investigators follow a specific plan to identify all the relevant studies to be included in the review. Key words to be used and the databases to be searched (e.g., Sport Discus, Pubmed) are identified beforehand. Similarly, specific inclusion and exclusion criteria for which studies to include in the review are decided before the literature review begins. The investigators may explicitly attempt to include unpublished studies that may be relevant. This helps to correct for "file drawer effect" (also known as publication bias), which refers to the phenomenon in which studies that show a statistically significant treatment effect are more likely to be published than studies that fail to show a statistically significant effect.

Meta-analysis extends the process of a systematic review to include a pooled statistical analysis. Recall from chapter 7 that statistical power is directly related to sample size. By pooling the results of multiple studies, the effective sample size increases. This increases statistical power and narrows confidence intervals. In fact, by pooling the results it is possible that individual studies that fail to show a statistically significant effect could collectively show a significant effect. This may help investigators explain discrepancies in the research literature. For example, some studies may show that a treatment works and others studies may show that it does not. These discrepancies may be due to some studies having too little statistical power. Further, by pooling the results from multiple studies, we can get a better (more accurate) estimate of the effect of an independent variable.

The essential statistical pieces needed to understand meta-analysis are presented in earlier chapters. First, recall the idea of effect size. For example, one effect size index was presented in equation 10.18 on page 165:

$$ES = \frac{\bar{X}_1 - \bar{X}_2}{SD_{control}},$$

where the numerator reflects the differences between mean values and the denominator reflects the standard deviation of the control group or, alternatively, an estimate of a pooled standard deviation. The process of meta-analysis takes the relevant information from the included studies and calculates an effect size for each study. For our purposes, we focus on an effect size based on mean differences and use the situation in which we compare a treatment group with a control group. However, a variety of effect size indices exist. These include the Pearson r, proportions, and odds ratios. The essence of the process is to create a pooled effect size that reflects the best estimate of the population effect size and to create a confidence interval around the pooled effect size.

In order to pool the effect sizes from different studies, the meta-analyst must weight the effect sizes from the studies. The studies included in the meta-analysis will very likely have different sample sizes, and it would be inappropriate to treat a study with, for example, 20 subjects the same as a study with 200 subjects because, if all else is equal, the larger study carries more information and is likely a closer approximation to results we would see in the population. The weights that are used in meta-analysis are calculated based on the standard errors of the individual effect sizes of each study. Remember that standard errors are influenced by the sample size, so studies that have larger sample sizes have larger weights.

Once the effect sizes and weights are calculated, a mean effect size and a confidence interval about the mean effect size can be calculated. The mean effect size is a weighted mean of the effect sizes from each study. The confidence interval about the mean effect size is based on the Z distribution and the standard error of the mean effect size.

If the meta-analysis compares mean differences between experimental and control groups, then the null hypothesis value is zero. That is, if no effect exists of the independent variable on the dependent variable, then the mean effect size should be zero. But as with all inferential statistical tests, sampling error would likely cause the calculated mean effect size to be some non-zero value even if the null hypothesis is true. Imagine that a meta-analyst wishes to examine the effect of creatine supplementation on anaerobic power as determined from the Wingate test. If the mean effect size is 0.50 and the 95% CI is 0.20 to 0.80, then we would interpret this result as indicating that creatine supplementation improves anaerobic power by about one-half of a standard deviation. We could state that the effect is statistically significant at $p < .05$ because the 95% CI excludes zero. Further, we are 95% confident that the true effect in the population is somewhere between two-tenths and eight-tenths of a standard deviation.

Assumptions and Cautions

The statistical treatment of meta-analysis is reasonably straightforward. The primary concerns about meta-analysis center around how well the review process has winnowed the various studies into the pool of studies to be analyzed. Meta-analysis is open to the criticism of "garbage in, garbage out." That is, the inclusion of studies that are poorly designed or executed may bias the meta-analysis. Further, as noted before, publication bias may influence the studies that are readily available for analysis. A vast body of methodological literature exists to help researchers improve the meta-analysis process and detect and correct for potential bias. Despite these methodological difficulties, meta-analysis is an important tool in evidence-based practice and tens of thousands of meta-analyses have been published.

Multiple Analysis of Variance

In the *t* test, simple ANOVA, factorial ANOVA, and ANCOVA, only one dependent variable is considered. For example, the effect of dietary fat levels (independent variable) on body fat (dependent variable) may be studied. This design could be analyzed with a *t* test (comparing two groups), with simple ANOVA (comparing three or more groups), with a 3×2 factorial ANOVA (comparing three or more groups with an additional factor of activity level; e.g., active vs. inactive), or ANCOVA (comparing two or more groups whose mean values are adjusted to account for the effects of a covariate, such as blood cholesterol levels).

Each of these research designs may be expanded to determine the simultaneous effect of treatment on more than one dependent variable. For example, the effects of dietary fat levels on fat mass, lean body mass, body mass index, and aerobic capacity may be of interest. In this case, multiple (four) dependent variables are analyzed simultaneously using **multiple ANOVA** (MANOVA; also known as multivariate ANOVA).

MANOVA is useful for several reasons:

1. It helps to protect against type I errors. In MANOVA, a new dependent variable is formed from the linear composite of the several dependent variables using regression techniques. This new virtual dependent variable is then analyzed with ANOVA to determine whether differences exist among the treatment groups. This is sometimes referred to as the omnibus *F* value. If omnibus *F* is not significant, the analysis stops.

If differences are found for the newly formed dependent variable, further analysis of the original dependent variables using ANOVA may be justified. If ANOVA is significant for any of the dependent variables, post hoc tests to identify group mean differences may be conducted. This step-by-step process helps to reduce the familywise error rate (discussed in chapter 11).

MANOVA adds one more step to the step-down process previously described for ANOVA (see chapters 11 and 14). At each step, a nonsignificant result usually terminates the analysis. If omnibus MANOVA is significant, this may indicate that one or more of the several dependent variables differs among the treatment groups. A univariate ANOVA is then performed on each of the separate dependent variables to determine which one(s) contain the effect. If univariate ANOVA is not significant for a given dependent variable, the analysis stops for that dependent variable. When univariate ANOVA is significant for one of the dependent variables, post hoc tests may be conducted to identify individual group differences.

Wagoner (1994), however, cautions against this procedure. If the researcher wishes to analyze all of the dependent variables with MANOVA, Wagoner suggests a Bonferroni adjustment to the original α (see chapter 11). He asserts that MANOVA is most useful for determining the underlying factors, which may be represented by two or more of the dependent variables. To identify these factors, he recommends that MANOVA be followed by a descriptive discriminant analysis. The choice of whether to use ANOVA or MANOVA thus depends on the research question being asked. If the underlying constructs (factors) among the dependent variables are of interest, use MANOVA. If each of the dependent variables is of autonomous interest, use ANOVA with a Bonferroni adjustment.

2. With more than one dependent variable, MANOVA offers a greater chance of determining the effects of treatment. The independent variable may affect one dependent variable but not another. Researchers usually collect data on several dependent variables while the subjects are being tested. This is much more efficient than testing the subjects several times. With the additional data available from several dependent variables, more information about the phenomenon being studied may be analyzed with MANOVA. However, indiscriminate use of multiple dependent variables is not an appropriate technique. Before collecting data, the researcher should plan the dependent variables to be studied based on the logic of the study and the theory behind it. To simply add multiple dependent variables into the analysis in hopes of finding something significant is poor research and is an open invitation to a type I error.

3. Under certain conditions, MANOVA may be more powerful than ANOVA. However, this is usually not the case. MANOVA is frequently less powerful (i.e., more conservative) than ANOVA. When several of the dependent variables are not significant and one is just barely significant, MANOVA is

less powerful. The insignificant dependent variables may mask the effects of the one significant dependent variable producing an insignificant omnibus F for MANOVA. This may cause you to accept the null hypothesis for all of the dependent variables when it is false for at least one (a type II error). This error may be avoided by using Wagoner's suggestion as described earlier.

Although MANOVA seems to be a logical extension of ANOVA, which should be used whenever possible, it may not always be advantageous. Tabachnick and Fidell (1996, p. 376) state that

There are no free lunches in statistics, just as there are none in life. MANOVA is a substantially more complicated analysis than ANOVA. There are several important assumptions to consider, and there is often some ambiguity in interpretation of the effects of IVs on any single DV. Further, the situations in which MANOVA is more powerful than ANOVA are quite limited; often MANOVA is considerably less powerful than ANOVA. Thus, our recommendation is to think very carefully about the need for more than one DV in light of the added complexity and ambiguity of analysis.

Assumptions and Cautions

The assumptions that underlie the t test, ANOVA, and repeated measures designs also apply to MANOVA. As in any design, MANOVA may be generalized to a population only if the subjects in the study have been randomly selected from that population. Other assumptions and cautions must be observed when MANOVA is used.

Common Sense

The finding of a significant omnibus F value for MANOVA (as with any other statistical computation) does not ensure causality. Causality is a logical determination based on review of literature, proper research design, control of all extraneous variables, and finally, confirmation by statistical analysis. Beginning researchers tend to accept whatever comes from the computer without serious consideration of what went in. This is especially true of very complicated processes like MANOVA. Before the results of the MANOVA are confirmed, a review of the conditions that prompted the research and the design of the study must be conducted to determine that the results make sense from both logical and a discipline-based perspectives.

Relationships Among Dependent Variables

MANOVA uses multiple regression techniques to create a new dependent variable from a combination of the several dependent variables in the study. As was discussed in chapter 9, multiple regression is most effective if the variables in the prediction are not related. In MANOVA, highly related dependent variables are not useful because they essentially measure the same variance twice. For the most effective use of MANOVA, select dependent variables that are all pertinent to the research design but that measure independent factors of interest. Because the number of dependent variables that can be used is limited, it is not wise to waste degrees of freedom on two or more related variables, only one of which contributes significantly to the analysis.

Ratio of Subjects to Dependent Variable

To maintain an appropriate level of degrees of freedom, there must be more subjects per group than dependent variables. Some statisticians feel that the ratio of subjects per group to dependent variables should be at least three to one. As the ratio approaches one to one, the power of MANOVA becomes severely limited. In addition, with only one or two subjects per dependent variable, the assumption of sphericity in repeated measures designs (homogeneity of variance–covariance) is likely to be rejected (Tabachnick and Fidell, 1996, p. 413). This is a major concern in MANOVA. When univariate analysis is used, one may find studies with two or three subjects per group. This may be acceptable (if there is sufficient power to conduct the analysis) when only one dependent variable is analyzed. But in MANOVA, with two or more dependent variables, more subjects are required per group.

If you are faced with a low ratio of subjects per group to dependent variables (with no way to measure more subjects or reduce dependent variables), do not use MANOVA. Instead, follow the suggestion of Wagoner (1994) and perform ANOVA on each dependent variable, then adjust the α level with a Bonferroni adjustment based on the number of ANOVAs performed. This is a reasonable solution to the problem of low ratio of subjects per group to dependent variables.

Outliers

Outliers, values that are significantly beyond the range of typical scores in the data set, can seriously affect the results of MANOVA (see chapters 1 and 4 for a review of the problem of outliers). Outliers should be identified and corrected or eliminated prior to the analysis. Outliers may be univariate or multivariate. Multivariate outliers are particularly hard to find in a large database but have significant effects on MANOVA.

Outliers may result from measurement or recording errors, or they may represent a case that is not typical of the population from which the samples were drawn. If the outliers are the result of measurement or recording errors, they must be found and corrected. If they result from real data points that lie well beyond the other values in the study, you must evaluate each one to determine whether it is representative of the subjects to be studied. Failure to remove or correct significant outliers may invalidate the results of a MANOVA analysis.

Repeated Measures Designs

When repeated measures are used with multiple dependent variables, the chances of violating the assumptions of sphericity or circularity are dramatically increased. If the Greenhouse-Geisser or Huynh-Feldt adjustments to the p values do not adequately correct for the violation, an alternative solution using MANOVA in a doubly multivariate design is possible. Doubly multivariate means that the usual dependent variables serve as one set of multiple variables and the repeated measures serve as a second set of multiple variables. Under these conditions, the assumptions of sphericity and circularity are not required. See Schutz and Gessaroli (1987) for an in-depth discussion of this procedure.

Interpreting the Results

MANOVA may be applied to any research design where ANOVA is appropriate but where there is more than one dependent variable. The computation of MANOVA is complicated, and MANOVA is always performed on a computer. The student who understands the basics of ANOVA, repeated measures ANOVA, and factorial ANOVA will be able to apply this knowledge to interpret MANOVA.

Factor Analysis

When data are collected on many variables using the same set of subjects, it is possible, indeed probable, that some of the variables are related. For example, if we want to determine the overall motor skill and fitness components of a group of people, we might collect data on the following variables: chin-ups, curl-ups in 1 minute, bench presses, leg presses, 50-yard dash, 440-yard dash, 1-mile run, sit-and-reach performance, skinfolds, softball throw for distance, softball throw for accuracy, standing long jump, vertical jump, hip flexibility with a goniometer, push-ups, step test for aerobic capacity, $\dot{V}O_2$max on a treadmill, body mass index, shuttle run, stork stand for balance, Bass stick test, reaction time, and perhaps many others.

You may immediately think that we don't need to use all of these tests because some of them test the same component of skill or fitness. What you really mean is that correlations exist between or among many of the measured variables. When two items are highly correlated, we can surmise that they essentially measure the same factor. For example, the sit-and-reach test and the hip flexibility with a goniometer test would be expected to be highly related to each other because the sit-and-reach test measures hip and low-back flexibility and the goniometer test measures hip flexibility. The common factor of hip flexibility produces a correlation between the variables. Likewise, bench presses, push-ups, and chin-ups are all measures of upper body strength but in slightly different ways. The 1-mile run and $\dot{V}O_2$max on a treadmill are also highly related in that they both measure factors related to aerobic capacity.

One could discover these relationships by producing an intercorrelation matrix among all the variables and looking for Pearson r values in excess of some predetermined level; for example, ± 80. This may work fairly well with a few variables, but when many variables are measured the intercorrelation matrix grows rapidly and soon becomes too cumbersome to deal with manually.

To resolve this issue, a statistical technique called **factor analysis,** which identifies the common components among multiple dependent variables and labels that component as a factor, has been developed. The factor is derived (virtual); that is, it does not actually exist as a stand-alone measured value. It is simply the result of the common component between or among two or more actually measured variables. The factor has no name until the researcher gives it one. For example, the relationship among bench presses, push-ups, and chin-ups produces a factor that the researcher might label upper body strength.

Notice that we have not actually measured upper body strength; we have measured the number of bench presses, push-ups, and chin-ups a person can do. To determine the upper body strength of the subjects, we do not need to measure all three variables. Because the correlation among the variables is high (e.g., $r > .80$), if a person can do many bench presses, we can predict that he or she will also be able to do many chin-ups and push-ups. Hence, we need to measure only one of the variables to estimate the subject's upper body strength. Using this technique, researchers have identified the most common components of physical fitness: muscular strength, muscular endurance, aerobic capacity, body composition, and flexibility.

The same argument could be used for measures of speed. Perhaps we would find a correlation among the 50-meter dash, the 100-meter dash, and the shuttle run. The common factor here is sprinting speed. However, because each variable has unique characteristics (e.g., the shuttle run also requires the ability to stop and change directions quickly), the correlation among them is not perfect. But a common factor exists among them that we could label sprint speed.

Factor analysis is therefore simply a method of reducing a large data set to its common components. It removes the redundancy among the measured variables and produces a set of derived variables called factors. We address these ideas in our study of multiple regression. To illustrate, examine figure 18.1. Variable Y shares common variance with X_1, X_2, X_3, and X_4 but very little with X_5. Variables X_3 and X_5 are highly correlated (collinear) with each other, so they contain a common factor. Variable X_5 is redundant and does not need to be measured because X_3 can represent it in their common relationship with Y.

Computer programs that perform factor analysis list the original variables that group together into common derived factors. These factors are then given a number called an **eigenvalue.** The eigenvalue indicates the number of original variables that are associated with that factor (Kachigan, 1986, p. 246). If we are dealing with 10 original variables, then each one contributes an average of 10% of the total variance in the problem. If factor 1 had an eigenvalue of 4.26, then that factor would represent the same amount of variance as 4.26 of the original variables. Because each variable contributes an average of 10%, factor 1 would contribute 4.26×10 or 42.6% of the total variance.

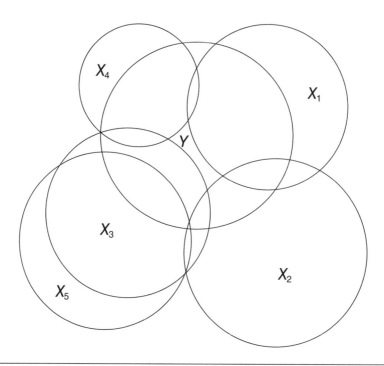

Figure 18.1 Example of shared variance among numerous variables.

The more variables that load on a given factor (i.e., its eigenvalue), the more important that factor is in determining the total variance of the entire set of data. By comparing eigenvalues, we can determine the relative contribution of each factor. This method usually identifies the two or three most important factors in a data set. Because the goal of factor analysis is to reduce the number of variables to be measured by grouping them into factors, we are looking for the smallest number of factors that will adequately explain the data. Generally speaking, when the eigenvalue for a factor decreases below the average variance explained by the original variables (in our example, 10%), it is not used. It is rare to identify more than three or four factors.

After the factors are identified, the computer can perform a procedure called **factor rotation.** This is a mathematical technique in which the factors are theoretically rotated in three-dimensional space in an attempt to maximize the correlation among the original variables and the factor on which they load. It also attempts to reduce the relationships among the factors. After all, if two factors are highly correlated to each other, we do not need both of them. Orthogonal rotation is the method in which the computer attempts to adjust or rotate the factors that have been selected as most representative of the data (i.e., the ones with the highest eigenvalues) so that the correlation among the factors is orthogonal (as close to zero as possible) and the correlation between the variables that load on a factor and the factor is maximized.

Assumptions and Cautions

Sample size is critical in factor analysis. Tabachnick and Fidell (1996, p. 579) suggest that problems may occur with too few cases. Sample size is less critical when the factors are independent and distinct. Variables should be reasonably normal; one should check them with a skewness and kurtosis test. Factor analysis assumes that the correlation among the variables is linear. If any correlation is curvilinear, the analysis will be less effective. Transformations of data can sometimes reduce curvilinearity. Multicollinearity or singularity is usually not a problem because we are actually trying to identify these characteristics of the data. If two variables are perfectly related (singular), both are not needed; one can be eliminated. Finally, outliers must be identified and eliminated or transformed. Outliers reduce the ability of the computer to find the appropriate factors.

Discriminant Analysis

With the *t* test and ANOVA, we test group means to determine whether significant differences exist between or among groups. Discriminant analysis is somewhat the opposite procedure. Using regression equations, **discriminant analysis** attempts to classify subjects into groups (the dependent variables) based on certain characteristics of the subjects (the independent variables). In other words, we attempt to discriminate among the subjects by classifying them into certain groups. For example, we may want to measure subjects using a questionnaire of their lifestyle habits in order to classify them as fit or not fit. To evaluate our group assignments, we could perform a chi-square test to determine whether a nonparametric classification is better than chance, or we could perform a *t* test on an acceptable parametric measure of fitness (e.g., $\dot{V}O_2max$) to determine whether the group means differ significantly. If they do, this would provide evidence to validate the classification produced by the discriminant analysis.

This procedure is similar to regression analysis and is used for basically the same purpose. However, in regression one attempts to predict the value of a continuous dependent variable and in discriminant analysis one tries to predict discrete group membership. Fitness consultants could distinguish who among a group of people should receive certain exercise prescriptions, medical researchers could determine who is most likely to be at risk for certain diseases, and employers could determine who among the applicants for a job would be the most promising employee.

Sometimes the independent variables (subject characteristics) are called predictor variables, and the dependent (grouping) variables are called criterion variables. The criterion variable must have at least two values, but it may have more. For example, to discriminate more precisely among subjects regarding their fitness levels, one might classify them as highly fit, moderately fit, and unfit.

Discriminant analysis is sometimes incorrectly used when regression analysis should be used. If the data are dichotomous or have several categories but are not continuous, discriminant analysis is appropriate. However, if the data are continuous, then regression analysis should be used. If discriminant analysis is inappropriately used on a continuous variable, it will force the dichotomization of the continuous variable (i.e., consider it as nominal data) when it should be analyzed as interval or ratio data.

Assumptions and Cautions

As with other parametric procedures, certain assumptions apply to discriminant analysis. First, it must be assumed that random selection of subjects from the population has taken place and that the distribution of the predictor variables and dependent variables in the population is normal. In addition, the variance of the predictor variables must be the same in the two or more populations into which the subjects will be classified. Second, the correlation between any two predictor variables must be the same in the two populations into which the subjects will be classified (Kachigan, 1986, p. 219). Finally, it is assumed that the relationships among the predictor variables and dependent variables are linear. To improve the results, outliers should be identified and transformed or eliminated before analysis and one should check for multicollinearity and singularity among the predictor variables (Tabachnick and Fidell, 1996, pp. 513–514).

Summary

Because advanced techniques are always performed on a computer, it is not necessary that you understand all of the formulas and mathematical calculations that produce the result. But it is critical that you understand what the answer means, how to interpret it, and the assumptions that underlie it. This chapter discusses four advanced statistical techniques: meta-analysis, MANOVA, factor analysis, and discriminant analysis. You should be aware of their existence, be able to describe the conditions under which they should be applied, and be able to consult statistical software programs and advanced statistical texts to make full use of these procedures.

Key Words

discriminant analysis
eigenvalue
factor analysis

factor rotation
meta-analysis
multiple analysis of variance

Appendix A

Statistical Tables

Area Between the Mean and Z (in Percent) in the Normal Distribution

Z	Area (%)	Z	Area (%)	Z	Area (%)
0.00	0.000	0.35	13.683	0.70	25.803
0.01	0.399	0.36	14.057	0.71	26.114
0.02	0.798	0.37	14.430	0.72	26.423
0.03	1.197	0.38	14.802	0.73	26.730
0.04	1.596	0.39	15.172	0.74	27.034
0.05	1.995	0.40	15.541	0.75	27.337
0.06	2.393	0.41	15.909	0.76	27.637
0.07	2.791	0.42	16.275	0.77	27.935
0.08	3.189	0.43	16.639	0.78	28.230
0.09	3.587	0.44	17.002	0.79	28.523
0.10	3.984	0.45	17.363	0.80	28.814
0.11	4.381	0.46	17.723	0.81	29.103
0.12	4.777	0.47	18.081	0.82	29.389
0.13	5.173	0.48	18.438	0.83	29.673
0.14	5.568	0.49	18.792	0.84	29.955
0.15	5.963	0.50	19.145	0.85	30.234
0.16	6.357	0.51	19.496	0.86	30.511
0.17	6.750	0.52	19.846	0.87	30.785
0.18	7.143	0.53	20.193	0.88	31.057
0.19	7.535	0.54	20.539	0.89	31.327
0.20	7.927	0.55	20.883	0.90	31.594
0.21	8.317	0.56	21.225	0.91	31.859
0.22	8.707	0.57	21.565	0.92	32.122
0.23	9.096	0.58	21.903	0.93	32.382
0.24	9.484	0.59	22.239	0.94	32.640
0.25	9.871	0.60	22.574	0.95	32.895
0.26	10.257	0.61	22.906	0.96	33.148
0.27	10.642	0.62	23.236	0.97	33.398
0.28	11.026	0.63	23.564	0.98	33.646
0.29	11.409	0.64	23.890	0.99	33.892
0.30	11.791	0.65	24.214	1.00	34.135
0.31	12.172	0.66	24.536		
0.32	12.551	0.67	24.856		
0.33	12.930	0.68	25.174		
0.34	13.307	0.69	25.490		

Z	Area (%)	Z	Area (%)	Z	Area (%)
1.01	34.376	1.36	41.309	1.71	45.637
1.02	34.614	1.37	41.467	1.72	45.728
1.03	34.850	1.38	41.622	1.73	45.818
1.04	35.084	1.39	41.774	1.74	45.907
1.05	35.315	1.40	41.925	1.75	45.994
1.06	35.544	1.41	42.074	1.76	46.079
1.07	35.770	1.42	42.220	1.77	46.163
1.08	35.994	1.43	42.365	1.78	46.246
1.09	36.215	1.44	42.507	1.79	46.327
1.10	36.434	1.45	42.648	1.80	46.407
1.11	36.651	1.46	42.786	1.81	46.485
1.12	36.865	1.47	42.923	1.82	46.562
1.13	37.077	1.48	43.057	1.83	46.637
1.14	37.287	1.49	43.189	1.84	46.711
1.15	37.494	1.50	43.320	1.85	46.784
1.16	37.699	1.51	43.448	1.86	46.855
1.17	37.901	1.52	43.575	1.87	46.925
1.18	38.101	1.53	43.700	1.88	46.994
1.19	38.299	1.54	43.822	1.89	47.061
1.20	38.494	1.55	43.943	1.90	47.128
1.21	38.687	1.56	44.062	1.91	47.193
1.22	38.878	1.57	44.18	1.92	47.256
1.23	39.066	1.58	44.295	1.93	47.319
1.24	39.252	1.59	44.409	1.94	47.380
1.25	39.436	1.60	44.520	1.95	47.440
1.26	39.618	1.61	44.630	1.96	47.499
1.27	39.797	1.62	44.739	1.97	47.557
1.28	39.974	1.63	44.845	1.98	47.614
1.29	40.148	1.64	44.950	1.99	47.670
1.30	40.321	1.65	45.053	2.00	47.724
1.31	40.491	1.66	45.154		
1.32	40.659	1.67	45.254		
1.33	40.825	1.68	45.352		
1.34	40.989	1.69	45.449		
1.35	41.150	1.70	45.543		

(continued)

Z	Area (%)	Z	Area (%)	Z	Area (%)
2.01	47.778	2.36	49.085	2.71	49.663
2.02	47.830	2.37	49.110	2.72	49.673
2.03	47.881	2.38	49.133	2.73	49.683
2.04	47.932	2.39	49.157	2.74	49.692
2.05	47.981	2.40	49.179	2.75	49.701
2.06	48.029	2.41	49.201	2.76	49.710
2.07	48.076	2.42	49.223	2.77	49.719
2.08	48.123	2.43	49.244	2.78	49.727
2.09	48.168	2.44	49.265	2.79	49.736
2.10	48.213	2.45	49.285	2.81	49.744
2.11	48.256	2.46	49.304	2.81	49.752
2.12	48.299	2.47	49.323	2.82	49.759
2.13	48.340	2.48	49.342	2.83	49.767
2.14	48.381	2.49	49.360	2.84	49.774
2.15	48.421	2.50	49.378	2.85	49.781
2.16	48.460	2.51	49.395	2.86	49.788
2.17	48.499	2.52	49.412	2.87	49.794
2.18	48.536	2.53	49.429	2.88	49.801
2.19	48.573	2.54	49.445	2.89	49.807
2.20	48.609	2.55	49.460	2.91	49.813
2.21	48.644	2.56	49.476	2.91	49.819
2.22	48.678	2.57	49.491	2.92	49.824
2.23	48.712	2.58	49.505	2.93	49.830
2.24	48.744	2.59	49.519	2.94	49.835
2.25	48.776	2.60	49.533	2.95	49.841
2.26	48.808	2.61	49.546	2.96	49.846
2.27	48.839	2.62	49.559	2.97	49.851
2.28	48.869	2.63	49.572	2.98	49.855
2.29	48.898	2.64	49.585	2.99	49.860
2.30	48.926	2.65	49.597	3.00	49.865
2.31	48.954	2.66	49.608	3.10	49.903
2.32	48.982	2.67	49.620	3.20	49.931
2.33	49.009	2.68	49.631	3.30	49.951
2.34	49.035	2.69	49.642	3.40	49.966
2.35	49.060	2.70	49.652	3.50	49.977
				4.00	49.997
				4.50	49.998
				5.00	50.000

Tabled values were generated using LabVIEW programming software.

Critical Values of the Correlation Coefficient (r)

df	$\alpha = .10$	$\alpha = .05$	$\alpha = .025$	$\alpha = .01$
1	0.988	0.997	0.999	1
2	0.900	0.950	0.975	0.990
3	0.805	0.878	0.924	0.959
4	0.729	0.811	0.868	0.917
5	0.669	0.754	0.817	0.875
6	0.621	0.707	0.771	0.834
7	0.582	0.666	0.732	0.798
8	0.549	0.632	0.697	0.765
9	0.521	0.602	0.667	0.735
10	0.497	0.576	0.640	0.708
11	0.476	0.553	0.616	0.684
12	0.458	0.532	0.594	0.661
13	0.441	0.514	0.575	0.641
14	0.426	0.497	0.557	0.623
15	0.412	0.482	0.541	0.606
16	0.400	0.468	0.526	0.590
17	0.389	0.456	0.512	0.575
18	0.378	0.444	0.499	0.561
19	0.369	0.433	0.487	0.549
20	0.360	0.423	0.476	0.537
21	0.352	0.413	0.466	0.526
22	0.344	0.404	0.456	0.515
23	0.337	0.396	0.447	0.505
24	0.330	0.388	0.439	0.496
25	0.323	0.381	0.430	0.487
26	0.317	0.374	0.423	0.479
27	0.312	0.367	0.415	0.471
28	0.306	0.361	0.409	0.463
29	0.301	0.355	0.402	0.456
30	0.296	0.349	0.396	0.449
31	0.291	0.344	0.390	0.442
32	0.287	0.339	0.384	0.436
33	0.283	0.334	0.378	0.430
34	0.279	0.329	0.373	0.424
35	0.275	0.325	0.368	0.418
36	0.271	0.320	0.363	0.413
37	0.267	0.316	0.359	0.408
38	0.264	0.312	0.354	0.403
39	0.260	0.308	0.350	0.398
40	0.257	0.304	0.346	0.393
41	0.254	0.301	0.342	0.389
42	0.251	0.297	0.338	0.384
43	0.248	0.294	0.334	0.380
44	0.246	0.291	0.330	0.376
45	0.243	0.288	0.327	0.372
46	0.240	0.285	0.323	0.368
47	0.238	0.282	0.320	0.365
48	0.235	0.279	0.317	0.361
49	0.233	0.276	0.314	0.358

(continued)

df	$\alpha = .10$	$\alpha = .05$	$\alpha = .025$	$\alpha = .01$
50	0.231	0.273	0.311	0.354
51	0.228	0.271	0.308	0.351
52	0.226	0.268	0.305	0.348
53	0.224	0.266	0.302	0.345
54	0.222	0.263	0.299	0.341
55	0.220	0.261	0.297	0.339
56	0.218	0.259	0.294	0.336
57	0.216	0.256	0.292	0.333
58	0.214	0.254	0.289	0.330
59	0.213	0.252	0.287	0.327
60	0.211	0.250	0.285	0.325
61	0.209	0.248	0.282	0.322
62	0.207	0.246	0.280	0.320
63	0.206	0.244	0.278	0.317
64	0.204	0.242	0.276	0.315
65	0.203	0.240	0.274	0.313
66	0.201	0.239	0.272	0.310
67	0.200	0.237	0.270	0.308
68	0.198	0.235	0.268	0.306
69	0.197	0.234	0.266	0.304
70	0.195	0.232	0.264	0.302
71	0.194	0.230	0.262	0.300
72	0.193	0.229	0.260	0.298
73	0.191	0.227	0.259	0.296
74	0.190	0.226	0.257	0.294
75	0.189	0.224	0.255	0.292
76	0.188	0.223	0.254	0.290
77	0.186	0.221	0.252	0.288
78	0.185	0.220	0.251	0.286
79	0.184	0.219	0.249	0.285
80	0.183	0.217	0.247	0.283
81	0.182	0.216	0.246	0.281
82	0.181	0.215	0.244	0.280
83	0.180	0.213	0.243	0.278
84	0.179	0.212	0.242	0.276
85	0.178	0.211	0.240	0.275
86	0.176	0.210	0.239	0.273
87	0.175	0.208	0.238	0.272
88	0.174	0.207	0.236	0.270
89	0.174	0.206	0.235	0.269
90	0.173	0.205	0.234	0.267
91	0.172	0.204	0.232	0.266
92	0.171	0.203	0.231	0.264
93	0.170	0.202	0.230	0.263
94	0.169	0.201	0.229	0.262
95	0.168	0.200	0.228	0.260
96	0.167	0.199	0.226	0.259
97	0.166	0.198	0.225	0.258
98	0.165	0.197	0.224	0.256
99	0.165	0.196	0.223	0.255
∞	0.016	0.020	0.022	0.026

Tabled values were generated using LabVIEW programming software.

Critical Values of the Student's *t* Distribution
for Two-Tailed Tests

df	Critical *t* (α = .10)	Critical *t* (α = .05)	Critical *t* (α = .025)	Critical *t* (α = .01)
1	6.314	12.706	25.452	63.657
2	2.920	4.303	6.205	9.925
3	2.353	3.183	4.177	5.841
4	2.132	2.776	3.495	4.604
5	2.015	2.571	3.163	4.032
6	1.943	2.447	2.969	3.707
7	1.895	2.365	2.841	3.500
8	1.860	2.306	2.752	3.355
9	1.833	2.262	2.685	3.250
10	1.813	2.228	2.634	3.169
11	1.796	2.201	2.593	3.106
12	1.782	2.179	2.560	3.055
13	1.771	2.160	2.533	3.012
14	1.761	2.145	2.510	2.977
15	1.753	2.131	2.490	2.947
16	1.746	2.120	2.473	2.921
17	1.740	2.110	2.458	2.898
18	1.734	2.101	2.445	2.878
19	1.729	2.093	2.434	2.861
20	1.725	2.086	2.423	2.845
21	1.721	2.080	2.414	2.831
22	1.717	2.074	2.405	2.819
23	1.714	2.069	2.398	2.807
24	1.711	2.064	2.391	2.797
25	1.708	2.060	2.385	2.787
26	1.706	2.056	2.379	2.779
27	1.703	2.052	2.373	2.771
28	1.701	2.048	2.368	2.763
29	1.699	2.045	2.364	2.756
30	1.697	2.042	2.360	2.750
31	1.696	2.040	2.356	2.744
32	1.694	2.037	2.352	2.739
33	1.692	2.035	2.348	2.733
34	1.691	2.032	2.345	2.728
35	1.690	2.030	2.342	2.724
36	1.688	2.028	2.339	2.720
37	1.687	2.026	2.336	2.715
38	1.686	2.024	2.334	2.712
39	1.685	2.023	2.331	2.708
40	1.684	2.021	2.329	2.704
41	1.683	2.020	2.327	2.701
42	1.682	2.018	2.325	2.698
43	1.681	2.017	2.323	2.695
44	1.680	2.015	2.321	2.692
45	1.679	2.014	2.319	2.690
46	1.679	2.013	2.317	2.687
47	1.678	2.012	2.315	2.685
48	1.677	2.011	2.314	2.682
49	1.677	2.010	2.312	2.680

(continued)

df	Critical t ($\alpha = .10$)	Critical t ($\alpha = .05$)	Critical t ($\alpha = .025$)	Critical t ($\alpha = .01$)
50	1.676	2.009	2.311	2.678
51	1.675	2.008	2.310	2.676
52	1.675	2.007	2.308	2.674
53	1.674	2.006	2.307	2.672
54	1.674	2.005	2.306	2.670
55	1.673	2.004	2.304	2.668
56	1.673	2.003	2.303	2.667
57	1.672	2.002	2.302	2.665
58	1.672	2.002	2.301	2.663
59	1.671	2.001	2.300	2.662
60	1.671	2.000	2.299	2.660
61	1.670	2.000	2.298	2.659
62	1.670	1.999	2.297	2.658
63	1.669	1.998	2.296	2.656
64	1.669	1.998	2.295	2.655
65	1.669	1.997	2.295	2.654
66	1.668	1.997	2.294	2.652
67	1.668	1.996	2.293	2.651
68	1.668	1.995	2.292	2.650
69	1.667	1.995	2.291	2.649
70	1.667	1.994	2.291	2.648
71	1.667	1.994	2.29	2.647
72	1.666	1.993	2.289	2.646
73	1.666	1.993	2.289	2.645
74	1.666	1.993	2.288	2.644
75	1.665	1.992	2.287	2.643
76	1.665	1.992	2.287	2.642
77	1.665	1.991	2.286	2.641
78	1.665	1.991	2.286	2.640
79	1.664	1.990	2.285	2.640
80	1.664	1.990	2.284	2.639
81	1.664	1.990	2.284	2.638
82	1.664	1.989	2.283	2.637
83	1.663	1.989	2.283	2.636
84	1.663	1.989	2.282	2.636
85	1.663	1.988	2.282	2.635
86	1.663	1.988	2.281	2.634
87	1.663	1.988	2.281	2.634
88	1.662	1.987	2.280	2.633
89	1.662	1.987	2.280	2.632
90	1.662	1.987	2.280	2.632
91	1.662	1.986	2.279	2.631
92	1.662	1.986	2.279	2.630
93	1.661	1.986	2.278	2.630
94	1.661	1.986	2.278	2.629
95	1.661	1.985	2.277	2.629
96	1.661	1.985	2.277	2.628
97	1.661	1.985	2.277	2.628
98	1.661	1.985	2.276	2.627
99	1.660	1.984	2.276	2.626
∞	1.645	1.960	2.242	2.576

Tabled values were generated using LabVIEW programming software.

Critical Values of the Student's *t* Distribution
for One-Tailed Tests

df	Critical *t* (α = .10)	Critical *t* (α =.05)	Critical *t* (α =.025)	Critical *t* (α =.01)
1	3.078	6.314	12.706	31.821
2	1.886	2.92	4.303	6.965
3	1.638	2.353	3.183	4.541
4	1.533	2.132	2.776	3.747
5	1.476	2.015	2.571	3.365
6	1.440	1.943	2.447	3.143
7	1.415	1.895	2.365	2.998
8	1.397	1.860	2.306	2.896
9	1.383	1.833	2.262	2.821
10	1.372	1.813	2.228	2.764
11	1.363	1.796	2.201	2.718
12	1.356	1.782	2.179	2.681
13	1.350	1.771	2.160	2.650
14	1.345	1.761	2.145	2.625
15	1.341	1.753	2.131	2.602
16	1.337	1.746	2.120	2.583
17	1.333	1.740	2.110	2.567
18	1.330	1.734	2.101	2.552
19	1.328	1.729	2.093	2.540
20	1.325	1.725	2.086	2.528
21	1.323	1.721	2.080	2.518
22	1.321	1.717	2.074	2.508
23	1.319	1.714	2.069	2.500
24	1.318	1.711	2.064	2.492
25	1.316	1.708	2.060	2.485
26	1.315	1.706	2.056	2.479
27	1.314	1.703	2.052	2.473
28	1.313	1.701	2.048	2.467
29	1.311	1.699	2.045	2.462
30	1.310	1.697	2.042	2.457
31	1.310	1.696	2.040	2.453
32	1.309	1.694	2.037	2.449
33	1.308	1.692	2.035	2.445
34	1.307	1.691	2.032	2.441
35	1.306	1.690	2.030	2.438
36	1.306	1.688	2.028	2.435
37	1.305	1.687	2.026	2.431
38	1.304	1.686	2.024	2.429
39	1.304	1.685	2.023	2.426
40	1.303	1.684	2.021	2.423
41	1.303	1.683	2.020	2.421
42	1.302	1.682	2.018	2.419
43	1.302	1.681	2.017	2.416
44	1.301	1.680	2.015	2.414
45	1.301	1.679	2.014	2.412
46	1.300	1.679	2.013	2.410
47	1.300	1.678	2.012	2.408
48	1.299	1.677	2.011	2.407
49	1.299	1.677	2.010	2.405

(continued)

df	Critical t (α = .10)	Critical t (α = .05)	Critical t (α = .025)	Critical t (α = .01)
50	1.299	1.676	2.009	2.403
51	1.298	1.675	2.008	2.402
52	1.298	1.675	2.007	2.400
53	1.298	1.674	2.006	2.399
54	1.297	1.674	2.005	2.397
55	1.297	1.673	2.004	2.396
56	1.297	1.673	2.003	2.395
57	1.297	1.672	2.002	2.394
58	1.296	1.672	2.002	2.392
59	1.296	1.671	2.001	2.391
60	1.296	1.671	2.000	2.390
61	1.296	1.670	2.000	2.389
62	1.295	1.670	1.999	2.388
63	1.295	1.669	1.998	2.387
64	1.295	1.669	1.998	2.386
65	1.295	1.669	1.997	2.385
66	1.295	1.668	1.997	2.384
67	1.294	1.668	1.996	2.383
68	1.294	1.668	1.995	2.382
69	1.294	1.667	1.995	2.382
70	1.294	1.667	1.994	2.381
71	1.294	1.667	1.994	2.380
72	1.293	1.666	1.993	2.379
73	1.293	1.666	1.993	2.379
74	1.293	1.666	1.993	2.378
75	1.293	1.665	1.992	2.377
76	1.293	1.665	1.992	2.376
77	1.293	1.665	1.991	2.376
78	1.293	1.665	1.991	2.375
79	1.292	1.664	1.990	2.375
80	1.292	1.664	1.990	2.374
81	1.292	1.664	1.990	2.373
82	1.292	1.664	1.989	2.373
83	1.292	1.663	1.989	2.372
84	1.292	1.663	1.989	2.372
85	1.292	1.663	1.988	2.371
86	1.291	1.663	1.988	2.371
87	1.291	1.663	1.988	2.370
88	1.291	1.662	1.987	2.370
89	1.291	1.662	1.987	2.369
90	1.291	1.662	1.987	2.369
91	1.291	1.662	1.986	2.368
92	1.291	1.662	1.986	2.368
93	1.291	1.661	1.986	2.367
94	1.291	1.661	1.986	2.367
95	1.291	1.661	1.985	2.366
96	1.290	1.661	1.985	2.366
97	1.290	1.661	1.985	2.365
98	1.290	1.661	1.985	2.365
99	1.290	1.660	1.984	2.365
∞	1.282	1.645	1.960	2.327

Tabled values were generated using LabVIEW programming software.

Critical Values of the F Distribution for $\alpha = .10$

Denominator df	Numerator df									
	1	2	3	4	5	6	7	8	9	10
1	39.863	49.5	53.593	55.833	57.24	58.204	58.906	59.439	59.858	60.195
2	8.526	9.000	9.162	9.243	9.293	9.326	9.349	9.367	9.381	9.392
3	5.538	5.462	5.391	5.343	5.309	5.285	5.266	5.252	5.24	5.230
4	4.545	4.325	4.191	4.107	4.051	4.010	3.979	3.955	3.936	3.920
5	4.060	3.780	3.619	3.520	3.453	3.404	3.368	3.339	3.316	3.297
6	3.776	3.463	3.289	3.181	3.107	3.055	3.014	2.983	2.958	2.937
7	3.589	3.257	3.074	2.960	2.883	2.827	2.785	2.752	2.725	2.702
8	3.458	3.113	2.924	2.806	2.726	2.668	2.624	2.589	2.561	2.538
9	3.360	3.006	2.813	2.693	2.611	2.551	2.505	2.469	2.44	2.416
10	3.285	2.924	2.728	2.605	2.522	2.461	2.414	2.377	2.347	2.323
11	3.225	2.859	2.660	2.536	2.451	2.389	2.341	2.304	2.273	2.248
12	3.176	2.807	2.605	2.480	2.394	2.331	2.283	2.245	2.213	2.188
13	3.136	2.763	2.560	2.434	2.347	2.283	2.234	2.195	2.164	2.138
14	3.102	2.726	2.522	2.395	2.307	2.243	2.193	2.154	2.122	2.095
15	3.073	2.695	2.490	2.361	2.273	2.208	2.158	2.118	2.086	2.059
16	3.048	2.668	2.462	2.333	2.244	2.178	2.128	2.088	2.055	2.028
17	3.026	2.645	2.437	2.308	2.218	2.152	2.102	2.061	2.028	2.001
18	3.007	2.624	2.416	2.286	2.196	2.130	2.078	2.038	2.005	1.977
19	2.990	2.606	2.397	2.266	2.176	2.109	2.058	2.017	1.984	1.956
20	2.975	2.589	2.380	2.249	2.158	2.091	2.040	1.998	1.965	1.937
21	2.961	2.575	2.365	2.233	2.142	2.075	2.023	1.982	1.948	1.920
22	2.949	2.561	2.351	2.219	2.128	2.060	2.008	1.967	1.933	1.904
23	2.937	2.549	2.339	2.206	2.115	2.047	1.995	1.953	1.919	1.890
24	2.927	2.538	2.327	2.195	2.103	2.035	1.983	1.941	1.906	1.877
25	2.918	2.528	2.317	2.184	2.092	2.024	1.971	1.929	1.895	1.866
26	2.909	2.519	2.307	2.174	2.082	2.014	1.961	1.919	1.884	1.855
27	2.901	2.511	2.299	2.165	2.073	2.005	1.951	1.909	1.874	1.845
28	2.894	2.503	2.291	2.157	2.064	1.996	1.943	1.900	1.865	1.836
29	2.887	2.495	2.283	2.149	2.057	1.988	1.934	1.892	1.857	1.827
30	2.881	2.489	2.276	2.142	2.049	1.980	1.927	1.884	1.849	1.819
31	2.875	2.482	2.270	2.135	2.042	1.973	1.920	1.877	1.842	1.812
32	2.869	2.476	2.263	2.129	2.036	1.967	1.913	1.870	1.835	1.805
33	2.864	2.471	2.258	2.123	2.030	1.961	1.907	1.864	1.828	1.799
34	2.859	2.466	2.252	2.118	2.024	1.955	1.901	1.858	1.822	1.793
35	2.855	2.461	2.247	2.113	2.019	1.950	1.896	1.852	1.817	1.787
36	2.850	2.456	2.243	2.108	2.014	1.945	1.890	1.847	1.811	1.781
37	2.846	2.452	2.238	2.103	2.009	1.940	1.886	1.842	1.806	1.776
38	2.842	2.448	2.234	2.099	2.005	1.935	1.881	1.837	1.802	1.772
39	2.839	2.444	2.230	2.095	2.001	1.931	1.877	1.833	1.797	1.767
40	2.835	2.440	2.226	2.091	1.997	1.927	1.872	1.829	1.793	1.763
41	2.832	2.437	2.222	2.087	1.993	1.923	1.869	1.825	1.789	1.759
42	2.829	2.434	2.219	2.084	1.989	1.919	1.865	1.821	1.785	1.755
43	2.826	2.43	2.216	2.080	1.986	1.916	1.861	1.817	1.781	1.751
44	2.823	2.427	2.213	2.077	1.983	1.912	1.858	1.814	1.778	1.747
45	2.820	2.424	2.210	2.074	1.980	1.909	1.855	1.811	1.774	1.744
46	2.818	2.422	2.207	2.071	1.977	1.906	1.852	1.808	1.771	1.741
47	2.815	2.419	2.204	2.068	1.974	1.903	1.849	1.805	1.768	1.738
48	2.813	2.417	2.202	2.066	1.971	1.901	1.846	1.802	1.765	1.735
49	2.811	2.414	2.199	2.063	1.968	1.898	1.843	1.799	1.763	1.732
50	2.809	2.412	2.197	2.061	1.966	1.895	1.840	1.796	1.760	1.729
51	2.807	2.410	2.194	2.058	1.964	1.893	1.838	1.794	1.757	1.727

(continued)

Denominator df	Numerator df									
	1	2	3	4	5	6	7	8	9	10
52	2.805	2.408	2.192	2.056	1.961	1.891	1.836	1.791	1.755	1.724
53	2.803	2.406	2.190	2.054	1.959	1.888	1.833	1.789	1.752	1.722
54	2.801	2.404	2.188	2.052	1.957	1.886	1.831	1.787	1.750	1.719
55	2.799	2.402	2.186	2.05	1.955	1.884	1.829	1.785	1.748	1.717
56	2.797	2.400	2.184	2.048	1.953	1.882	1.827	1.782	1.746	1.715
57	2.796	2.398	2.182	2.046	1.951	1.880	1.825	1.780	1.744	1.713
58	2.794	2.396	2.181	2.044	1.949	1.878	1.823	1.778	1.742	1.711
59	2.792	2.395	2.179	2.043	1.947	1.876	1.821	1.777	1.74	1.709
60	2.791	2.393	2.177	2.041	1.946	1.875	1.819	1.775	1.738	1.707
61	2.790	2.392	2.176	2.039	1.944	1.873	1.818	1.773	1.736	1.705
62	2.788	2.390	2.174	2.038	1.942	1.871	1.816	1.771	1.734	1.703
63	2.787	2.389	2.173	2.036	1.941	1.870	1.814	1.770	1.733	1.702
64	2.786	2.387	2.171	2.035	1.939	1.868	1.813	1.768	1.731	1.700
65	2.784	2.386	2.170	2.033	1.938	1.867	1.811	1.767	1.730	1.699
66	2.783	2.385	2.169	2.032	1.936	1.865	1.810	1.765	1.728	1.697
67	2.782	2.384	2.167	2.031	1.935	1.864	1.808	1.764	1.727	1.695
68	2.781	2.382	2.166	2.029	1.934	1.863	1.807	1.762	1.725	1.694
69	2.780	2.381	2.165	2.028	1.933	1.861	1.806	1.761	1.724	1.693
70	2.779	2.38	2.164	2.027	1.931	1.860	1.804	1.760	1.723	1.691
71	2.778	2.379	2.163	2.026	1.930	1.859	1.803	1.758	1.721	1.690
72	2.777	2.378	2.161	2.025	1.929	1.858	1.802	1.757	1.720	1.689
73	2.775	2.377	2.160	2.023	1.928	1.856	1.801	1.756	1.719	1.687
74	2.775	2.376	2.159	2.022	1.927	1.855	1.800	1.755	1.718	1.686
75	2.774	2.375	2.158	2.021	1.926	1.854	1.798	1.753	1.716	1.685
76	2.773	2.374	2.157	2.020	1.925	1.853	1.797	1.752	1.715	1.684
77	2.772	2.373	2.156	2.019	1.924	1.852	1.796	1.751	1.714	1.683
78	2.771	2.372	2.155	2.018	1.923	1.851	1.795	1.750	1.713	1.682
79	2.770	2.371	2.154	2.017	1.922	1.850	1.794	1.749	1.712	1.681
80	2.769	2.37	2.154	2.016	1.921	1.849	1.793	1.748	1.711	1.68
81	2.768	2.369	2.153	2.016	1.920	1.848	1.792	1.747	1.710	1.679
82	2.768	2.368	2.152	2.015	1.919	1.847	1.791	1.746	1.709	1.677
83	2.767	2.368	2.151	2.014	1.918	1.846	1.790	1.745	1.708	1.677
84	2.766	2.367	2.150	2.013	1.917	1.845	1.790	1.744	1.707	1.676
85	2.765	2.366	2.149	2.012	1.916	1.845	1.789	1.744	1.706	1.675
86	2.765	2.365	2.149	2.011	1.915	1.844	1.788	1.743	1.705	1.674
87	2.764	2.365	2.148	2.011	1.915	1.843	1.787	1.742	1.704	1.673
88	2.763	2.364	2.147	2.010	1.914	1.842	1.786	1.741	1.704	1.672
89	2.763	2.363	2.146	2.009	1.913	1.841	1.785	1.740	1.703	1.671
90	2.762	2.363	2.146	2.008	1.912	1.841	1.785	1.739	1.702	1.670
91	2.761	2.362	2.145	2.008	1.912	1.840	1.784	1.739	1.701	1.670
92	2.761	2.361	2.144	2.007	1.911	1.839	1.783	1.738	1.700	1.669
93	2.760	2.360	2.144	2.006	1.910	1.838	1.782	1.736	1.699	1.668
94	2.760	2.360	2.143	2.006	1.910	1.838	1.782	1.736	1.699	1.667
95	2.759	2.359	2.142	2.005	1.909	1.837	1.781	1.736	1.698	1.667
96	2.759	2.359	2.142	2.004	1.908	1.836	1.780	1.735	1.698	1.666
97	2.758	2.358	2.141	2.004	1.908	1.836	1.780	1.734	1.697	1.665
98	2.757	2.358	2.140	2.003	1.907	1.835	1.779	1.734	1.696	1.665
99	2.757	2.357	2.140	2.002	1.906	1.834	1.778	1.733	1.696	1.664
100	2.756	2.356	2.139	2.002	1.906	1.834	1.778	1.732	1.695	1.663
120	2.748	2.347	2.130	1.992	1.896	1.824	1.767	1.722	1.684	1.652
150	2.739	2.338	2.121	1.983	1.886	1.814	1.757	1.711	1.674	1.642
200	2.731	2.329	2.111	1.973	1.876	1.804	1.747	1.701	1.663	1.631
∞	2.710	2.300	2.080	1.940	1.850	1.770	1.720	1.670	1.630	1.600

Tabled values were generated using LabVIEW programming software.

TABLE A.5

Critical Values of the F Distribution for $\alpha = .05$

Denominator df	Numerator df									
	1	2	3	4	5	6	7	8	9	10
1	161.44	199.5	215.707	224.583	230.162	233.986	236.768	238.883	240.543	241.88
2	18.513	19.00	19.164	19.247	19.296	19.329	19.353	19.371	19.385	19.396
3	10.128	9.552	9.277	9.117	9.013	8.941	8.887	8.845	8.812	8.786
4	7.709	6.944	6.591	6.388	6.256	6.163	6.094	6.041	5.999	5.964
5	6.608	5.786	5.409	5.192	5.050	4.950	4.876	4.818	4.772	4.735
6	5.987	5.143	4.757	4.534	4.387	4.284	4.207	4.147	4.099	4.060
7	5.591	4.737	4.347	4.120	3.971	3.866	3.787	3.726	3.677	3.636
8	5.318	4.459	4.066	3.838	3.687	3.581	3.500	3.438	3.388	3.347
9	5.117	4.256	3.863	3.633	3.482	3.374	3.293	3.230	3.179	3.137
10	4.965	4.103	3.708	3.478	3.326	3.217	3.135	3.072	3.020	2.978
11	4.844	3.982	3.587	3.357	3.204	3.095	3.012	2.948	2.896	2.854
12	4.747	3.885	3.490	3.259	3.106	2.996	2.913	2.849	2.796	2.753
13	4.667	3.806	3.411	3.179	3.025	2.915	2.832	2.767	2.714	2.671
14	4.600	3.739	3.344	3.112	2.958	2.848	2.764	2.699	2.646	2.602
15	4.543	3.682	3.287	3.056	2.901	2.790	2.707	2.641	2.588	2.544
16	4.494	3.634	3.239	3.007	2.852	2.741	2.657	2.591	2.538	2.493
17	4.451	3.591	3.197	2.965	2.810	2.699	2.614	2.548	2.494	2.450
18	4.414	3.555	3.160	2.928	2.773	2.661	2.577	2.510	2.456	2.412
19	4.381	3.522	3.127	2.895	2.740	2.628	2.543	2.477	2.423	2.378
20	4.351	3.493	3.098	2.866	2.711	2.599	2.514	2.447	2.393	2.348
21	4.325	3.467	3.072	2.840	2.685	2.573	2.488	2.420	2.366	2.321
22	4.301	3.443	3.049	2.817	2.661	2.549	2.464	2.396	2.342	2.297
23	4.279	3.422	3.028	2.796	2.640	2.528	2.442	2.375	2.320	2.275
24	4.260	3.403	3.009	2.776	2.621	2.508	2.423	2.355	2.300	2.255

(continued)

Numerator df

Denominator df	1	2	3	4	5	6	7	8	9	10
25	4.242	3.385	2.991	2.759	2.603	2.490	2.405	2.337	2.282	2.236
26	4.225	3.369	2.975	2.743	2.587	2.474	2.388	2.320	2.265	2.220
27	4.210	3.354	2.960	2.728	2.572	2.459	2.373	2.305	2.250	2.204
28	4.196	3.340	2.947	2.714	2.558	2.445	2.359	2.291	2.236	2.190
29	4.183	3.328	2.934	2.701	2.545	2.432	2.346	2.278	2.223	2.177
30	4.171	3.316	2.922	2.690	2.533	2.420	2.334	2.266	2.211	2.165
31	4.160	3.305	2.911	2.679	2.522	2.409	2.323	2.255	2.199	2.153
32	4.149	3.294	2.901	2.668	2.512	2.399	2.313	2.244	2.189	2.142
33	4.139	3.285	2.891	2.659	2.503	2.389	2.303	2.235	2.179	2.132
34	4.130	3.276	2.883	2.650	2.494	2.380	2.294	2.225	2.170	2.123
35	4.121	3.267	2.874	2.641	2.485	2.372	2.285	2.217	2.161	2.114
36	4.113	3.259	2.866	2.633	2.477	2.364	2.277	2.209	2.153	2.106
37	4.105	3.252	2.859	2.626	2.470	2.356	2.269	2.201	2.145	2.098
38	4.098	3.245	2.852	2.619	2.462	2.349	2.262	2.194	2.137	2.091
39	4.091	3.238	2.845	2.612	2.456	2.342	2.255	2.187	2.131	2.084
40	4.085	3.232	2.839	2.606	2.449	2.336	2.249	2.180	2.124	2.077
41	4.079	3.226	2.833	2.6	2.443	2.330	2.243	2.174	2.118	2.071
42	4.073	3.220	2.827	2.594	2.438	2.324	2.237	2.168	2.112	2.065
43	4.067	3.214	2.822	2.589	2.432	2.318	2.232	2.162	2.106	2.059
44	4.062	3.209	2.816	2.584	2.427	2.313	2.226	2.157	2.101	2.054
45	4.057	3.204	2.811	2.579	2.422	2.308	2.221	2.152	2.096	2.049
46	4.052	3.200	2.807	2.574	2.417	2.303	2.216	2.147	2.091	2.044
47	4.047	3.195	2.802	2.569	2.413	2.299	2.212	2.143	2.086	2.039
48	4.043	3.191	2.798	2.565	2.408	2.295	2.207	2.138	2.082	2.035
49	4.038	3.187	2.794	2.561	2.404	2.290	2.203	2.134	2.077	2.030
50	4.034	3.183	2.790	2.557	2.400	2.286	2.199	2.130	2.073	2.026
51	4.030	3.179	2.786	2.553	2.397	2.283	2.195	2.126	2.069	2.022

52	4.027	3.175	2.783	2.550	2.393	2.279	2.192	2.122	2.066	2.018
53	4.023	3.172	2.779	2.546	2.389	2.275	2.188	2.119	2.062	2.015
54	4.020	3.168	2.776	2.543	2.386	2.272	2.185	2.115	2.058	2.011
55	4.016	3.165	2.773	2.540	2.383	2.269	2.181	2.112	2.055	2.008
56	4.013	3.162	2.769	2.537	2.380	2.265	2.178	2.109	2.052	2.005
57	4.010	3.159	2.766	2.534	2.377	2.263	2.175	2.106	2.049	2.001
58	4.007	3.156	2.763	2.531	2.374	2.260	2.172	2.103	2.046	1.998
59	4.004	3.153	2.761	2.528	2.371	2.257	2.169	2.100	2.043	1.995
60	4.001	3.150	2.758	2.525	2.368	2.254	2.166	2.097	2.040	1.993
61	3.998	3.148	2.755	2.523	2.366	2.251	2.164	2.094	2.037	1.990
62	3.996	3.145	2.753	2.52	2.363	2.249	2.161	2.092	2.035	1.987
63	3.993	3.143	2.751	2.518	2.361	2.246	2.159	2.089	2.032	1.985
64	3.991	3.140	2.748	2.515	2.358	2.244	2.156	2.087	2.030	1.982
65	3.989	3.138	2.746	2.513	2.356	2.242	2.154	2.084	2.027	1.980
66	3.986	3.136	2.744	2.511	2.354	2.239	2.152	2.082	2.025	1.977
67	3.984	3.134	2.742	2.509	2.352	2.237	2.150	2.080	2.023	1.975
68	3.982	3.132	2.739	2.507	2.350	2.235	2.147	2.078	2.021	1.973
69	3.980	3.130	2.737	2.505	2.348	2.233	2.145	2.076	2.019	1.971
70	3.978	3.128	2.735	2.503	2.346	2.231	2.143	2.074	2.017	1.969
71	3.976	3.126	2.734	2.501	2.344	2.229	2.142	2.072	2.015	1.967
72	3.974	3.124	2.732	2.499	2.342	2.227	2.140	2.070	2.013	1.965
73	3.972	3.122	2.730	2.497	2.340	2.226	2.138	2.068	2.011	1.963
74	3.970	3.120	2.728	2.495	2.338	2.224	2.136	2.066	2.009	1.961
75	3.968	3.119	2.727	2.494	2.337	2.222	2.134	2.064	2.007	1.959
76	3.967	3.117	2.725	2.492	2.335	2.220	2.133	2.063	2.005	1.958
77	3.965	3.115	2.723	2.490	2.333	2.219	2.131	2.061	2.004	1.956
78	3.963	3.114	2.722	2.489	2.332	2.217	2.129	2.059	2.002	1.954
79	3.962	3.112	2.720	2.487	2.330	2.216	2.128	2.058	2.001	1.953
80	3.960	3.111	2.719	2.486	2.329	2.214	2.126	2.056	1.999	1.951
81	3.959	3.109	2.717	2.484	2.327	2.213	2.125	2.055	1.998	1.950
82	3.957	3.108	2.716	2.483	2.326	2.211	2.123	2.053	1.996	1.948
83	3.956	3.106	2.715	2.482	2.324	2.210	2.122	2.052	1.995	1.947

(continued)

	Numerator df									
Denominator df	1	2	3	4	5	6	7	8	9	10
84	3.955	3.105	2.713	2.480	2.323	2.209	2.121	2.051	1.993	1.945
85	3.953	3.104	2.712	2.479	2.322	2.207	2.119	2.049	1.992	1.944
86	3.952	3.102	2.711	2.478	2.320	2.206	2.118	2.048	1.991	1.943
87	3.951	3.101	2.709	2.476	2.319	2.205	2.117	2.047	1.989	1.941
88	3.949	3.100	2.708	2.475	2.318	2.203	2.115	2.045	1.988	1.940
89	3.948	3.099	2.707	2.474	2.317	2.202	2.114	2.044	1.987	1.939
90	3.947	3.098	2.706	2.473	2.316	2.201	2.113	2.043	1.986	1.938
91	3.946	3.097	2.705	2.472	2.315	2.200	2.112	2.042	1.984	1.936
92	3.944	3.095	2.704	2.471	2.313	2.199	2.111	2.041	1.983	1.935
93	3.943	3.094	2.702	2.470	2.312	2.198	2.110	2.040	1.982	1.934
94	3.942	3.093	2.701	2.469	2.311	2.197	2.109	2.038	1.981	1.933
95	3.941	3.092	2.700	2.467	2.31	2.196	2.107	2.037	1.980	1.932
96	3.940	3.091	2.699	2.466	2.309	2.194	2.106	2.036	1.979	1.931
97	3.939	3.090	2.698	2.465	2.308	2.193	2.105	2.035	1.978	1.930
98	3.938	3.089	2.697	2.465	2.307	2.192	2.104	2.034	1.977	1.929
99	3.937	3.088	2.696	2.464	2.306	2.192	2.103	2.033	1.976	1.928
100	3.936	3.087	2.695	2.463	2.305	2.191	2.102	2.032	1.975	1.927
120	3.920	3.072	2.680	2.447	2.290	2.175	2.087	2.016	1.959	1.910
150	3.904	3.056	2.665	2.432	2.274	2.159	2.071	2.001	1.943	1.894
200	3.888	3.041	2.650	2.417	2.259	2.144	2.056	1.985	1.927	1.878
∞	3.840	3.000	2.600	2.370	2.210	2.100	2.010	1.940	1.880	1.830

Tabled values were generated using LabVIEW programming software.

TABLE A.6

Critical Values of the F Distribution for $\alpha = .01$

Denominator df	Numerator df									
	1	2	3	4	5	6	7	8	9	10
1	4052	5000	5403	5625	5764	5859	5928	5981	6022	6056
2	98.502	99.000	99.166	99.249	99.299	99.333	99.356	99.374	99.388	99.399
3	34.116	30.816	29.457	28.71	28.237	27.911	27.672	27.489	27.345	27.229
4	21.198	18.000	16.694	15.977	15.522	15.207	14.976	14.799	14.659	14.546
5	16.258	13.274	12.06	11.392	10.967	10.672	10.455	10.289	10.158	10.051
6	13.745	10.925	9.779	9.148	8.746	8.466	8.260	8.102	7.976	7.874
7	12.246	9.547	8.451	7.847	7.460	7.191	6.993	6.840	6.719	6.620
8	11.259	8.649	7.591	7.006	6.632	6.371	6.178	6.029	5.911	5.814
9	10.561	8.021	6.992	6.422	6.057	5.802	5.613	5.467	5.351	5.257
10	10.044	7.559	6.552	5.994	5.636	5.386	5.20	5.057	4.942	4.849
11	9.646	7.206	6.217	5.668	5.316	5.069	4.886	4.744	4.632	4.539
12	9.330	6.927	5.952	5.412	5.064	4.821	4.639	4.499	4.387	4.296
13	9.074	6.701	5.739	5.205	4.862	4.620	4.441	4.302	4.191	4.100
14	8.862	6.515	5.564	5.035	4.695	4.456	4.278	4.140	4.030	3.939
15	8.683	6.359	5.417	4.893	4.556	4.318	4.141	4.004	3.895	3.805
16	8.531	6.226	5.292	4.773	4.437	4.202	4.026	3.890	3.780	3.691
17	8.400	6.112	5.185	4.669	4.336	4.101	3.927	3.791	3.682	3.593
18	8.285	6.013	5.092	4.579	4.248	4.015	3.841	3.705	3.597	3.508
19	8.185	5.926	5.010	4.500	4.171	3.939	3.765	3.631	3.522	3.434
20	8.096	5.849	4.938	4.431	4.103	3.871	3.699	3.564	3.457	3.368
21	8.017	5.780	4.874	4.369	4.042	3.812	3.640	3.506	3.398	3.310
22	7.945	5.719	4.817	4.313	3.988	3.758	3.587	3.453	3.346	3.258
23	7.881	5.664	4.765	4.264	3.939	3.710	3.539	3.406	3.299	3.211

(continued)

Numerator df

Denominator df	1	2	3	4	5	6	7	8	9	10
24	7.823	5.614	4.718	4.218	3.895	3.667	3.496	3.363	3.256	3.168
25	7.770	5.568	4.675	4.177	3.855	3.627	3.457	3.324	3.217	3.129
26	7.721	5.526	4.636	4.140	3.818	3.591	3.421	3.288	3.182	3.094
27	7.677	5.488	4.601	4.106	3.785	3.558	3.388	3.256	3.149	3.062
28	7.636	5.453	4.568	4.074	3.754	3.528	3.358	3.226	3.119	3.032
29	7.598	5.420	4.538	4.045	3.725	3.499	3.330	3.198	3.092	3.005
30	7.562	5.390	4.510	4.018	3.699	3.473	3.304	3.173	3.067	2.979
31	7.530	5.362	4.484	3.993	3.674	3.449	3.281	3.149	3.043	2.955
32	7.499	5.336	4.459	3.969	3.652	3.427	3.258	3.127	3.021	2.933
33	7.471	5.312	4.437	3.948	3.630	3.406	3.238	3.106	3.000	2.913
34	7.444	5.289	4.416	3.927	3.611	3.386	3.218	3.087	2.981	2.894
35	7.419	5.268	4.396	3.908	3.592	3.368	3.200	3.069	2.963	2.876
36	7.396	5.248	4.377	3.890	3.574	3.351	3.183	3.052	2.946	2.859
37	7.373	5.229	4.359	3.873	3.558	3.334	3.167	3.036	2.930	2.843
38	7.353	5.211	4.343	3.857	3.542	3.319	3.152	3.021	2.915	2.828
39	7.333	5.194	4.327	3.842	3.528	3.305	3.137	3.006	2.901	2.814
40	7.314	5.178	4.313	3.828	3.514	3.291	3.124	2.993	2.888	2.801
41	7.296	5.163	4.299	3.815	3.501	3.278	3.111	2.980	2.875	2.788
42	7.28	5.149	4.285	3.802	3.488	3.266	3.099	2.968	2.863	2.776
43	7.264	5.136	4.273	3.790	3.476	3.254	3.087	2.957	2.851	2.764
44	7.248	5.123	4.261	3.778	3.465	3.243	3.076	2.946	2.840	2.754
45	7.234	5.110	4.249	3.767	3.454	3.232	3.066	2.935	2.830	2.743
46	7.220	5.099	4.238	3.757	3.444	3.222	3.056	2.925	2.820	2.733
47	7.207	5.087	4.228	3.747	3.434	3.213	3.046	2.916	2.811	2.724
48	7.194	5.077	4.218	3.737	3.425	3.204	3.037	2.907	2.802	2.715
49	7.182	5.066	4.208	3.728	3.416	3.195	3.028	2.898	2.793	2.706
50	7.171	5.057	4.199	3.719	3.408	3.186	3.020	2.89	2.785	2.698

51	7.159	5.047	4.191	3.711	3.399	3.178	3.012	2.882	2.777	2.690
52	7.149	5.038	4.182	3.703	3.392	3.171	3.005	2.874	2.769	2.683
53	7.139	5.030	4.174	3.695	3.384	3.163	2.997	2.867	2.762	2.675
54	7.129	5.021	4.166	3.688	3.377	3.156	2.990	2.860	2.755	2.668
55	7.119	5.013	4.159	3.681	3.370	3.149	2.983	2.853	2.748	2.662
56	7.110	5.005	4.152	3.674	3.363	3.143	2.977	2.847	2.742	2.655
57	7.102	4.998	4.145	3.667	3.357	3.136	2.971	2.841	2.736	2.649
58	7.093	4.991	4.138	3.661	3.351	3.130	2.965	2.835	2.730	2.643
59	7.085	4.984	4.132	3.655	3.345	3.124	2.959	2.829	2.724	2.637
60	7.077	4.977	4.126	3.649	3.339	3.119	2.953	2.823	2.718	2.632
61	7.069	4.971	4.120	3.643	3.333	3.113	2.948	2.818	2.713	2.626
62	7.062	4.965	4.114	3.638	3.328	3.108	2.942	2.813	2.708	2.621
63	7.055	4.959	4.109	3.632	3.323	3.103	2.937	2.808	2.703	2.616
64	7.048	4.953	4.103	3.627	3.318	3.098	2.932	2.803	2.698	2.611
65	7.042	4.947	4.098	3.622	3.313	3.093	2.928	2.798	2.693	2.607
66	7.035	4.942	4.093	3.617	3.308	3.088	2.923	2.793	2.689	2.602
67	7.029	4.937	4.088	3.613	3.304	3.084	2.919	2.789	2.684	2.598
68	7.023	4.932	4.083	3.608	3.299	3.079	2.914	2.785	2.680	2.593
69	7.017	4.927	4.079	3.604	3.295	3.075	2.910	2.781	2.676	2.589
70	7.011	4.922	4.074	3.600	3.291	3.071	2.906	2.776	2.672	2.585
71	7.006	4.917	4.070	3.596	3.287	3.067	2.902	2.773	2.668	2.581
72	7.000	4.913	4.066	3.591	3.283	3.063	2.898	2.769	2.664	2.578
73	6.995	4.908	4.062	3.588	3.279	3.060	2.895	2.765	2.660	2.574
74	6.990	4.904	4.058	3.584	3.275	3.056	2.891	2.761	2.657	2.570
75	6.985	4.900	4.054	3.580	3.272	3.052	2.887	2.758	2.653	2.567
76	6.981	4.896	4.050	3.577	3.268	3.049	2.884	2.755	2.650	2.563
77	6.976	4.892	4.047	3.573	3.265	3.046	2.881	2.751	2.647	2.560
78	6.971	4.888	4.043	3.570	3.261	3.042	2.877	2.748	2.644	2.557
79	6.967	4.884	4.040	3.566	3.258	3.039	2.874	2.745	2.640	2.554
80	6.963	4.881	4.036	3.563	3.255	3.036	2.871	2.742	2.637	2.551
81	6.958	4.877	4.033	3.560	3.252	3.033	2.868	2.739	2.634	2.548
82	6.954	4.874	4.030	3.557	3.249	3.030	2.865	2.736	2.632	2.545

(continued)

Denominator df / Numerator df

df	1	2	3	4	5	6	7	8	9	10
83	6.950	4.870	4.027	3.554	3.246	3.027	2.863	2.733	2.629	2.542
84	6.947	4.867	4.024	3.551	3.243	3.024	2.860	2.730	2.626	2.539
85	6.943	4.864	4.021	3.548	3.240	3.022	2.857	2.728	2.623	2.537
86	6.939	4.861	4.018	3.545	3.238	3.019	2.854	2.725	2.621	2.534
87	6.935	4.858	4.015	3.543	3.235	3.017	2.852	2.723	2.618	2.532
88	6.932	4.855	4.012	3.540	3.233	3.014	2.849	2.720	2.616	2.529
89	6.928	4.852	4.010	3.537	3.230	3.011	2.847	2.718	2.613	2.527
90	6.925	4.849	4.007	3.535	3.228	3.009	2.844	2.715	2.611	2.524
91	6.922	4.846	4.004	3.532	3.225	3.007	2.842	2.713	2.609	2.522
92	6.919	4.844	4.002	3.530	3.223	3.004	2.840	2.711	2.606	2.520
93	6.915	4.841	3.999	3.528	3.221	3.002	2.838	2.709	2.604	2.517
94	6.912	4.838	3.997	3.525	3.218	3.000	2.835	2.706	2.602	2.515
95	6.909	4.836	3.995	3.523	3.216	2.998	2.833	2.704	2.600	2.513
96	6.906	4.833	3.992	3.521	3.214	2.996	2.831	2.702	2.598	2.511
97	6.904	4.831	3.990	3.519	3.212	2.994	2.829	2.700	2.596	2.509
98	6.901	4.828	3.988	3.517	3.210	2.992	2.827	2.698	2.594	2.507
99	6.898	4.826	3.986	3.515	3.208	2.990	2.825	2.696	2.592	2.505
100	6.895	4.824	3.984	3.513	3.206	2.988	2.823	2.694	2.590	2.503
120	6.851	4.786	3.949	3.479	3.174	2.956	2.792	2.663	2.559	2.472
150	6.807	4.749	3.915	3.447	3.142	2.924	2.761	2.632	2.528	2.441
200	6.763	4.713	3.881	3.414	3.110	2.893	2.730	2.601	2.497	2.411
∞	6.630	4.610	3.780	3.320	3.020	2.800	2.640	2.510	2.410	2.320

Tabled values were generated using LabVIEW programming software.

TABLE A.7

Values of the Studentized Range for α = .10

df_E	Number of groups (k)								
	2	3	4	5	6	7	8	9	10
2	4.128	5.733	6.773	7.538	8.139	8.633	9.049	9.409	9.725
3	3.328	4.467	5.199	5.738	6.162	6.511	6.806	7.062	7.267
4	3.015	3.976	4.586	5.035	5.388	5.679	5.926	6.139	6.327
5	2.850	3717	4.264	4.664	4.979	5.238	5.458	5.648	5.816
6	2.748	3.558	4.065	4.435	4.726	4.966	5.168	5.344	5.499
7	2.679	3.451	3.931	4.280	4.555	4.780	4.971	5.137	5.283
8	2.630	3.374	3.834	4.169	4.431	4.646	4.829	4.987	5.126
9	2.592	3.316	3.761	4.084	4.337	4.545	4.721	4.873	5.007
10	2.563	3.270	3.704	4.018	4.264	4.465	4.636	4.783	4.913
11	2.540	3.234	3.658	3.965	4.205	4.401	4.567	4.711	4.838
12	2.521	3.204	3.621	3.921	4.156	4.349	4.511	4.652	4.776
13	2.504	3.179	3.589	3.885	4.116	4.304	4.464	4.602	4.724
14	2.491	3.158	3.563	3.854	4.081	4.267	4.424	4.560	4.679
15	2.479	3.140	3.540	3.828	4.052	4.235	4.390	4.524	4.641
16	2.469	3.124	3.520	3.804	4.026	4.207	4.360	4.492	4.608
17	2.460	3.110	3.503	3.784	4.003	4.182	4.334	4.464	4.579
18	2.452	3.098	3.487	3.766	3.984	4.161	4.310	4.440	4.553
19	2.445	3.087	3.474	3.751	3.966	4.142	4.290	4.418	4.530
20	2.439	3.077	3.462	3.736	3.950	4.124	4.271	4.398	4.510
21	2.433	3.069	3.451	3.724	3.936	4.109	4.255	4.380	4.491
22	2.428	3.061	3.441	3.712	3.923	4.095	4.239	4.364	4.474
23	2.424	3.054	3.432	3.701	3.911	4.082	4.226	4.350	4.459
24	2.420	3.047	3.423	3.692	3.900	4.070	4.213	4.336	4.445
25	2.416	3.041	3.416	3.683	3.890	4.059	4.201	4.324	4.432
26	2.412	3.036	3.409	3.675	3.881	4.049	4.191	4.313	4.420
27	2.409	3.030	3.402	3.667	3.873	4.040	4.181	4.302	4.409
28	2.406	3.026	3.396	3.660	3.865	4.032	4.172	4.293	4.399
29	2.403	3.021	3.391	3.654	3.858	4.024	4.163	4.284	4.389
30	2.400	3.017	3.386	3.648	3.851	4.016	4.155	4.275	4.381
40	2.381	2.988	3.348	3.605	3.802	3.963	4.099	4.215	4.317
50	2.370	2.970	3.326	3.579	3.774	3.932	4.065	4.179	4.279
60	2.362	2.959	3.312	3.562	3.755	3.911	4.042	4.155	4.254
70	2.357	2.951	3.302	3.550	3.741	3.896	4.026	4.138	4.236
80	2.353	2.945	3.294	3.541	3.731	3.885	4.014	4.125	4.223
90	2.350	2.940	3.288	3.534	3.723	3.876	4.005	4.115	4.212
100	2.348	2.936	3.283	3.528	3.717	3.870	3.998	4.108	4.204
120	2.344	2.930	3.276	3.520	3.707	3.859	3.986	4.096	4.191
∞	2.326	2.903	3.241	3.479	3.661	3.809	3.932	4.038	4.130

Tabled values were generated using the qtukey command in R statistical software (www.r-project.org). We are indebted to Dr. Simon Geletta for his assistance with R.

Values of the Studentized Range for $\alpha = .05$

df_E	Number of groups (k)								
	2	3	4	5	6	7	8	9	10
2	6.080	8.331	9.799	10.881	11.734	12.435	13.028	13.542	13.994
3	4.501	5.910	6.825	7.502	8.037	8.478	8.852	9.177	9.462
4	3.927	5.040	5.757	6.287	6.706	7.053	7.347	7.602	7.826
5	3.635	4.602	5.218	5.673	6.033	6.330	6.582	6.801	6.995
6	3.460	4.339	4.896	5.305	5.628	5.895	6.122	6.319	6.493
7	3.344	4.165	4.681	5.060	5.359	5.606	5.815	5.997	6.158
8	3.261	4.041	4.529	4.886	5.167	5.399	5.596	5.767	5.918
9	3.199	3.948	4.415	4.755	5.024	5.244	5.432	5.595	5.738
10	3.151	3.877	4.397	4.654	4.912	5.124	5.304	5.460	5.598
11	3.113	3.820	4.256	4.574	4.823	5.028	5.202	5.353	5.486
12	3.081	3.773	4.199	4.508	4.750	4.950	5.119	5.265	5.395
13	3.055	3.734	4.151	4.453	4.690	4.884	5.049	5.192	5.318
14	3.033	3.701	4.111	4.407	4.639	4.829	4.990	5.130	5.253
15	3.014	3.673	4.076	4.367	4.595	4.782	4.940	5.077	5.198
16	2.998	3.649	4.046	4.333	4.557	4.741	4.896	5.031	5.150
17	2.984	3.628	4.020	4.303	4.524	4.705	4.858	4.991	5.108
18	2.971	3.609	3.997	4.276	4.494	4.673	4.824	4.955	5.071
19	2.960	3.593	3.977	4.253	4.468	4.645	4.794	4.924	5.037
20	2.950	3.578	3.958	4.232	4.445	4.620	4.768	4.895	5.008
21	2.941	3.565	3.942	4.213	4.424	4.597	4.743	4.870	4.981
22	2.933	3.553	3.927	4.196	4.405	4.577	4.722	4.847	4.957
23	2.926	3.542	3.914	4.180	4.388	4.558	4.702	4.826	4.935
24	2.919	3.532	3.901	4.166	4.373	4.541	4.684	4.807	4.915
25	2.913	3.523	3.890	4.153	4.358	4.526	4.667	4.789	4.897
26	2.910	3.514	3.880	4.141	4.345	4.511	4.652	4.773	4.880
27	2.902	3.506	3.870	4.130	4.333	4.498	4.638	4.758	4.864
28	2.897	3.499	3.861	4.120	4.322	4.486	4.625	4.745	4.850
29	2.892	3.493	3.853	4.111	4.311	4.475	4.613	4.732	4.837
30	2.888	3.486	3.845	4.102	4.301	4.464	4.601	4.720	4.824
40	2.858	3.442	3.791	4.039	4.232	4.388	4.521	4.634	4.735
50	2.841	3.416	3.758	4.002	4.190	4.344	4.473	4.584	4.681
60	2.829	3.399	3.737	3.977	4.163	4.314	4.441	4.550	4.646
70	2.821	3.386	3.722	3.960	4.144	4.293	4.419	4.527	4.621
80	2.814	3.377	3.711	3.947	4.129	4.277	4.402	4.509	4.603
90	2.810	3.370	3.702	3.937	4.118	4.265	4.389	4.495	4.588
100	2.806	3.365	3.695	3.929	4.109	4.256	4.379	4.484	4.577
120	2.800	3.356	3.685	3.917	4.096	4.241	4.363	4.468	4.560
∞	2.772	3.315	3.634	3.858	4.031	4.170	4.287	4.387	4.475

Tabled values were generated using the qtukey command in R statistical software (www.r-project.org). We are indebted to Dr. Simon Geletta for his assistance with R.

TABLE A.9

Values of the Studentized Range for $\alpha = .01$

df_E	2	3	4	5	6	7	8	9	10
				Number of groups (k)					
2	13.902	19.016	22.564	25.372	27.757	29.856	31.730	33.412	34.926
3	8.260	10.620	12.170	13.322	14.239	14.998	15.646	16.212	16.713
4	6.511	8.120	9.173	9.958	10.583	11.101	11.542	11.925	12.263
5	5.702	6.976	7.804	8.421	8.913	9.321	9.669	9.971	10.239
6	5.243	6.331	7.033	7.556	7.972	8.318	8.612	8.869	9.097
7	4.949	5.919	6.542	7.005	7.373	7.678	7.939	8.166	8.367
8	4.745	5.635	6.204	6.625	7.959	7.237	7.474	7.680	7.863
9	4.596	5.428	5.957	6.347	6.657	6.915	7.134	7.325	7.494
10	4.482	5.270	5.769	6.136	6.428	6.669	6.875	7.054	7.213
11	4.392	5.146	5.621	5.970	6.247	6.476	6.671	6.841	6.992
12	4.320	5.046	5.502	5.836	6.101	6.320	6.507	6.670	6.814
13	4.260	4.964	5.404	5.726	5.981	6.192	6.372	6.528	6.666
14	4.210	4.895	5.322	5.634	5.881	6.085	6.258	6.409	6.543
15	4.167	4.435	5.252	5.556	5.796	5.994	6.162	6.309	6.438
16	4.131	4.786	5.192	5.489	5.722	5.915	6.079	6.222	6.348
17	4.099	4.742	5.140	5.430	5.659	5.847	6.007	6.147	6.270
18	4.071	4.703	5.094	5.379	5.603	5.787	5.944	6.081	6.201
19	4.046	4.669	5.054	5.334	5.553	5.735	5.889	6.022	6.141
20	4.024	4.639	5.018	5.293	5.510	5.688	5.839	5.970	6.086
21	4.004	4.612	4.986	5.257	5.470	5.646	5.794	5.924	6.038
22	3.986	4.588	4.957	5.225	5.435	5.608	5.754	5.882	5.994
23	3.970	4.566	4.931	5.195	5.403	5.573	5.718	5.844	5.955
24	3.955	4.546	4.907	5.168	5.373	5.542	5.685	5.809	5.919
25	3.942	4.527	4.885	5.144	5.347	5.513	5.655	5.778	5.886
26	3.930	4.510	4.865	5.121	5.322	5.487	5.627	5.749	5.856
27	3.918	4.495	4.847	5.101	5.300	5.463	5.602	5.722	5.828
28	3.908	4.481	4.830	5.082	5.279	5.441	5.578	5.697	5.802
29	3.898	4.467	4.814	5.064	5.260	5.420	5.556	5.674	5.778
30	3.889	4.455	4.799	5.048	5.242	5.401	5.536	5.653	5.756
40	3.825	4.367	4.695	4.931	5.114	5.265	5.392	5.502	5.599
50	3.787	4.316	4.634	4.863	5.040	5.185	5.308	5.414	5.507
60	3.762	4.282	4.594	4.818	4.991	5.133	5.253	5.356	5.447
70	3.745	4.258	4.566	4.786	4.957	5.096	5.214	5.315	5.404
80	3.732	4.241	4.545	4.763	4.931	5.069	5.185	5.284	5.372
90	3.722	4.227	4.529	4.745	4.911	5.048	5.162	5.261	5.348
100	3.714	4.216	4.516	4.730	4.896	5.031	5.144	5.242	5.328
120	3.702	4.200	4.497	4.709	4.872	5.005	5.118	5.214	5.300
∞	3.643	4.121	4.404	4.604	4.758	4.883	4.988	5.079	5.158

Tabled values were generated using the qtukey command in R statistical software (www.r-project.org). We are indebted to Dr. Simon Geletta for his assistance with R.

Critical Values for the Bryant Paulson Procedure
With One Covariate and $\alpha = .05$

df error	Number of groups			
	3	4	5	6
2	11.000	13.009	14.462	15.61
3	7.181	8.320	9.170	9.840
4	5.833	6.690	7.319	7.819
5	5.180	5.879	6.401	6.822
6	4.784	5.402	5.861	6.230
7	4.522	5.089	5.507	5.841
8	4.336	4.869	5.259	5.567
9	4.198	4.706	5.074	5.365
10	4.094	4.681	4.932	5.208
11	4.011	4.481	4.821	5.084
12	3.943	4.401	4.729	4.983
13	3.888	4.334	4.653	4.900
14	3.842	4.279	4.590	4.831
15	3.803	4.230	4.535	4.771
16	3.770	4.189	4.489	4.719
17	3.741	4.153	4.448	4.674
18	3.716	4.121	4.411	4.634
19	3.694	4.094	4.380	4.599
20	3.674	4.068	4.351	4.568
21	3.657	4.046	4.326	4.540
22	3.641	4.025	4.303	4.515
23	3.627	4.008	4.281	4.492
24	3.614	3.990	4.262	4.473
25	3.602	3.975	4.245	4.453
26	3.591	3.961	4.228	4.436
27	3.580	3.948	4.214	4.420
28	3.572	3.936	4.200	4.405
29	3.564	3.925	4.188	4.391
30	3.555	3.914	4.176	4.378
40	3.501	3.841	4.092	4.288
50	3.469	3.797	4.043	4.233
60	3.449	3.769	4.009	4.198
70	3.434	3.749	3.987	4.174
80	3.423	3.734	3.970	4.155
90	3.415	3.722	3.956	4.141
100	3.409	3.713	3.946	4.129
120	3.399	3.699	3.930	4.113

Tabled values were generated using LabVIEW programming software. For more extensive tabled values for the Bryant-Paulson procedure, see table 1 in Bryant and Paulson (1976).

Values of the Chi-Square Distribution

df	$\alpha = .10$	$\alpha = .05$	$\alpha = .01$
1	2.706	3.841	6.635
2	4.605	5.991	9.210
3	6.251	7.815	11.345
4	7.779	9.488	13.277
5	9.236	11.070	15.086
6	10.645	12.592	16.812
7	12.017	14.067	18.475
8	13.361	15.507	20.090
9	14.684	16.919	21.666
10	15.987	18.307	23.209
11	17.275	19.675	24.725
12	18.549	21.026	26.217
13	19.812	22.362	27.688
14	21.064	23.685	29.141
15	22.307	24.996	30.578
16	23.542	26.296	32.000
17	24.769	27.587	33.409
18	25.989	28.869	34.805
18	27.204	30.143	36.191
20	28.412	31.410	37.566
21	29.615	32.671	38.932
22	30.813	33.924	40.289
23	32.007	35.172	41.638
24	33.196	36.415	42.980
25	34.382	37.652	44.314
26	35.563	38.885	45.642
27	36.741	40.113	46.963
28	37.916	41.337	48.278
29	39.087	42.557	49.588
30	40.256	43.773	50.892
31	41.422	44.985	52.191
32	42.585	46.194	53.486
33	43.745	47.400	54.775
34	44.903	48.602	56.061
35	46.059	49.802	57.342
36	47.212	50.998	58.619
37	48.363	52.192	59.892
38	49.513	53.383	61.162
39	50.660	54.572	62.428
40	51.805	55.758	63.691
41	52.948	56.942	64.950
42	54.090	58.124	66.206
43	55.230	59.303	67.459
44	56.368	60.481	68.709
45	57.505	61.656	69.957

(continued)

df	$\alpha = .10$	$\alpha = .05$	$\alpha = .01$
46	58.640	62.830	71.201
47	59.774	64.001	72.443
48	60.907	65.171	73.683
49	62.037	66.339	74.919
50	63.167	67.505	76.154
60	74.397	79.082	88.379
70	85.527	90.531	100.425
80	96.578	101.879	112.329
90	107.565	113.145	124.116
100	118.498	124.342	135.807
∞	226.021	233.994	249.445

Tabled values were generated using LabVIEW programming software.

Critical Values of the Spearman
Rank Order Correlation Coefficient

| | Two-tailed alpha level | | |
df	.10	.05	.01
5	0.696	0.786	0.911
6	0.631	0.726	0.869
7	0.592	0.683	0.825
8	0.558	0.642	0.788
9	0.527	0.609	0.755
10	0.500	0.584	0.724
11	0.478	0.558	0.698
12	0.459	0.536	0.675
13	0.443	0.518	0.654
14	0.428	0.501	0.632
15	0.413	0.485	0.615
16	0.401	0.471	0.599
17	0.389	0.458	0.582
18	0.379	0.446	0.568
19	0.369	0.435	0.555
20	0.360	0.425	0.543
21	0.352	0.415	0.531
22	0.344	0.406	0.520
23	0.337	0.397	0.510
24	0.330	0.389	0.501
25	0.324	0.382	0.491
26	0.317	0.375	0.483
27	0.312	0.368	0.474
28	0.306	0.362	0.467
29	0.301	0.356	0.459
30	0.296	0.350	0.452
31	0.291	0.345	0.445
32	0.287	0.339	0.439
33	0.283	0.335	0.432
34	0.279	0.330	0.427
35	0.275	0.325	0.421
36	0.271	0.321	0.415
37	0.267	0.317	0.410
38	0.264	0.313	0.405
39	0.260	0.309	0.400
40	0.257	0.305	0.395
41	0.254	0.301	0.391
42	0.251	0.298	0.386
43	0.248	0.294	0.382
44	0.246	0.291	0.378
45	0.243	0.288	0.374
46	0.240	0.285	0.370
47	0.238	0.282	0.366

(continued)

df	Two-tailed alpha level		
	.10	.05	.01
48	0.235	0.279	0.363
49	0.233	0.276	0.359
50	0.231	0.274	0.356
51	0.228	0.271	0.352
52	0.226	0.268	0.349
53	0.224	0.266	0.346
54	0.222	0.263	0.343
55	0.220	0.261	0.340
56	0.218	0.259	0.337
57	0.216	0.257	0.334
58	0.214	0.254	0.331
59	0.213	0.252	0.329
60	0.211	0.250	0.326
61	0.209	0.248	0.323
62	0.207	0.246	0.321
63	0.206	0.244	0.318
64	0.204	0.243	0.316
65	0.203	0.241	0.314
66	0.201	0.239	0.311
67	0.200	0.237	0.309
68	0.198	0.235	0.307
69	0.197	0.234	0.305
70	0.195	0.232	0.303
∞	0.165	0.197	0.257

Tabled values were generated using LabVIEW programming software using the calculations described by Olds (1938) and Zar (1972).

Critical Values of the Mann-Whitney U Distribution
for $\alpha = .05$ (One-Tailed) and $\alpha = .10$ (Two-Tailed)

N_1						N_2					
	10	11	12	13	14	15	16	17	18	19	20
5	11	12	13	15	16	18	19	20	22	23	25
6	14	16	17	19	21	23	25	26	28	30	32
7	17	19	21	24	26	28	30	33	35	37	39
8	20	23	26	28	31	33	36	39	41	44	47
9	24	27	30	33	36	39	42	45	48	51	54
10	27	31	34	37	41	44	48	51	55	58	62
11	31	34	38	42	46	50	54	57	61	65	69
12	34	38	42	47	51	55	60	64	68	72	77
13	37	42	47	51	56	61	65	70	75	80	84
14	41	46	51	56	61	66	71	77	82	87	92
15	44	50	55	61	66	72	77	83	88	94	100
16	48	54	60	65	71	77	83	89	95	101	107
17	51	57	64	70	77	83	89	96	102	109	115
18	55	61	68	75	82	88	95	102	109	116	123
19	58	65	72	80	87	94	101	109	116	123	130
20	62	69	77	84	92	100	107	115	123	130	138

Tabled values calculated using utest software (http://elegans.swmed.edu/~leon/stats/utest.html).

Critical Values of the Mann-Whitney U Distribution for $\alpha = .025$ (One-Tailed) and $\alpha = .05$ (Two-Tailed)

N_1						N_2					
	10	11	12	13	14	15	16	17	18	19	20
5	8	9	11	12	13	14	15	17	18	19	20
6	11	13	14	16	17	19	21	22	24	25	27
7	15	16	18	20	22	24	26	28	30	32	34
8	17	19	22	24	26	29	31	34	36	38	41
9	20	23	26	28	31	34	37	39	42	45	48
10	23	26	29	33	36	39	42	45	48	52	55
11	26	30	33	37	40	44	47	51	55	58	62
12	29	33	37	41	45	49	53	57	62	65	69
13	33	37	41	45	50	54	59	63	67	72	76
14	36	40	45	50	55	59	64	69	74	78	83
15	39	44	49	54	59	64	70	75	80	85	90
16	42	47	53	59	64	70	75	81	86	92	97
17	45	51	57	63	69	75	81	87	93	99	105
18	48	55	62	67	74	80	86	93	99	106	112
19	52	58	65	72	78	85	92	99	106	113	119
20	55	62	69	76	83	90	97	105	112	119	127

Tabled values calculated using utest software (http://elegans.swmed.edu/~leon/stats/utest.html).

Critical Values of the Mann-Whitney U Distribution
for α = .01 (One-Tailed) and α = .02 (Two-Tailed)

N_1	N_2										
	10	11	12	13	14	15	16	17	18	19	20
5	6	7	8	9	10	11	12	13	14	15	16
6	8	9	11	12	13	15	16	18	19	20	22
7	11	12	14	16	17	19	21	23	24	26	28
8	13	15	17	20	22	24	26	28	30	32	34
9	16	18	21	23	26	28	31	33	36	38	40
10	19	22	24	27	30	33	36	38	41	44	47
11	22	25	28	31	34	37	41	44	47	50	53
12	24	28	31	35	38	42	46	49	53	56	60
13	27	31	35	39	43	47	51	55	59	63	67
14	30	34	38	43	47	51	56	60	65	69	73
15	33	37	42	47	51	56	61	66	70	75	80
16	36	41	46	51	56	61	66	71	76	82	87
17	38	44	49	55	60	66	71	77	82	88	93
18	41	47	53	59	65	70	76	82	88	94	100
19	44	50	56	63	69	75	82	88	94	101	107
20	47	53	60	67	73	80	87	93	100	107	114

Tabled values calculated using utest software (http://elegans.swmed.edu/~leon/stats/utest.html).

Appendix B

Raw Data

A researcher randomly divided 110 men into two groups. For 6 weeks, the control group ate normally and the experimental group followed a low-fat diet. At the end of the 6-week period, the weights of all of the men were recorded. Four subjects did not complete the diet.

TABLE B.1

Weight in Pounds

		Control group		
161	155	164	167	154
168	163	176	158	170
150	171	155	149	186
171	175	169	184	161
164	165	181	168	188
173	172	163	191	183
177	168	173	169	152
154	145	167	157	160
166	159	175	166	179
170	164	147	203	157
162	170	156	164	166
		Experimental group		
179	160	161	173	181
165	152	150	154	169
204	174	186	161	157
159	156	148	145	150
152	151	144	139	146
164	149	167	188	165
153	198	158	142	154
177	159	147	155	160
155	148	132	140	145
151	150	146	126	136
168				

A physical education teacher tested two classes on the softball throw for distance. Each student threw the ball twice (trial 1 and trial 2). Scores were recorded to the nearest foot.

TABLE B.2

Softball Throw in Feet

Subject	Class 1 trial		Subject	Class 2 trial	
	1	2		1	2
1	121	136	1	132	138
	143	149		101	115
	102	96		114	121
	141	162		141	136
5	99	97	5	78	76
	134	99		123	96
	129	135		159	162
	148	141		95	119
	111	106		125	129
10	155	162	10	114	146
	118	105		123	117
	127	110		114	108
	72	79		128	122
	137	161		148	133
15	139	144	15	87	93
	113	119		143	116
	152	126		119	115
	126	126		127	109
	106	84		145	152
20	149	153	20	109	84
	123	92		134	138
	177	175		115	82
	133	126		168	154
	119	87		126	122
25	133	138	25	75	103
	147	116		131	139
	94	105		112	125
	139	135		121	119
	125	143		147	153
30	158	162	30	97	115
	116	146		153	159
	136	132		129	150
	121	118		105	110
	164	159		181	167
35	103	109	35	123	126
	176	181		118	100
	125	108		126	115
	145	160		134	134
	119	122		83	126

(continued)

Subject	Class 1 trial		Subject	Class 2 trial	
	1	2		1	2
40	156	155	40	137	142
	129	110		119	99
	186	179		149	135
	86	84		103	109
	134	127		122	136
45	117	138	45	138	129
	143	145		106	101
	128	150		128	143
	109	98		91	102
	137	134			
50	113	120			

Appendix C

Answers to Problems

Chapter 1

1. Measurement is the process of comparing with a standard. This process produces data, which are disorganized. The data are organized, treated, and presented for evaluation through a procedure called statistics. Evaluation determines the worth or value of the data.

2. A variable is a characteristic of a person or object that can assume more than one value. If a characteristic can assume only one value, it is a constant. Independent variables are unrelated to each other (e.g., height and intelligence), whereas dependent variables are related to each other (e.g., body fat and weight).

3. Nominal: places values into mutually exclusive categories, such as male or female, without qualitative value differences.

Ordinal: ordered values such as tallest, next tallest, to shortest.

Interval: Values on a scale on which negative scores are possible, such as temperature scales or judges' scores.

Ratio: values on a scale on which negative scores are not possible, such as time (races), distance (height), force (weight), or counting events (heart rate).

4. Statistical inference means to infer characteristics of an unmeasured population based on a sample taken from that population. A sample is a certain portion or fraction of the population. A population is any group of persons, places, or objects that have at least one characteristic in common. To infer from sample statistics to population parameters, the sample must be random. To be random, everyone or everything in the population must have an equal chance of being selected in the sample. If the population contains subgroups that are of interest, each subgroup should be sampled so that the total sample has the same representation of subgroups as the population; this process is called stratified random sampling.

5. A parameter is a characteristic of a population, whereas a statistic is a characteristic of a sample taken from a population. Statistics are used to estimate parameters.

6. A theory is a belief about a concept or series of concepts. It is neither right nor wrong when it is conceived; it is just an opinion. However, theories produce hypotheses, which can be tested. When a hypothesis is tested and found to be true, it supports the theory. For example, suppose you believed that distributed practice was better than massed practice for learning to make free throws in basketball. One hypothesis that logically results from the theory is that a group of players who shot 100 free throws in a row during practice would have a lower free throw percentage in the games than a group who shot 100 free throws, 10 at a time with a 5-minute break between each set of 10 shots. If the massed group has a lower percentage in games than the distributed group, the hypothesis is true and the theory is credible.

7. Answers will vary based on individual experience.

8. Some examples are *Research Quarterly for Exercise and Sport, Journal of Applied Biomechanics, Medicine and Science in Sports and Exercise, Journal of Strength and Conditioning Research, Journal of Athletic Training, International Journal of Sports Medicine, Journal of Motor Behavior, Sports Medicine Training and Rehabilitation, Perceptual Motor Skill, Journal of Teaching Physical Education*, and *Journal of Applied Sport Psychology*.

Chapter 2

1. $H = 15.9$, $L = 9.1$, $R = 15.9 - 9.1 = 6.8$, $N = 72$.

2. $i = 6.8/15 = 0.453$; i is rounded to 0.5.

3.

X	f	Cum. f
15.5-15.9	1	72
15.0-15.4	2	71
14.5-14.9	3	69
14.0-14.4	2	66
13.5-13.9	4	64
13.0-13.4	8	60
12.5-12.9	12	52
12.0-12.4	10	40
11.5-11.9	11	30
11.0-11.4	5	19
10.5-10.9	5	14
10.0-10.4	5	9
9.5-9.9	2	4
9.0-9.4	2	2
	$N = 72$	

4. Real limits of group 12.0 to 12.4 are 11.95 to 12.449 . . . .

5. Histogram:

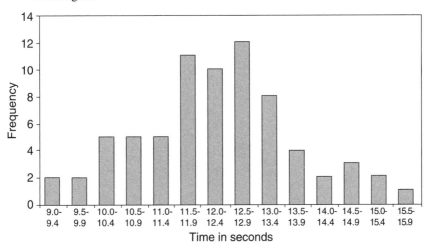

6. Frequency polygon:

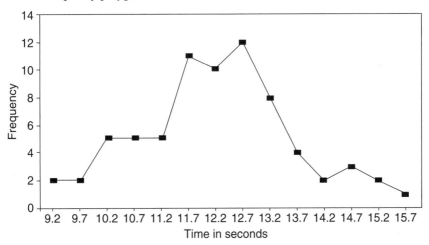

The data appear to be reasonably normal.

    7. With more cases, the curve would probably smooth out and approach a more normal shape.

8. Cumulative frequency curve:

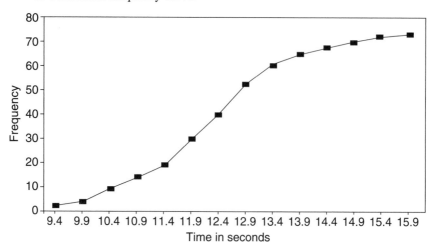

9. Computer answers should be the same as hand-calculated answers, except for small differences due to rounding during hand calculations. Computerized graphs may differ slightly in the size of the bars and the labeling of the axes, but the shape should be similar.

Chapter 3

1. (a) $4/11 \times 100 = 36.4\%$; (b) $8/11 \times 100 = 72.7\%$.
2. (a) $.7 \times 11 = 7.7$ from bottom—nearest score is 10; (b) $.45 \times 11 = 4.95$ from bottom—nearest score is 8.
3. (a) $67/70 \times 100 = 95.7\%$; (b) $4/70 \times 100 = 5.7\%$.
4. (a) $.47 \times 70 = 32.9$ from bottom—score is 10; (b) $.80 \times 70 = 56$ from bottom—score is 13.
5. (a) 13.9%; (b) 30%; (c) 74.6%.
6. (a) 37 sit-ups; (b) 46 sit-ups; (c) 55 sit-ups.
7. Computer answers should be the same as hand-calculated answers, except for small differences due to rounding during hand calculations.

Chapter 4

1. (a) Mode = 8, median = 9, mean = 9.1. (b) Mode = 11, median = 12, mean = 12.1.
2. Computer answers should be the same as hand-calculated answers, except for small differences due to rounding during hand calculations.

Chapter 5

1. 4.3.
2. 4.6.
3. $(4.3)^2 = 18.49$.
4. $N = 49$; $SD = 1.97$. Mode = 24, Median = 24, Mean = 24.16.
5. Computer answers should be the same as hand-calculated answers, except for small differences due to rounding during hand calculations.

Chapter 6

1. 13.68.
2. (a) –0.79; (b) 0.62; (c) –0.23. All are within 1 SD of the mean; a and c are below, and b is above.
3. (a) 21.48%; (b) 73.24%; (c) 40.90%.
4. (a) 42.1; (b) 56.2; (c) 47.7.
5. (a) 3; (b) 6; (c) 5.
6. (a) 1.0 or 1 to 1; (b) .333 or 1 to 3; (c) .111 or 1 to 9; (d) .053 or 1 to 19; (e) .0101 or 1 to 99.

Chapter 7

1. (a) 3.57; (b) .20; (c) 13.01. As standard deviation increases, the standard error of the mean increases. As N increases, the standard error of the mean decreases.
2. $SE_M = 0.79$; $\mu = 38.5 \pm 1.96\,(0.79)$; $\mu = 38.5 \pm 1.55$, $p = .05$. This could be written as $36.95 \le \mu \le 40.05$.
3. Answers will vary based on class data.

Chapter 8

1. (a) +; (b) +; (c) +; (d) –; (e) +; (f) 0; (g) –; (h) –; (i) –; (j) –.
2. (a) $p < .05$. (b) $p < .01$. (c) not significant.
3. (a) $\bar{X} = 24.23$ dips, $\bar{Y} = 13.47$ pull-ups; (b) $r = .776$; (c) better than 99%, $p < .01$; (d) 11.72 pull-ups; (e) $SE_E = 2.96$; (f) $11.72 \pm 1.96\,(2.96) = 11.72 \pm 5.80$, or 5.92 to 17.52.
4. Relationship between RT and MT is not significant. $r = -.034$; RT mean = 240.6; $SD_{RT} = 58.19$; MT mean = 340.6; $SD_{MT} = 94.74$.

5. (a) Mile run: mean = 535.72, SD = 133.44; $\dot{V}O_2$max: mean = 41.26, SD = 10.31, $r = -.94$ (b) $p < .001$. The correlation is negative because high mile run–walk times are associated with low $\dot{V}O_2$max values. (c) Slope = $-.073$, Y-intercept = 80.17, SE_E = 3.63. (d) Y_P = 80.17 – 0.073 (540) = 40.75; 95% CI = 40.75 ± (1.96)(3.63) = 40.75 ± 7.11 = 33.64 to 47.86 milliliters per kilogram per minute.

6. Answers will differ based on various data sets. Residual is equal to the difference between the actual and predicted weight.

Chapter 9

1. Step 1: X_1 = age (years); Y-intercept (a) = -56.23 Newton meters; slope (b) = 7.16 Newton meters per year, R^2 =.65, SE_E = 11.56 Newton meters.

Step 2: X_1 = age (years); X_2 = fat-free weight (kilograms); $a = -51.39$; b_1 = 4.95 Newton meters per year; b_2 = 0.55 Newton meters per kilogram; R^2 =.71; SE_E = 10.69 Newton meters. The increment in R^2 from step 1 to step 2 was significant ($F = 4.74$, $p = .041$).

Step 3: X_1 = age (years); X_2 = fat-free weight (kilograms); X_3 = height (m); $a = -28.09$; b_1 = 5.46 Newton meters per year; b_2 = 0.72 Newton meters per kilogram; b_3 = -0.24 Newton meters per centimeter; R^2 =.72; SE_E = 10.77 Newton meters. The increment in R^2 from step 2 to step 3 was not significant ($F = 0.67$, $p = .42$).

2. Step 1: standard error for b (SE_b) = 1.13; $df = N - k - 1 = 24 - 1 - 1 = 22$; critical $t = 2.074$; 95% CI = $b \pm t(SE_b)$ = 7.16 ± 2.074 (1.13) = 4.82 to 9.50 Newton meters per year.

Step 2: For age, SE_b = 1.46, df = 21, critical $t = 2.08$, 95% CI = 4.95 ± 2.08 (1.46) = 1.91 to 7.99 Newton meters per year.

For fat-free weight, SE_b = 0.25, df = 21, critical $t = 2.08$, 95% CI = 0.55 ± 2.08 (0.25) = 0.03 to 1.07 Newton meters per kilogram.

Step 3: For age, SE_b = 1.59, df = 20, critical $t = 2.086$, 95% CI = 5.46 ± 2.086 (1.59) = 2.14 to 8.78 Newton meters per year.

For fat-free weight, SE_b = 0.33, df = 20, critical $t = 2.086$, 95% CI = 0.72 ± 2.086 (0.33) = 0.03 to 1.41 Newton meters per kilogram.

For height, SE_b = 0.29, df = 20, critical $t = 2.086$, 95% CI = -0.24 ± 2.086 (0.29) = -0.84 to 0.36 Newton meters per centimeter. Notice that zero is included in the 95% CI.

3. Reversing the order of entry for age and fat-free weight has no effect on the equation or the interpretation of the data.

Chapter 10

1. (a) 3.81; (b) 1.19.

2. (a) $t = -.80$, not significant; (b) $t = 2.27$, $df = 23$, $p < .05$.

3. The standard error of the difference is the standard deviation of an infinite set of scores, each of which was derived by computing the difference between the means of two samples randomly drawn from the same population. It is the amount of difference expected between two randomly drawn sample means due to chance alone.

4. (a) No significant difference, accept H_0, $t = .511$. Mean active = 2.282 ($SD = 1.244$), mean passive = 1.966 ($SD = 1.506$), $SE_D = 0.618$. (b) Computer answers should be the same as hand-calculated answers, except for small differences due to rounding during hand calculations.

5. $SE_D = 0.066$, $df = 35$, $t = 3.03$, $p < .01$, $\omega^2 = .18$, $ES = 0.87$. The difference is significant, the effect of being on the cross-country team on stride length is large, and 18% of the variability in stride length can be explained by participation on a cross-country ski team.

6. $SE_{M1} = 2.84$, $SE_{M2} = 3.18$, $SE_D = 1.78$, $t = 2.81$. The research hypothesis could be used because it is logical to assume that the dominant hand would be stronger. For $df = 19$, a one-tailed test is significant at $p < .01$ and a two-tailed test is significant at $p < .05$.

7. (a) Males: mean = 20.0, $SD = 3.16$. Females: mean = 21.17, SD = 2.32. $SE_D = 1.65$, $t = -.707$, not significant. (b) Accept. (c) Computer answers should be the same as hand-calculated answers, except for small differences due to rounding during hand calculations.

8. (a) Yes. First: mean = 39.8, $SD = 11.86$. Second: mean = 47.8, $SD = 9.83$. $SE_D = 1.50$, $r = .921$, $t = -5.34$. (b) $p < .01$, reject null. (c) Computer answers should be the same as hand-calculated answers, except for small differences due to rounding during hand calculations.

9. Area for $Z_b = 40\%$, $Z_b = 1.28$. The value of N needed for 90% power at $p = .05$ is approximately 9 subjects per group.

Chapter 11

1. (a) Means: poor = 6.00, average = 7.83, good = 11.00. (b–d) ANOVA table:

Source	SS	df	MS	F	p
Between	76.78	2	38.39	15.61	< .01
Within	36.83	15	2.46		
Total	113.61	17			

(e) Mean difference table:

	1 (Good)	2 (Average)	3 (Poor)
1 (Good)	0.00	1.83	5.00*
2 (Average)		0.00	3.17**
3 (Poor)			0.00

* $p < .01$ for both Scheffé and Tukey.
** $p < .05$ for Scheffé and $p < .01$ for Tukey.

(f) $I_{.05} = 2.46$; $I_{.01} = 3.23$; 1 vs. 3, 2 vs. 3 differ. (g) $HSD_{.05} = 2.35$; $HSD_{.05} = 3.09$; 1 vs. 3, 2 vs. 3 differ. (h) Scheffé and Tukey agree on groups 1 and 3, $p < .01$. Scheffé finds 2 and 3 different at $p < .05$ but Tukey finds 2 and 3 different at $p < .01$. (i) $R^2 = .68$, $\omega^2 = .62$, effect is large.

2. (a) Control: mean = 26.5, SD = 4.53; weight: mean = 28.8, SD = 3.52; jump: mean = 32.6, SD = 5.50. (b) $F = 6.303$, yes, $p = .006$. (c) Control vs. jump, $p = .004$; control vs. weight, not significant; jump vs. weight, not significant. (d) Reject H_0 for control vs. jump only.

3. (a) independent variable = activity level; dependent variable = body fat. (b)

	Inactive	Semi	Normal	Active	Very Active
Means	27.13	24.97	22.22	16.12	12.35
SD	5.80	4.37	2.73	3.73	1.95

(c) Yes, $F = 14.68$, $p < .001$. (d) Mean difference table:

	Inactive	Semi	Normal	Active	Very Active
Inactive	0.00	2.16	4.91	11.01*	14.78*
Semi		0.00	2.75	8.85*	12.62*
Normal			0.00	6.10	9.87*
Active				0.00	3.77
Very Active					0.00

$HSD_{.01} = 8.33$, *$p < .01$

Chapter 12

1. (a) Means = 16.6, 20.8, 25.0, 30.6, 35.4. (b–e) ANOVA repeated measures table:

Source	SS	df	MS	F	p
Columns	1127.04	4	281.76	98.52	< .01
Rows	220.64	4	55.16	19.29	< .01
Error	45.76	16	2.86		
Total	1393.44	24			

(f) Yes. $F = 98.52$, $p < .01$, indicates that at least two of the means are significantly different. (g) With ten pairwise comparisons, $\alpha_{FW} = .005$ for Bonferroni correction; all mean differences are significant except minute 2 versus minute 4.

	Machine 1	Machine 2	Machine 3	Machine 4	Control
Means	623.60	597.70	568.30	586.30	620.50
SD	367.15	392.00	333.73	448.57	395.25

2. (a)

(b) No, $F = .492$, not significant. (c) $p = .742$ by computer analysis. (d) No, when F is not significant, further mean difference comparisons are not justified.

Chapter 13

Source	SS	df	MS	F	p
Between (or Subjects)	3,962.6	19	208.56 (MS_B 1-way) (MS_S: 2-way)	32.92	
Within	248.0	40	6.2 (MS_W)		
Trials	7.2	2	3.62 (MS_T)	0.57	0.57
Error	240.8	38	6.34 (MS_E)		
Total	4,210.6				

1.
The nonsignificant effect for trials from the ANOVA suggests that systematic error is not present in the data.

2. ICC 2,1 = 0.92; ICC 3,1 = 0.91.

3. $SEM = \sqrt{MS_E} = \sqrt{6.34} = 2.52°$.

4. $T = 35 \pm 1.96\ (2.52) = 30.1$ to $39.9°$.

5. $MD = 2.52° \times 1.96 \times \sqrt{2} = 6.99°$.

Chapter 15

Source	SS	df	MS	F	p
Covariate	35,475.2	1	35,475.2	42.85	<.001
Group	6,856.4	2	3,428.2	4.14	.028
Error	21,524.8	26	827.9		
Total	57,500.0				

1. The statistically significant effect for group indicates that a differential response occurred over the 8-week period among the three groups.

2. The test for the homogeneity of regression assumption yielded an F value of 0.18 with 2 and 24 degrees of freedom, $p = .84$. This indicates that the homogeneity of regression assumption was not violated.

3. The adjusted means for the three groups are as follows: (1) control = 157.57; (2) low intensity = 170.76; (3) high intensity = 196.68. From an ANOVA on the covariate we find that $MS_{Bcov} = 3,520.8$ and $SS_{Wcov} = 42,875.0$, and from the previous ANCOVA analysis we find $MS_{EANOVA} = 827.9$ and $n = 10$. Using equation 15.02, the Q is as follows: (1) control vs. low intensity = 1.39; (2) control vs. high intensity = 4.13; (3) low intensity vs. high intensity = 2.74. From table A.10, with df $= N - J - 1 = 30 - 3 - 1 = 26$, the critical Q = 3.591. The only difference that is statistically significant is high intensity > control.

Chapter 16

1. No significant difference. Chi-square = 1.46, not significant. $df = (2 - 1)$ $(2 - 1) = 1$. See following tables for calculations.

	Yes		No		
	O	E	O	E	Total
Boys	27	30.6	42	38.5	69
Girls	35	31.4	36	39.5	71
Total	62	62	78	78	140
% expected	44.3		55.7		

O	E	O - E	(O - E)²	(O - E)²/E
27	30.6	−3.6	12.96	0.42
35	31.4	3.6	12.96	0.41
42	38.5	3.5	12.25	0.32
36	39.5	−3.5	12.25	0.31
	Chi-square = 1.46			

2. $\rho = .88$; $p < .01$.

3. No; difference is not significant.

$$\Sigma R_{West} = 134; \Sigma R_{East} = 191.$$

$$U_{West} = (9)(16) + \left[\frac{(9)(9+1)}{2} \right] - 134 = 55$$

$$U_{East} = (9)(16) - 55 = 89$$

$$Z_{West} = \frac{55 - \frac{(9)(16)}{2}}{\sqrt{\frac{(9)(16)\ (9+16+1)}{12}}} = -.96, \text{ N.S.}$$

4. Yes; live is better than control.

$H = (.05) (1,115.2) - 48 = 7.76; df = 3 - 1 = 2; p < .05.$

Mean difference table:

	Control	Video	Live
Control	0.0	5.6	7.6
Video		0.0	2.0
Live			0.0

$SE = 2.83$; critical difference $= 2.83 \times 2.39 = \pm6.67$
* $p = .017$

5. Yes; the students like baseball best. *Note:* data are fabricated.

Chapter 17

1. (a) The relative risk is ~0.33. This means that those who wore the knee braces had about one-third the risk of an ACL tear compared with those who did not wear the brace. This suggests the knee braces are protective against ACL tears. (b) The attributable risk reduction ~0.22. (c) The number needed to treat ~4.5. Therefore, we estimate that we will prevent one ACL tear for every four to five linemen who wear a knee brace.

2. The 2×2 contingency table is as follows:

	Cases	Controls	
Knee brace	A = 21	B = 37	60
No knee brace	C = 29	D = 13	45
	50	50	Total = 100

The odds ratio is ~0.25. Therefore the odds that linemen with torn ACLs wore a knee brace are about one fourth of the odds that noninjured linemen wore a knee brace.

3. (a) 83.6%; (b) 85.1%; (c) 80.6%; (d) 87.5%; (e) 5.6; (f) 0.2.

Glossary

actual mean difference—The difference between the raw score mean values in the numerator of a t test.

adjusted means—In analysis of covariance, where group means are modified based on the regression of the covariate on the dependent variable.

alpha (α)—(1) the area under the normal curve for rejection of H_0; (2) the probability of chance occurrence; (3) the probability of a type I error if the null hypothesis is true.

alternate hypothesis—The statistical hypothesis that is the logical alternative to the null hypothesis; also known as research hypothesis.

analysis of covariance (ANCOVA)—The adjustment of dependent variable mean values to account for the influence of one or more covariates that are not controlled by the research design.

analysis of variance (ANOVA)—An F value that represents the ratio of between-group and within-group variance.

apparent limit—The integer or discrete value listed for each group in a grouped frequency distribution.

backward elimination—A process of multiple regression where a computer algorithm sequentially removes independent variables from the regression model.

best fit line—(1) a line on a scatter plot that best indicates the relationship between the plotted values; (2) a line on a scatter plot that balances the positive and negative residual values so that they sum to zero.

beta (β)—The probability of a type II error if the null hypothesis is false.

beta weight—A coefficient in multiple regression indicating the weight to be assigned to each independent variable in the prediction equation. Beta weights are in the form of Z scores when the independent variables have been converted to standard scores.

between–between—A factorial analysis of variance where factor A is a between analysis of independent groups and factor B is a between analysis on a second set of independent groups.

between-group variance—In analysis of variance, the deviation of a set of group means from the grand mean.

between–within—A factorial analysis of variance where factor A is a between analysis of independent groups and factor B is a within (repeated measures) analysis on the same subject. This is sometimes called a mixed model.

bias—The factors operating on a sample so it is not representative of the population from which it was drawn. Error introduced into the research model by investigators.

bimodal—A distribution of values with more than one mode.

bivariate regression—Regression analysis applied to one independent variable (X) and one dependent variable (Y).

Bonferroni adjustment—An adjustment of the p value (probability of error) to correct for a familywise error rate when making multiple comparisons on the same set of subjects.

Bryant-Paulson procedure—In analysis of covariance, a modification of Tukey's honestly significant difference procedure for conducting pairwise comparisons.

ceiling effect—The phenomenon that makes it more difficult to improve as a raw score approaches the maximum possible score.

central limit theorem—A theorem stating that the means of a series of samples randomly selected from a population will be normally distributed even if the population from which they were taken is not normal.

central tendency—Values that describe the middle, or central, characteristics of a set of data. The three values of central tendency are the mode, median, and mean.

chi-square—A nonparametric statistical technique for determining the significance of the difference between frequency counts on nominal data.

circularity—The assumption of sphericity applied to the pooled data (across all groups) in a between–within or within–within factorial analysis of variance.

coefficient—A known numerical quantity used to explain or modify a variable.

coefficient of determination—The amount of shared variance between two variables; typically quantified by squaring the correlation coefficient (r^2).

coefficient of nondetermination—The amount of variance that is not shared between variables; typically quantified as $1 - r^2$.

confidence interval—The range of values between which the true or population value is believed to lie; usually defined by a given probability value (e.g., 95% confidence interval).

constant—A characteristic of a person, place, or thing that can assume only one value.

continuous variable—A variable that can theoretically assume any level on a continuum of data depending only on the accuracy of the instrument used to measure the variable (i.e., it can be measured to the nearest 10th, 100th, 1,000th, and so on and no gaps exist in the range of the data).

correlation—(1) a numerical coefficient between +1.00 and –1.00 that indicates the extent to which two variables are related or associated; (2) the extent to which the direction and size of deviations from the mean in one variable are related to the direction and size of deviations from the mean in another variable.

covariance—An estimate of the degree to which two variables vary together.

covariate—In analysis of covariance, the variable that is used to adjust the scores on the dependent variable.

criterion variable—Another name for the dependent variable.

critical ratio—The ratio or numerical result of a *t* test that must be met to reach a given level of confidence when rejecting the null hypothesis.

cross validation—A confirmation of the accuracy of a bi- or multivariate regression equation by applying the equation to a second independent sample taken from the same population.

cumulative frequency graph—A line graph of ordered scores plotted against the frequency of the scores that fall at or below a given score.

curve—A line on an *X-Y* plot representing the succession of change on *Y* as values of *X* are altered.

curvilinear—A plot of bivariate data where the best fit line is not straight.

data—Information gathered by measurement.

decile—One-tenth of the range of values.

degrees of freedom—The number of values in a data set that are free to vary when restrictions are imposed on the set.

dependent sample—A sample related to another sample and dependent on it. Dependent samples usually contain the same people measured more than once (i.e., pre–post comparisons).

dependent *t* test (aka paired *t* test)—A *t* test comparing means from two related samples; typically the means come from the same subjects tested twice; also known as paired *t* test.

dependent variable—A variable whose value is partially determined by the effects of other variables. It is not free to assume any value. It is usually the variable that is measured in the research design.

descriptive research—An attempt to determine the current state of events or conditions.

descriptive statistics—Numerical values that describe a current event or condition.

discrete variable—A variable that is limited in its assessment to certain values, usually integers (i.e., the data is not continuous; gaps exist between values in the range of the data).

discriminant analysis—Classification of subjects into like groups based on certain measureable characteristics of the subjects.

effect size—(1) an estimate of the percent of total variance between means that can be attributed to the result of treatment; (2) a measurement of the magnitude of the difference between two mean values independent of sample size; (3) a measurement of the magnitude of association between variables.

eigenvalue—A coefficient used in factor analysis that indicates the number of original variables that are associated with a single factor.

eta squared—A measure of effect size in analysis of variance; same as R^2.

evaluation—The philosophical process of determining the worth, or value, of the data.

Excel—A computer program that can be used in creating spreadsheets.

expected mean difference—Another term for standard error of the difference.

experimental design—A research process that involves manipulating and controlling certain events or conditions to solve a problem.

external validity—The ability to generalize the results of an experiment to the population from which the samples were drawn.

F ratio—In analysis of variance, the ratio of the mean squares.

factor—(1) a component in the design of a study that is combined with other components to answer multiple questions about the data; (2) a virtual variable that is the result of a combination of two or more variables in a factor analysis design.

factor analysis—A statistical technique used to identify the common components (factors) among multiple variables.

factorial analysis of variance—Analysis of variance performed on more than one factor simultaneously [e.g., the effects of gender (factor A) and treatment (factor B) on a dependent variable]. Factorial analysis of variance permits evaluation of the interaction of the factors on the dependent variable.

factor rotation—A mathematical technique in which the factors are theoretically rotated in three-dimensional space in an attempt to maximize the correlation among the original variables and the factor on which they load.

familywise error rate—An inflation of the error rate when making a series or family of comparisons on the same set of subjects; sometimes called inflated alpha.

floor effect—The phenomenon that can occur when there is a limit on how low scores can go and scores tend to bunch around that limit.

forward selection—A process of multiple regression in which a computer algorithm sequentially adds independent variables to the regression model.

frequency data—(1) values associated with raw scores indicating the number of subjects who received each raw score; (2) values associated with nominal data indicating the number of subjects classified into each category.

frequency polygon—A line graph of scores plotted against frequency.

Friedman's two-way analysis of variance—Nonparametric test similar to repeated measures analysis of variance used to determine the significance of the differences among groups of ranked data collected as repeated measures.

graph—(1) a diagrammatic representation of quantities designed to show their relative values; (2) a visual representation of data.

Greenhouse-Geisser adjustment—A conservative adjustment to the degrees of freedom in repeated measures analysis of variance to correct for violation of the assumption of sphericity.

grouped frequency distribution—An ordered listing of the values of a variable organized into groups with a frequency column indicating the number of cases included in each group.

heteroscedasticity—A condition in which the variances of the residuals of the independent variables in multiple regression are not equal.

hierarchical multiple regression—A multiple regression technique in which the investigator specifies the order of inclusion of the independent variables in the solution.

histogram—A graph plotting blocked scores against frequency; commonly known as a bar graph.

historical research—A search of the records of the past in an attempt to determine what happened and why.

homogeneity of regression—In analysis of covariance, the assumption that the relationship between the covariate and the dependent variable is the same across all levels of the independent variable.

homogeneity of variance—Equality of the variances among a set of two or more measures.

homoscedasticity—A condition in which the variance of the residuals of each of the independent variables in multiple regression is equal or nearly so.

Huynh-Feldt adjustment—A liberal adjustment to the degrees of freedom in repeated measures analysis of variance to correct for violation of the assumption of sphericity.

hypothesis—A prediction or assumption that can be tested to determine whether it is correct.

independent sample—A sample that is unrelated to a second sample. Independent samples usually contain different subjects in each sample.

independent *t* test—A t test performed to compare the means from two independent samples; typically each sample is from a different group of subjects.

independent variable—(1) variable that is free to vary and that is not dependent on the influence of another variable; (2) a variable in the research design that is permitted to exert influence over other variables (i.e., the dependent variable) in the study. The independent variable is usually controlled by the research design.

instrument error—Bias or error in the data produced by inaccurate or improperly calibrated instruments.

interaction—The combined effect of two or more factors on the dependent variable.

interindividual variability—The amount of variability in measurements that can be attributed to differences between two or more different subjects (people).

internal validity—The ability of the research design to determine that the results are due to the treatment applied.

interquartile range—The range in raw score units from the 25th percentile (Q_1) to the 75th percentile (Q_3).

inter-rater reliability—The reliability of different raters testing the same subjects.

interval scale—A parametric scale of measurement with equal units or intervals between data points, but that has no absolute zero point (e.g., the Fahrenheit scale of temperature where negative values are possible). Ratio comparisons among the data points are not appropriate.

interval size—The numerical size of each group in a group frequency distribution (i.e., the number of data points in the group).

intervening variable—An extraneous variable not controlled by the research design that has an unintended or unknown effect on the dependent variable.

intraclass correlation coefficient—A method of determining relative reliability (unit-less) using mean square terms from repeated measures analysis of variance; it may be used on two or more repeated measures.

intraindividual variability—The amount of variability in measurements that can be attributed to changes within the individual subjects (people).

intrarater reliability—The reliability of the same rater testing subjects on multiple occasions.

investigator error—Bias or error in the data produced by the human limitations of the investigator.

J-curve—A curve that results when frequency is high at one end of the X axis, decreases rapidly, and flattens at the other end of the scale.

kinesiology—The study of the art and science of human movement.

Kruskal-Wallis analysis of variance—A nonparametric test similar to analysis of variance used to determine the significance of the differences among groups of ranked data.

kurtosis—A measure of the vertical deviation from normality (amount of peakedness or flatness) in the plot of a data set.

leptokurtic—A curve that is more peaked than a mesokurtic curve.

level of confidence—(1) the amount of confidence that can be placed in a conclusion; (2) a value expressed as a percentage that establishes the probability that a statement is correct.

main effects—The F value based on marginal means in a factorial analysis of variance.

Mann-Whitney U test—A nonparametric statistical technique for determining the significance of the difference between rankings of two groups of subjects who have been ranked on the same variable.

marginal mean—The average score across all groups or all trials in a factorial analysis of variance.

mean—The arithmetic average.

mean square—In analysis of variance, the average of the squared deviations from the mean of a set of scores.

measurement—The process of comparing with a standard.

measurement error—The difference between an observed score and the true score.

median—The 50th percentile, or the score that falls midway in the range of ordered values.

mesokurtic—A typical, bell-shaped, normal curve.

meta-analysis—A process whereby a systematic review of the relevant research literature is conducted and data from the multiple studies are statistically analyzed together.

minimal difference (*MD*)—A value equal to the smallest difference on repeated testing that cannot reasonably be attributed to chance or noise in the measurement; calculated from the standard error of measurement in reliability assessments. Also known as minimal detectable difference.

mixed design—A between–within research design in factorial analysis of variance; also known as mixed model.

mode—The score that occurs the most often in a distribution of values.

motor learning—A study of how people learn motor skills including the neurological and muscular components of the process.

multicollinearity—A condition in which two or more independent variables in multiple regression are highly correlated with each other.

multiple analysis of variance (MANOVA)—The simultaneous analysis of two or more dependent variables in a research design using analysis of variance.

multiple correlation—The simultaneous correlation between multiple independent variables and the dependent variable.

multiple regression—Regression analysis applied to more than one independent variable (X_1, X_2, X_3, and so on) and one dependent variable (Y).

multivariate—Data sets consisting of measurement of more than one variable per subject.

negative likelihood ratio—In a diagnostic test, a ratio of probabilities in which the numerator is the probability of a negative test in a person with the condition and the denominator is the probability of a negative test in a person without the condition.

negative predictive value—In a diagnostic test, the proportion of individuals with a negative test who do not have the condition.

nominal scale—A nonparametric classification of data based on names of categories and the frequency of occurrence within each category.

nonparametric—Data that do not meet the assumption of normality.

normal curve—A curve, which has known characteristics, formed by the bilaterally symmetrical, bell-shaped distribution of values around the mean.

null hypothesis—A hypothesis that predicts the absence of a relationship among subjects or no differences between or among groups of subjects; traditionally symbolized as H_0. It is typically the hypothesis that is tested statistically.

objectivity—An appraisal of the amount of bias in the measurement process.

odds—Ratio of the number of occurrences of an event happening to the number of occurrences of an event not happening.

odds ratio—In a case control study, the ratio of the odds of exposure to a purported risk or protective factor in cases to the odds of exposure in controls.

omega squared (ω^2)—A measure of effect size that determines the percent of total variance that may be attributed to treatment.

one-tailed test—A test of the research hypothesis wherein the difference between two mean values is predicted to be significant. It uses only one tail of the normal curve.

ordinal scale—A nonparametric listing of data based on order without consideration of the absolute value of each data point (e.g., a listing from highest to lowest; first, second, third).

outlier—A value in a data set that lies beyond the limits of the typical scores.

paired *t* test—See dependent *t* test.

parameter—A characteristic of a population.

parametric—Data that meet the assumptions of normality.

partial correlation—The correlation (common variance) between an independent variable and a dependent variable that remains after the influence of a second variable has been removed.

partial eta squared—A measure of effect size in analysis of variance; equal to eta squared in simple analysis of variance.

Pearson's product moment correlation coefficient—A correlation coefficient appropriate for use on parametric data. A measure of the relationship or association between two parametric variables symbolized by *r*.

percentile—A point or position on a continuous scale of 100 theoretical divisions such that a certain fraction of the population of scores lies at or below that point.

platykurtic—A curve that is more flat than a mesokurtic curve.

population—A group of people, places, or things that have at least one common characteristic.

positive likelihood ratio—In a diagnostic test, the ratio of probabilities in which the numerator is the probability of a positive test in a person with the condition and the denominator is the probability of a positive test in a person without the condition.

positive predictive value—In a diagnostic test the proportion of individuals with a positive test who have the condition.

post hoc—After the fact.

power—The ability of a test to correctly reject a false null hypothesis.

predictor variable—Another name for the independent variable.

probability—The long-run proportion of a particular outcome; varies between zero and 1.0.

probability of error—(1) the probability that a statement is incorrect; (2) a value expressed as a decimal that establishes the probability that a statement is incorrect (i.e., the mean weight of college men is greater than the mean weight of college women; $p < .001$).

quartile—One-fourth of the range of values.

quintile—One-fifth of the range of values.

R squared (R^2)—(1) a measure of effect size in analysis of variance; (2) the ratio of the between-group variance and the total variance in analysis of variance; same as eta squared.

random error—Unpredictable error due to random fluctuations of observed scores about the true score (i.e., noise).

random sample—A sample taken from a population where every member of the population has an equal chance of being selected in the sample.

range—The numerical distance from the highest to the lowest score.

rank order correlation—Same as Spearman's rho.

rank order distribution—An ordered listing of data in a single column.

ratio scale—A parametric scale of measurement based on order, with equal units of measurement and an established zero point (i.e., data based on time, distance, force, or counting events).

real limits—The assumed upper and lower values for a group in a grouped frequency distribution that include all possible values on a continuous scale.

rectangular curve—A curve that results when the frequencies of the values in the middle of the data are the same.

region of rejection—The area of a sampling distribution that contains scores that would result in the rejection of the null hypothesis.

regression—(1) a method of predicting values on the Y-variable based on a value on one or more X-variables and the relationship between the variables; (2) a statistical term meaning prediction.

relative risk—A ratio of the rate of exposure to a purported risk (or beneficial factor) in individuals who have a condition divided by the rate of exposure in individuals who do not have the condition.

reliability—A measure of the consistency of the data when measurements are taken more than once under the same conditions. See intraclass reliability.

repeated measures—Measuring the same set of subjects more than once, as in a pre–post comparison; same as within-subjects design.

research—A special technique for solving problems.

research hypothesis—A hypothesis that typically prompts research. It is usually a prediction that significant relationships exist among subjects or significant differences exist between or among groups of subjects. It is traditionally symbolized as H_1.

residual—The vertical distance from any point in a scatter diagram to the line of best fit.

robust—Reasonably reliable results, even if assumptions are not met totally.

SAS—Statistical analysis system.

sample—A portion or fraction of a population.

sampling distribution of the mean—A frequency distribution of sample means.

sampling distribution of mean differences—A frequency distribution of mean differences.

sampling error—The amount of error in the estimate of a population parameter based on a sample statistic.

scatter plot—A plot of bivariate data with one variable on the X-axis and the other on the Y-axis that produces a visual picture of the relationship between the variables.

Scheffé's confidence interval—A post hoc test conducted after a significant analysis of variance to determine the significance of all possible combinations of cell contrasts.

semi-interquartile range—Half the interquartile range.

sensitivity—The true positive rate of a diagnostic test.

shrinkage—The reduction in the multiple correlation coefficient observed in the cross validation process.

significant—A statistical term meaning that a relationship or a mean difference is not due solely to chance.

simple effect—The F value based on a single group or a single trial in factorial analysis of variance.

simple frequency distribution—An ordered listing of the values of the variable with a frequency column that indicates the number of cases for each value.

singularity—A condition in which two or more independent variables in multiple regression are perfectly correlated ($r = 1.00$) with each other.

skewed—A plot of values that is not normal (i.e., a disproportionate number of subjects fall toward one end of the scale—the curve is not bilaterally symmetrical).

skewness—A measure of lateral deviation from normality (bilateral symmetry) in the plot of a data set.

Spearman's rho—A nonparametric correlation technique for determining the significance of the relationships among ordinal data sets; same as rank order correlation. Also known as Spearman's rank order correlation coefficient.

specificity—The true negative rate of a diagnostic test.

sphericity—An assumption that in repeated measures designs the variances of the difference scores are all equal.

spreadsheet—Computer term for table.

SPSS—Formerly known as Statistical Package for the Social Sciences, now referred to as PASW (Predictive Analytic SoftWare).

standard deviation—(1) a measure of the spread, or dispersion, of the values in a parametric data set standardized to the scale of the unit of measurement of the data; (2) the square root of the average of the squared deviations around the mean.

standard error of measurement—An absolute index of reliability (units are the same as the units of the variable); can be used to construct confidence intervals about individual scores and about individual change scores.

standard error of the difference—(1) the amount of difference between the means of two randomly drawn samples that may be attributed to chance alone; (2) the denominator in the t test. Also known as standard error of mean differences.

standard error of the estimate—A numerical value that indicates the amount of error in the prediction of a Y value in bivariate or multivariate regression.

standard error of the mean—The numeric value that indicates the amount of error in the prediction of a population mean based on a sample.

standard score—A score that is derived from raw data and that has a standard basis for comparison (i.e., it has a known central tendency and a known variability).

stanine—A standard score based on the division of the normal curve into nine sections, each of which is one-half of a standard deviation wide, with a mean of 5 and a range of 1 to 9.

statistic—A characteristic of a sample drawn randomly from a population.

statistical inference—To infer certain characteristics of a population based on random samples taken from that population.

statistics—A mathematical technique by which data are organized, treated, and presented for interpretation and evaluation.

step-down process—A step-by-step process comparing main effects first, simple effects second, and cell contrasts last in a factorial analysis of variance.

stepwise multiple regression—A multiple regression technique in which the order of inclusion of the independent variables in the solution is determined according to the amount of explained variance each independent variable can offer, and in which previously included variables can be subsequently removed.

stratified sample—A series of samples taken from various subgroups of a population so that the subgroups are represented in the total sample in the same proportion that they are found in the population.

sum of squares—The sum of the squares of the deviations of each score from the mean in a set of data.

systematic error—Differences between observed and true scores that follow a pattern or trend.

***T* score**—A standard score with a mean of 50 and a standard deviation of 10.

***t* test**—See dependent (paired) and independent *t* tests.

terminal statistic—A statistic that does not provide information that can be used in further analysis of the data.

test–retest reliability—The reliability determined when a test is administered to the same individuals on multiple occasions.

theory—A belief about a concept or series of concepts.

tolerance—The denominator of the variance inflation factor.

total variance—The sum of between-group and within-group variability.

Tukey's honestly significant difference (*HSD*)—A post hoc test conducted after a significant analysis of variance to determine the significance of pairwise cell contrasts.

two-tailed test—A test of the null hypothesis wherein a difference between two mean values is predicted to be zero. It uses both tails of the normal curve.

type I error—Rejection of the null hypothesis when it is really true.

type II error—Acceptance of the null hypothesis when it is really false.

U-shaped curve—A curve that results when the frequencies of the values at the extremes of the scale are higher than the frequency of values in the middle.

validity—The soundness or correctness of a test or instrument in measuring what it is designed to measure (i.e., the truthfulness of the test or instrument).

variability—A measure of the spread or dispersion of a set of data. The four most commonly used measures of variability are range, interquartile range, variance, and standard deviation.

variable—A characteristic of a person, place, or thing that can assume more than one value on subsequent measurements.

variance—(1) a measure of the spread, or dispersion, of the values in a parametric data set; (2) the average of the squared deviations around the mean.

variance inflation factor (VIF)—An index of the extent of multicollinearity in a data set.

Venn diagram—A diagram of overlapping circles illustrating the amount of shared variance between variables.

within-group variance—In analysis of variance, the deviation of a set of values from the mean of the group in which they are included.

within-subjects design—Measurements within the same set of subjects; same as a repeated measures design.

within–within—A factorial analysis of variance in which factor A is a within (repeated measures) analysis and factor B is a second within (repeated measures) analysis on the same subject.

Y-intercept—The point at which the extension of the best fit line intercepts the Y-axis.

Z score—A standard score with a mean of 0 and standard deviation of 1; a score in standard deviation units.

References

AndersonBell. (1989). *ABstat users manual.* Parker, CO: Author.

Armitage, P., Berry, G., & Matthews, J.N.S. (2002). *Statistical methods in medical research (4th ed.).* Oxford: Blackwell Science.

Benjaminse, A., Gokeler, A., & van der Schans, C.P. (2006). Clinical diagnosis of an anterior cruciate ligament rupture: A meta-analysis. *Journal of Orthopaedic and Sports Physical Therapy,* 36(5), 267-288.

Bruning, J.L., & Kintz, B.L. (1977). *Computational handbook of statistics (2nd ed.).* Glenview, IL: Scott Foresman.

Bryant, J.L., & Paulson, A.S. (1976). An extension of Tukey's method of multiple comparisons to experimental designs with random concomitant variables. *Biometrika,* 63(3), 631-638.

Christenfeld, N.J.S., Sloan, R.P., Carroll, D., & Greenland, S. (2004). Risk factors, confounding, and the illusion of statistical control. *Psychosomatic Medicine,* 66, 868-875.

Cohen, J. (1968). Multiple regression as a general data-analytic system. *Psychological Bulletin,* 70, 426-443.

Cohen, J. (1988). *Statistical power analysis for the behavioral sciences (2nd ed.).* Hillsdale, NJ: Erlbaum.

Cooperman, J.M., Riddle, D.L., & Rothstein, J.M. (1990). Reliability and validity of judgments of the integrity of the anterior cruciate ligament of the knee using the Lachman's test. *Physical Therapy,* 70, 225-233.

Dixon, W.J. (Ed.) (1990). *BMDP statistical software manual.* Berkeley, CA: University of California Press.

Duoos, B.A. (1984). Fatigue and diagonal stride in cross-country skiing. In *Sport Biomechanics: Proceedings of the International Symposium of Biomechanics in Sports* (p. 219). Del Mar, CA: The Research Center for Sports, and Academic Publishers.

Egstrom, G. (1964). Effects of an emphasis on conceptualizing techniques during early learning of a gross motor skill. *Research Quarterly,* 35, 472-481.

Finney, C. (1985). Further evidence: Employee recreation and improved performance. *Journal of Employee Recreation, Health, and Education,* 28, 8-10.

Franks, B.D., & Huck, S.W. (1986). Why does everyone use the .05 significance level? *Research Quarterly for Exercise and Sport,* 57, 245-249. [Also see *Research Quarterly for Exercise and Sport* (1987), 58, 81-89, for comments and responses.]

Friedman, M. (1937). The use of ranks to avoid the assumption of normality implicit in the analysis of variance. *Journal of the American Statistical Association,* 32, 675-701.

Fritz, J.M., & Wainner, R.S. (2001). Examining diagnostic tests: An evidence-based perspective. *Physical Therapy,* 81, 1546-1564.

Girden, E.R. (1991). ANOVA. Repeated Measures. Thousand Oaks, CA: Sage Publications.

Huitema, B. (1980). *The Analysis of Covariance and Alternatives.* New York: John Wiley and Sons.

Jones, J.G. (1965). Motor learning without demonstration of physical practice under two conditions of mental practice. *Research Quarterly*, 36, 270.

Kachigan, S. (1986). *Statistical analysis*. New York: Radius Press.

Keppel, G. (1991). *Design and analysis: A researcher's handbook (3rd ed.)*. Englewood Cliffs, NJ: Prentice Hall.

Kotz, S., & Johnson, N.L. (Eds.) (1982). *Encyclopedia of statistical sciences (Vol. 3)*. New York: Wiley.

Leedy, P. (1980). *Practical research planning and design (2nd ed.)*. New York: Macmillan.

Lipsey, M.W., & Wilson, D.B. (2001). *Practical meta-analysis*. Applied Social Science Methods Series Volume 49, Thousand Oaks, CA: Sage Publications.

Looney, M.A. (2000). When is the intraclass correlation coefficient misleading? *Measurement in Physical Education and Exercise Science*, 4, 73.

Maxwell, S.E. (1980). Pairwise multiple comparisons in repeated measures designs. *Journal of Educational Statistics*, 5(3), 269-287.

McGraw, K.O., & Wong, S.P. (1996). Forming inferences about some intraclass correlation coefficients. *Psychological Methods*, 1: 30-46, 1996.

Mokone, G.G., Schwellnus, M.P., Noakes, T.D., & Collins, M. (2006). The COL5A1 gene and Achilles tendon pathology. *Scandinavian Journal of Medicine and Science in Sports*, 16, 19-26.

Morrow, J.R., Jackson, A.W., Disch, J.G., & Mood, D.P. (2000). *Measurement and evaluation in human performance (2nd ed.)*. Champaign, IL: Human Kinetics.

Nunnally, J.C., & Bernstein, I.H. (1994). Psychometric Theory. (3rd Edition). New York: McGraw-Hill.

Olds, E.G. (1938). Distributions of sums of squares of rank differences for small numbers of individuals, *The Annals of Mathematical Statistics*, 7, 133-148.

Olsen, O.-E., Myklebust, G., Engebretsen, L., Holme, I., & Bahr, R. (2005). Exercises to prevent lower limb injuries in youth sports: Cluster randomized controlled trial. *British Medical Journal*, 330(7489), 449.

Oxendine, J.B. (1969). Effect of mental and physical practice on the learning of three motor skills. *Research Quarterly*, 40, 755.

Richardson, A. (1967). Mental practice: A review and discussion. *Research Quarterly*, 38, 95.

Schutz, R.W., & Gessaroli, M.E. (1987). The analysis of repeated measures designs involving multiple dependent variables. *Research Quarterly for Exercise and Sport*, 58, 132-149.

Shrout, P.E., & Fleiss, J.L. (1979). Intraclass correlations: Uses in assessing rater reliability. *Psychological Bulletin*, 86: 420-428.

Smith, S.W. (1997). *The scientist and engineer's guide to digital signal processing (1st ed.)*. San Diego: California Technical Publishing.

Spence, J.T., Underwood, B.J., Duncan, C.P., & Cotton, J.W. (1968). *Elementary statistics (2nd ed.)*. New York: Appleton-Century-Crofts.

Spiegel, M.R. (1961). *Theory and problems of statistics*. New York: Schaum.

Statistical Analysis Systems. (2004). *SAS Institute*. Cary, NC: Author.

Statistical Package for the Social Sciences. (2004). *SPSS Inc.* Chicago: Author.

Streiner, D.L. (2007). A shortcut to rejection: How not to write the results section. *Canadian Journal of Psychiatry*, 562(6), 385-389.

Tabachnick, B.G., & Fidell, L.S. (1996). *Using multivariate statistics (3rd ed.)*. New York: HarperCollins College Publishers.

Thomas, J.R., Nelson, J.K., & Silverman, S. (2010). *Research Methods in Physical Activity (6th ed)*. Champaign, IL: Human Kinetics.

Thomas, J.R., Salazar, W., & Landers, D.M. (1991). What is missing in p less than .05? Effect size. *Research Quarterly for Exercise and Sport*, 62, 344.

Thorland, W.G., Tipton, C.M., Lohman, T.G., Bowers, R.W., Housh, T.J., Johnson, G.O., Kelly, J.M., Oppliger, R.A., & Tceng, T.-K. (1991). Midwest wrestling study: Prediction of minimal weight for high school wrestlers. *Medicine and Science in Sports and Exercise*, 23(9), 1102-1110.

Tran, Z.V. (1997). Estimating sample size in repeated measures analysis of variance. *Measurement in Physical Education and Exercise Science*, 1, 89.

Twinning, W.E. (1949). Mental practice and physical practice in learning a motor skill. *Research Quarterly*, 20, 432.

Vincent, W.J. (1968). Transfer effects between motor skills judged similar in perceptual components. *Research Quarterly*, 39, 380.

Wagoner, K.D. (1994). Descriptive discriminant analysis: A follow-up procedure to a "significant" MANOVA. Monograph presented at the 1994 American Alliance for Health, Physical Education, Recreation and Dance Convention, Denver, CO, April 13.

Weir, J.P. (2005). Quantifying test-retest reliability using the intraclass correlation coefficient and the SEM. *Journal of Strength and Conditioning Research*, 19(1), 231-240.

Whiting, P.F., Sterne, J.A.C., Westwood, M.E., Bachmann, L.M., Harbord, R., Egger, M., & Deeks, J.M. (2008). Graphical presentation of diagnostic information. *BMC Medical Research Methodology*, 8, 20.

Winer, B.J., Brown, D.R., & Michels, K.M. (1991). *Statistical principles in experimental design (3rd ed.)* New York: McGraw-Hill.

Witte, R.S. (1985). *Statistics (2nd ed.)*. New York: Holt, Reinhart & Winston.

Zar, J.H. (1972). Significance testing of the Spearman rank correlation coefficient. *Journal of the American Statistical Association*, 67, 578-580.

Zilak, S.T., & McCloskey, D.N. (2008). *The cult of statistical significance: How the standard error costs us jobs, justice, and lives*. Ann Arbor: University of Michigan Press.

Index

Note: The italicized *f* and *t* following page numbers refer to figures and tables, respectively.

About the Authors

William J. Vincent, EdD, is currently an adjunct professor and is the former director of the general education wellness program in the department of exercise sciences at Brigham Young University in Provo, Utah. He is professor emeritus and former chair of the department of kinesiology at California State University at Northridge. He was employed at CSUN for 39 years and taught statistics and measurement theory for 35 of those years. In 1995 he received the University Distinguished Teaching Award.

William J. Vincent

Dr. Vincent has been a member of the American Alliance for Health, Physical Education, Recreation and Dance since 1964. In 2007, he received the AAHPERD National Honor Award for distinguished service to the profession. He has served as the president of the Southwest District of AAHPERD and was a member of the AAHPERD Board of Governors from 1993 to 1995. In 1988 he was named the Southwest District Scholar and delivered the keynote address titled "From Means to Manova" at the 1989 convention.

Dr. Vincent is the author or coauthor of 4 books and more than 70 professional articles. Fifty-one of those articles appeared in refereed journals, including *Research Quarterly for Exercise and Sport,* the *International Journal of Sports Medicine,* and the *Journal of Athletic Training.* He has a bachelor's degree in physical education (pedagogy), a master's degree in physical education (exercise physiology), and a doctorate in educational psychology (perception and learning), all from the University of California at Los Angeles.

Dr. Vincent and his wife, Diana, live in Lindon, Utah, and have 6 children and 20 grandchildren. In his free time, Dr. Vincent enjoys camping, snow skiing and water skiing, conducting genealogical research, and reading.

Joseph P. Weir, PhD, is a professor in the doctor of physical therapy program at Des Moines University in Iowa. He earned his doctorate in exercise physiology from the University of Nebraska at Lincoln.

Joseph P. Weir

Dr. Weir is a fellow of both the American College of Sports Medicine (ACSM) and the National Strength and Conditioning Association (NSCA). He was given the NSCA President's Award in 2007 and its William J. Kraemer Outstanding Sport Scientist Award in 2006. He served as president of the National Strength and Conditioning Association Foundation from 2006 to 2009, and he was cochair of the ACSM's Biostatistics Interest Group from 2001 to 2003.

Dr. Weir is the associate editor of the *Journal of Strength and Conditioning Research,* and he is a member of the editorial board of *Medicine and Science in Sports and Exercise.* He is the author of numerous research articles, which have appeared in journals including *European Journal of Applied Physiology, Physiological Measurement, American Journal of Physiology,* and the *Journal of Orthopaedic and Sports Physical Therapy.* He is coauthor of *Physical Fitness Laboratories on a Budget,* and he has contributed chapters to seven texts, including *NSCA's Essentials of Personal Training.*

Dr. Weir is originally from Glennallen, Alaska. He and his wife, Loree, live near Des Moines, Iowa, and have three children. Dr. Weir is an avid motorcyclist and fan of University of Nebraska football.